SELECTED TOPICS IN MINERAL PROCESSING

Managing Editor

A Ghosh

Editors

Pradip

Rakesh Kumar

Publishing Globally

NEW AGE INTERNATIONAL (P) LIMITED, PUBLISHERS

New Delhi • Bangalore • Chennai • Cochin • Guwahati

Hyderabad • Kolkata • Lucknow • Mumbai

Visit us at **www.newagepublishers.com**

Published by New Age International (P) Ltd., Publishers
First Edition: 1995
Reprint: 2015

BRANCHES

- **Bangalore** 37/10, 8th Cross (Near Hanuman Temple), Azad Nagar, Chamarajpet, Bangalore-560 018 Tel.: (080) 26756823, Telefax: 26756820 **E-mail: bangalore@newagepublishers.com**
- **Chennai** 26, Damodaran Street, T. Nagar, Chennai-600 017 Tel.: (044) 24353401, Telefax: 24351463 **E-mail: chennai@newagepublishers.com**
- **Cochin** CC-39/1016, Carrier Station Road, Ernakulam South, Cochin-682 016 Tel.: (0484) 2377303, Telefax: 4051303 **E-mail: cochin@newagepublishers.com**
- **Guwahati** Hemsen Complex, Mohd. Shah Road, Paltan Bazar, Near Starline Hotel, Guwahati-781 008 Tel.: (0361) 2513881, Telefax: 2543669 **E-mail: guwahati@newagepublishers.com**
- **Hyderabad** 105, 1st Floor, Madhiray Kaveri Tower, 3-2-19, Azam Jahi Road Near Kumar Theater Nimboliadda, Kachiguda, Hyderabad-500 027 Tel.: (040) 24652456, Telefax: 24652457 **E-mail: hyderabad@newagepublishers.com**
- **Kolkata** RDB Chambers (Formerly Lotus Cinema) 106A, 1st Floor, S.N. Banerjee Road, Kolkata-700 014 Tel.: (033) 22273773, Telefax: 22275247 **E-mail: kolkata@newagepublishers.com**
- **Lucknow** 16-A, Jopling Road, Lucknow-226 001 Tel.: (0522) 2209578, 4045297, Telefax: 2204098, **E-mail:lucknow@newagepublishers.com**
- **Mumbai** 142C, Victor House, Ground Floor, N.M. Joshi Marg, Lower Parel, Mumbai-400 013 • Tel.: (022) 24927869, Telefax: 24915415 **E-mail: mumbai@newagepublishers.com**
- **New Delhi** 22, Golden House, Daryaganj, New Delhi-110 002 Tel.: (011) 23262368, 23262370, Telefax: 43551305, **E-mail: sales@newagepublishers.com**

ISBN: 978-81-224-0745-7

₹ 695.00

C-14-11-8063

Printed in India at Glorious Printers, Delhi.
Typeset at Grover DTP Systems, Delhi

PUBLISHING GLOBALLY
NEW AGE INTERNATIONAL (P) LIMITED, PUBLISHERS
7/30 A, Daryaganj, New Delhi-110002
Visit us at **www.newagepublishers.com**

Foreword

Efforts for the publication of this multi-authored book on "Selected Topics in Mineral Processing" in honour of Professor Arun Kumar Biswas were started in 1992. The following committee was constituted.

Prof. A. Ghosh
Materials and Metallurgical Engineering
Indian Institute of Technology, Kanpur

Prof. R.K. Ray
Materials and Metallurgical Engineering
Indian Institute of Technology, Kanpur

Dr. Pradip
Tata Research Development and Design Centre, Pune

Prof. S.G. Dixit
Chemical Technology
University of Bombay, Bombay

Dr. Rakesh Kumar
NFP Division
National Metallurgical Laboratory, Jamshedpur

Prof. N.K. Khosla
Metallurgical Engineering
Indian Institute of Technology, Bombay

Dr. Pradip and Dr. Rakesh Kumar were also entrusted with the responsibility of editing and all technical matters concerned with this publication.

The financial requirements for this activity were partially met by generous contributions from several ex-students of Department of Materials and Metallurgical Engineering, Indian Institute of Technology, Kanpur. Many of them had carried out their research work under the supervision of Professor Biswas.

We are grateful to all the authors for their valuable contributions. Suggestions received from Professor Biswas were of considerable help. Secretarial and other assistance received by editors and myself at organizations with which we have been associated during this period, are also gratefully acknowledged.

I wish to thank all who contributed towards the success of this endeavour. Lastly, our thanks also go to Wiley Eastern Ltd., publishers, for bringing out the publication.

Department of Materials
and Metallurgical Engineering
Indian Institute of Technology
Kanpur 208 016, U.P., India

Ahindra Ghosh
Managing Editor and President
Professor Arun Kumar Biswas
60th Birthday Anniversary
Celebration Committee

Preface

This multi-authored book, dedicated to Prof. Arun Kumar Biswas, consists of invited contributions from all around the world. The authors have highlighted the recent innovations made in the areas of their own expertise and experience in these papers. The subject matter covered in the book deals with important facets of modern mineral processing science and technology such as the advances in characterization of mineral separation systems including the measurements of hydrophobic force and mineral association & liberation, modelling of ball mill grinding and fine coal treatment processes, optimum design of flotation circuits, applications of grinding aids, flotation of silicate minerals, beneficiation of complex sulfide ores and chemistry of sulfidization flotation, selective flocculation, biobeneficiation, electroflotation and high gradient magnetic separation. Most of these papers, providing an in-depth coverage of such a diverse set of subjects, are written by former students, associates and colleagues of Prof. Biswas at Indian Institute of Technology, Kanpur. The fact that several of his students are today established professional leaders is indicative of the kind of impact Prof. Biswas has made in the field through his inspiring teaching and research over the past three decades.

It is indeed heartening to put on record that this book will be released during a seminar on "Teaching and Research Needs in Mineral Processing" to be held at Indian Institute of Technology, Bombay in April 1995. Prof. Biswas will deliver the keynote address on this occasion.

This book is a result of the sincere efforts of many persons. We thank all the authors for their valuable contributions. We extend our sincere thanks to Prof. A. Ghosh for his timely advice and help in this effort. He has been the main guiding force behind this book. We are grateful to Prof. E.C. Subbarao and Prof. P. Ramachandra Rao for their kind help and encouragement in our efforts to bring out this book in time. The infrastructural support provided by Tata Research Development and Design Centre, Pune during the typing, editing and the preparation of the final versions of the various manuscripts is gratefully acknowledged. A special thank you to Mrs. Patricia Lobo without whose timely, painstaking and expert secretarial assistance we could not have met the publication deadlines. We also express our appreciation to the staff of Wiley Eastern Ltd. particularly Mr. Batra and Mr. Poplai.

March 1995

Pradip
Rakesh Kumar

Contents

Other Separation Techniques

Professor Arun Kumar Biswas – A Biographical Sketch

Professor Arun Kumar Biswas was born on the 6th of July, 1934 in Calcutta. After studying at the Model High School and the Saint Xaviers College, Calcutta, Prof. Biswas obtained his Masters and Doctorate degrees in Applied Chemistry from the Calcutta University in 1955 and 1959 respectively with research work in the area of surfactants. Subsequently, he proceeded to the Massachusetts Institute of Technology, USA in 1960 to pursue his studies in mineral engineering in order to extend his expertise on surface phenomena to applications in the field of minerals and metals. He carried out his research work under the guidance of Late Professor Antoine Marc Gaudin, the world renowned mineral engineer, on hysteresis of contact angle. It was a fruitful era of studentship (1960-63), when Prof. Biswas came into contact and had the opportunity to learn from several other noted scientists and professors at MIT.

Prof. Biswas returned to India and joined as an Assistant Professor in the Department of Metallurgical Engineering (now Materials and Metallurgical Engineering), Indian Institute of Technology, Kanpur in 1963. He was promoted as Professor in 1971. He married Sulekha in 1964, and their son, Sandeepan was born in 1973. During his distinguished career of 31 years (1963-94) at Indian Institute of Technology, Kanpur, Prof. Biswas developed and taught several courses related to mineral processing. He also established the Materials Separation and Surface Chemistry laboratories at IIT Kanpur.

Prof. Biswas has been a reputed and dedicated teacher. Besides teaching, he has guided many students in their research at Doctoral, Masters and Bachelors levels. Many of them are well-established professionals now in India and abroad. Prof. Biswas always believed that both basic and applied research are to be pursued. Soon after joining IIT Kanpur, he initiated industrial consultation between his institution and several mineral-based industries which he used to visit almost every summer. In tune with such activities, many of his research projects were devoted to beneficiation of low-grade mineral deposits of India such as monazite beach sand, zircon, molybdenite, diamond, phosphorites, zinc ferrite, zinc tailings and residues, alumina-rich iron ore, pyrite, calcite-fluorite systems, Kudremukh iron ore tailing, deep sea manganese nodules, ferro-tungsten deposits etc. Simultaneously, studies were made in his laboratory on unit operations such as comminution, froth flotation, selective flocculation, leaching and bacterial leaching.

In basic research, some studies by Prof. Biswas and his coworkers received international attention, such as:

- role of CO_2 in the flotation of calcite
- collector-frother interaction in flotation
- adsorption at three-phase interline
- tannin-fatty acid interactions in calcite and fluorite systems

In the eighties, assisted by his colleagues, specially Profs. T. R. Ramachandran and R. K. Ray of Department of Materials and Metallurgical Engineering, IIT Kanpur, Prof. Biswas made important contributions towards microstructural characterization of mineral separation systems using TEM, SEM / EDAX and EPMA. The systems studied included sea-bed manganese nodules from the Indian ocean, ancient zinc slag, and alumina-rich iron ores.

Prof. Biswas participated in several international get-togethers such as International Mineral Processing Congresses (Cannes, 1963; Prague, 1970; Cagliari, 1975; Warsaw, 1979) and International Symposium on Surfactants, Gainesville, Florida, 1990. He also made endeavours to strengthen the Mineral Engineering profession in India. Prof. Biswas is one of the founder members of the Indian Institute of Mineral Engineers, founded in the sixties under the leadership of Late Dr. K. K. Majumdar. He strongly advocated the setting up of a National Mineral Engineering Research and Training Institute in India, an institute devoted to pursuit of excellence in this area of major importance to a country like India.

Intellectual activities of Prof. Biswas extend far beyond mineral engineering. He is well-known nationally and internationally for his writings on archaeometallurgy, history of science, languages, literatures and religions - all in the Indian context.

In archaeometallurgy, he accumulated several archaeological and literary evidences with considerable efforts, and established the primacy of India in brass and zinc metallurgy (Indian J. of History of Science, Dec. 1993, pp. 309-330). Ably assisted by his wife, Sulekha, who holds a doctorate degree in Sanskrit, Prof. Biswas has delved deep into the ancient Indian texts and established the etymology of beryllium-containing minerals in the Indian and world literature - particularly of Beryl (Vaidurya) and emerald (Marakata) (Ind. J. of History of Science, June 1994, pp. 13-28). He is recognised as an expert on the antiquity of minerals and metals in India.

Prof. Biswas has written and edited several books on diverse topics such as history of science, religion, socialism and languages. The following list of books written and edited by him indicates the multifaceted personality of a man who transcended the barriers of narrow disciplines, a man who studied and contributed to several other areas of human knowledge.

- *Science in India*, Firma K. L. Mukhopadhyay, Calcutta
- *Profiles in Indian Languages and Literatures*, Indian Languages Society, IIT Kanpur (editor)

- *Swami Vivekananda and the Indian Quest for Socialism*, Firma KLM Pvt Ltd, Calcutta
- *A Pilgrimage to Khetri and the Sarasvati Valley*, Sujan Publications, Calcutta
- *Buddha and Bodhisattva – A Hindu View*, Cosmo Publications, New Delhi
- *Frontiers in Applied Chemistry*, Springer Verlag, Berlin and Narosa Publishing House, New Delhi (editor)

In addition, some more books are awaiting publication.

Prof. Biswas is also the Founder-President of the Indian Languages Society. These endeavours by him arise both from his academic interest as well as his conviction about the relevance of social sciences and humanities along with science and technology for human progress. He has made consistent efforts to utilize his background in science and scientific research in these areas also.

Hence as a fitting tribute to Professor Biswas, we are bringing out this multi-authored book. The biographical sketch would be incomplete without recognizing the invaluable support and cooperation of Mrs. Biswas which contributed in no small measure to the accomplishments of Professor Biswas.

I have been closely associated with Professor Biswas from 1960, firstly when I was a student at MIT and later as a colleague at IIT Kanpur. This association has been of great benefit to me intellectually, professionally and personally. I am happy that I have been able to assist towards publication of this book.

Kanpur, India
March, 1995

Ahindra Ghosh

Recent Applications of Some Microscopic, Diffraction and Spectroscopic Techniques in Mineralogy and Mineral Processing

RAKESH KUMAR[1] and R.K. RAY[2]
[1]*Nonferrous Process Division,*
National Metallurgical Laboratory, Jamshedpur-831007 (India)
[2]*Department of Materials and Metallurgy,*
Indian Institute of Technology, Kanpur-208016 (India)

ABSTRACT

An overview on the applications of selected microscopic, diffraction and spectroscopic techniques in mineralogy and mineral processing is presented. The selected techniques include : (a) transmission electron microscopy (TEM), scanning electron microscopy (SEM), energy dispersive X-ray spectrometry (EDS) and electron energy loss spectroscopy (EELS) for the bulk characterization of minerals and their aggregates; (b) secondary ion mass spectrometry (SIMS) and proton induced X-ray emission (PIXE) spectrometry for trace elements analysis; (c) Mössbauer spectroscopy and X-ray absorption spectroscopy (XAS) for local structure determination; and (d) Auger electron spectroscopy (AES), X-ray photoelectron spectroscopy (XPS), scanning tunneling and atomic force microscopy (STM and AFM), low energy electron diffraction (LEED), internal reflectance spectroscopy (IRS), SIMS and laser ionization mass spectrometry (LIMS) for surface characterization. Brief description of these techniques is given with emphasis on relative potential of different techniques available for a specific purpose. Applications of the techniques are found to cover wide range of materials, e.g. complex ores, slags, leach residues etc, and processes such as grinding, pelletization, flotation, flocculation, magnetic separation, and bio and chemical leaching. The mineralogical applications described in detail focus on complex ores. Specific examples of deep-sea nodules, an ore characterized by fine grain size and complex microstructure, and invisible gold in ores, i.e. gold present in the form of colloidal size inclusions or as trace element in solid solution, are discussed in detail. The problem of local structure determination is discussed in relation to the occurrence of Au in arsenopyrite and Ni, Co and Cu in manganese oxides. Applications in mineral processing are discussed highlighting leaching of minerals and ores, and surface characterization in the study of flotation. Besides flotation and leaching, several other pertinent applications are described in brief.

INTRODUCTION

Many scientific and technological problems emerging in mineral processing are associated with the processing of complex or rebellious ores [1], i.e. off grade ores with complex mineralogy and unique surface characteristics. The domain and complexity of problems have expanded in the past due to gradual increase in materials of interest which include not only ores but also other materials, such as roasted products, slags and leach residues [2-6]. Mineralogical factors which are generally considered important in mineral processing are identity (crystal structure and chemistry) and proportion of minerals, textural associations, variability in composition and particle size, shape and distribution with regards to individual minerals and in the bulk sample [7]. Minerals present in complex ores, leach residues and slags are often amorphous or crypto-crystalline and/or characterized by non-stoichiometry and presence of impurity elements [2-5, 8]. In some ores, e.g. sea nodules, laterites, refractory gold ores, valuable element occurs as part of major mineral(s), rather than as discrete mineral, and its behaviour is governed by its location, i.e. inside or outside the mineral lattice or at surface, and distribution [10-13]. Minerals react with their environment via surfaces. Chemical composition, atomic structure and fine scale morphology of surface, not the bulk characteristics, control the reactivity of minerals during interfacial reactions which are of relevance in various aspects of mineral processing, such as flotation, flocculation and bio and chemical leaching [14-16].

There has been growing realization in the recent past to include raw materials characterization as an essential component of any fundamental study and process development or improvement scheme [17, 18]. Accompanying this development, an increase in the convenience and efficiency of well established microscopic, diffraction and spectroscopic techniques has taken place with a simultaneous explosion of new techniques [19-23]. The increase in the number of techniques available for the characterization of materials is evident from the list of acronyms in Table 1. The list is not all inclusive. Besides, there are techniques which do not find acronyms, e.g. Mössbauer spectroscopy, neutron diffraction etc. This article profiles the applications in mineralogy and mineral processing of some techniques which are relatively common, such as XRD, SEM, TEM, Mössbauer spectroscopy annd XPS; and those which are emerging, such as SIMS, PIXE, LIMS and FTIR-IRS. The contents of the article are organized in two major sections. In the first section a brief description of the techniques is presented with focus not on the techniques per se but on the potential, limitations and comparative merit and demerit of the techniques. General reviews on the techniques can be found in the recent monographs [19-23]. Applications of selected techniques in mineralogy and mineral processing are presented in the second section, where the emphasis is on the nature of characterization problem vis-a-vis techniques available to tackle the same.

Table 1 : Microscopy, diffraction and spectroscopy acronyms revealed (compiled from reference [19-21])

MICROSCOPY	
AEM	Analytical Electron Microscopy
AFM	Atomic Force Microscopy
BEEM	Ballistic Electron Emission Microscopy
CCM	Channelling Contrast Microscopy
CFM	Columbic Force Microscopy
CSTIM	Channelling Scanning Transmission Ion Microscopy
FIM	Field Ion Microscopy
HM	High Resolution Optical Microscope
HREM	High Resolution Electron Microscope
MFM	Magnetic Force Microscopy
OM	Optical Microscope
OAM	Optical Absorption Microscope
PSTM	Photon Scanning Tunnelling Microscopy
QAM	Quantitative Acoustic Microscopy
SCPM	Scanning Chemical Potential Microscopy
SAM	Scanning Auger Microscopy
SCM	Scanning Chemical Potential Microscope
SEM	Scanning Electron Microscopy
SICM	Scanning Ion Conductance Microscopy
SNOM	Scanning Near Field Optical Microscopy
SPM	Scattering Proton Microscopy
STM	Scanning Tunnelling Microscopy
STEM	Scanning Transmission Electron Microscopy
TEM	Transmission Electron Microscopy

DIFFRACTION	
CBED	Convergent Beam Electron Diffraction
EDXRD	Energy Dispersive X-ray Diffraction
GXRD	Gazing X-ray Diffraction
LEED	Low Energy Electron Diffraction
RHEED	Reflective High Energy Electron Diffraction
SAED	Selected Area Electron Diffraction
TED	Transmission Electron Diffraction
XRD	X-Ray Diffraction

SPECTROSCOPY	
AES	Auger Electron Spectroscopy
AFS	Atomic Force Spectroscopy
AMS	Accelerator Mass Spectrometry
ATR	Attenuated Total Reflectance apectroscopy
BIS	Bremsstrahlung Isochromat Spectroscopy
CDS	Crystal Diffraction Spectrometry
CITS	Current Imaging Tunnelling Spectroscopy
DCEMS	Depth-resolved Conversion Electron Mössbauer Spectroscopy
DSIMS	Dynamic Secondary Ion Mass Spectroscopy
EDS	Energy Dispersive Spectrometry
EELS	Electron Energy Loss Spectroscopy
EPR	Electron Paramagnetic Resonance spectroscopy
ESCA	Electron Spectroscopy for Chemical Analysis (same as XPS)
ESR	Electron Spin Resonance spectroscopy
EXAFS	Extended X-ray Absorption Fine Structure spectroscopy
EXELFS	EXtended Electron (energy) Loss Fine Structure
FABS	Fast Atom Bombardment Spectroscopy
FRES	Forward Recoil Spectroscopy
FTIR	Fourier Transform InfraRed spectroscopy
HREELS	High Resolution Electron Energy Loss Spectroscopy
IETS	Inelastic Electron Tunnelling Spectroscopy
IPES	Inverse PhotoElectron Spectroscopy
IR	InfraRed spectroscopy
IRS	Internal Reflectance Spectroscopy
ISS	Ion Scattering Spectroscopy
LEAFS	Laser Excited Atomic Fluorescence Spectroscopy
LIMS	Laser Ionization Mass Spectrometry
LRS	Laser Raman Spectroscopy
MS	Mass Spectrometry
NMR	Nuclear Magnetic Resonance spectroscopy
NEXAFS	Near Edge X-ray Absorption Fine Structure spectroscopy
OAS	Optical Absorption Spectroscopy
PAS	PhotoAcoustic Spectroscopy
PDS	Photothermal Deflection Spectroscopy
PES	Photo Electron Spectroscopy
PIXE	Proton Induced X-ray Emission spectroscopy
QEXAFS	Quick scanning EXAFS spectroscopy
RBS	Rutherford Backscattering Spectroscopy
REELS	Reflection Electron Energy Loss Spectroscopy
SALI	Surface Analysis by Laser Ionization spectroscopy
SEXAFS	Surface Extended X-ray Absorption Fine Structure spectroscopy
SERS	Surface Enhanced Raman Spectroscopy
SIMS	Secondary Ion Mass Spectroscopy
SNMS	Secondary Neutrals Mass Spectrometry
SSIMS	Static Secondary Ion Mass Spectrometry
STS	Scanning Tunnelling Spectroscopy
TL	ThermoLuminescence spectroscopy
UPS	Ultraviolet Photoelectron Spectroscopy
VBXPS	Valence Band X-ray Photoelectron Spectroscopy
WDS	Wavelength Dispersive Spectrometry
XANES	X-ray Absorption Near Edge Structure spectroscopy
XAS	X-ray Absorption Spectroscopy
XES	X-ray Emission Spectroscopy
XPS	X-ray Photoelectron Spectroscopy (ESCA)
XRF	X-ray Fluorescence spectroscopy

TECHNIQUES

Probing of bulk, surface and near surface region of a mineral involves electrons, photons and ions which are altered with respect to the number, energy or types of particle emitted. Changes can also occur in the frequency or absorbance of the radiation transmitted through or reflected from the mineral [20, 23]. Each type of technique looks at these emissions in a different way to provide information about certain aspects of the sample, such as its structure, composition or chemistry. The potential and limitations of different techniques considered in this paper are summarized in Table 2 along with key references for recent reviews on the theoretical and practical aspects of the techniques.

Table 2 : Informations obtained and limitations of selected techniques

Name	Probe in/out	Informations obtained	Major limitations	Key references
XRD	X-ray/ X-ray	minerals, crystallinity, stress	mineral identification difficult for amorphous or crypto-crystalline minerals and minor minerals	[19, 24]
TEM	e^-/e^-	morphology, crystal structure, disorder	sample preparation, small area of examination	[19, 25-28]
SEM	e^-/e^-	morphology, crystallography	charging problem with insulating minerals	[19, 29-31]
EDS/ WDS (in EM)	e^-/ X-ray	elemental analysis	low atomic number (generally $Z < 11$ for EDS and $Z < 3$ for WDS, WDS time consuming	[19, 32-34]
EELS (in AEM)	e^-/e^-	elemental analysis, valency state	difficult sample preparation and sample damage	[19, 21, 35, 36]
SIMS	ion/ion	elemental analysis, surface image	sample damage, difficult quantification, spectrum very complicated	[20, 37-40]
PIXE	proton/ X-ray	elemental analysis, surface image	availability of proton source, difficult quantification	[20, 41, 42]
LIMS	laser/ ions	elemental analysis of bulk and surface	difficult quantification	[43, 44]
Mössbauer spectroscopy	γ-ray/ γ-ray	local structure and bonding	applicable to only selected elements such as Fe, Au, Sn etc.	[21]
EXAFS	X-ray/ X-ray	local structure and bonding	availability of tuneable X-ray source (synchrotron)	[19, 21, 45]
IRS	IR/IR	molecular structure	poor depth resolution in surface analysis	[20, 21, 46, 47]
AES/ SAM	e^-/e^-	surface composition and image	severe sample damage and charging for insulating samples, quantification difficult	[15, 20, 21, 48]
XPS/ESCA	X-ray/ e^-	surface composition and bonding	chemical information limited by instrument resolution, sample charging for mono-chromatized X-ray source	[15, 21, 48]
STM , AFM	e^-/e^-	surface micro-topography	STM can not image insulating minerals and sample damage possible	[15, 20, 49-52]
LEED	e^-/e^-	surface crystallography	sample damage, conducting surface required	[15, 20, 53-54]

Identification of Minerals

Traditional methods of optical microscopy and X-ray diffraction (XRD) continue to be of great practical importance for the characterization of minerals and their aggregates but many problems of current scientific thrust are not amenable to solution with these techniques. Scanning and transmission electron microscopes (TEM and SEM) offer distinct advantages over an optical microscope in terms of image resolution and mineral identification potential (Table 3). The selected area electron diffraction (SAED) mode in TEM makes identification of minor crystalline and/or extremely fine grained (crypto-crystalline) minerals, not identified by conventional XRD, possible because of the shorter wavelength of the electron beam and the option of choosing small particles of reasonable degree of crystallinity. Some distinct morphological features of the mineral observed in TEM can also be used occasionally as an aid in identification. The spatial resolution for SAED pattern in a conventional 100 kV TEM is about 500 nm. In a modern TEM, the spatial resolution for the convergent beam electron diffraction (CBED) may range between 50-100 nm [26].

Table 3 : Comparison of optical and electron microscope

Criterion	Optical microscopy		Electron microscopy	
	Reflected beam	Transmitted beam	SEM	TEM/HRTEM
Maximum magnification	2000 x	2000 x	~300000 x	> 300000 x
Spatial Resolution	1000 nm	1000 nm	1.5-7 nm	0.1-0.5 nm
Ability to yield information on				
Texture	limited	quantitative	limited	little e.g. coherency
Phase identification	limited	excellent : optical constants	limited : atomic number contrast	excellent with diffraction
Phase composition	none	limited : optical constants	many with EDS and WDS	many with STEM and EELS
Crystal structures	little	crystal system	limited	space groups
Grain boundaries	little	limited	limited	atomic structure
Sample				
Preparation	easy	medium	easy	difficult
Size	0.01–10 cm^2	0.01–10 cm^2	10^{-4}–10 cm^2	2–3 × 10^{-3} cm^2
Environment	air	air	vacuum	vacuum

The sample to be examined in an electron microscope should be vacuum compatible. In some of the recently available scanning electron microscopes the vacuum requirement in the sample chamber is relaxed to a great extent [55]. For example, "environmental" scanning electron microscope (ESEM) can operate in the working environment of 3 Pa to 3.5 kPa range [55]. These microscopes allow low vacuum examination of wet and non-conducting specimens without the need for tedious and time consuming specimen preparations, such as sample drying and gold coating.

Elemental Analysis and Microstructure

Elemental analysis capability in an electron microscope adds up to its potential enormously (Table 3). The analysis in a SEM is carried by X-ray spectrometry using an energy dispersive spectrometer (EDS) or crystal diffraction spectrometer (CDS or WDS) system. Modern scanning transmission electron microscope or analytical electron microscope (STEM or AEM) may have both an EDS and electron energy loss spectrometer (EELS) for elemental analysis. The EDS technique is quick and simple to use in comparison with WDS. With the advent of a windowless EDS detector, which makes it possible to analyze lighter elements between atomic number 4-9, EDS detectors are being used increasingly. Quantitative analysis for elements, having atomic number (Z) equal to or greater than 11, can be performed with relative accuracies of about 2% for concentration greater than 1%. Quantitative analysis of lighter elements is more difficult to carry out and relative accuracies of 2-4% are usually all that can be attained [20]. The limit of detection for an element varies inversely with its atomic number : it is typically about a few hundred parts per million with X-ray spectrometry [20]. The EELS can detect all the elements including H, He and Li. However, quantification of the EELS spectra is far more complex than X-ray spectra and is still in the early stage of development [33]. From a practical point of view, the SEM-EDS with back-scattered electron (BSE) imaging, X-ray mapping and advanced imaging system, such as QEM*SEM, SEM-IPS, has proved quite useful for determining the phase quantities, grain sizes, grain shapes, mineral associations and degree of liberation of principal ore constituents in mineral beneficiation operations.

Trace Elements Analysis

Secondary ion mass spectrometry (SIMS) and proton induced X-ray emission spectrometry with a microbeam (μ-PIXE or PIXE) are emerging as complementary powerful elemental analysis techniques for the concentration range which is beyond the capability of an electron microprobe. Typical comparison of different microanalysis techniques is presented in Table 4. SIMS is suitable for trace analysis down to sub-ppm levels for all the elements including the lighter elements (H-F). The technique is one to four orders of magnitude more sensitive than electron microprobe and one order of magnitude more sensitive than PIXE. Typically in the analysis of gold ores, minimum detectable limits of Au by SIMS, PIXE and electron microprobe are reported to be 0.1, 1-10 and > 100 ppm, respectively [12]. Chemical imaging is well established for the SIMS. Though the lateral resolution in SIMS is worse than in electron microprobe (typically around 10 μm compared to 1 μm), it is much better with respect to depth because the sputtered ions originate from 2-3 atomic layers [36]. SIMS is a destructive technique in which material is continuously sputtered off the surface. The dynamic mode (DSIMS) involves rapid erosion of surface material. The technique can be used to obtain depth profile of elements for over several hundred nanometer of the material.

Table 4 : Comparison of different microanalysis techniques

Criterion	EPMA	SEM		AEM		SIMS	μ-PIXE
		EDS	WDS	EDS	EELS		
Lightest detectable element	Be	Na*	Li	Be	H	H	Al
Lateral resolution for analysis, μm	1-2	1-2	1-2	0.01	< 0.01	> 10	~1
Depth resolution for analysis, μm	1-3	1-3	1-3	sample thickness		~ 0.01	10-70
Quantification	quant.	quant.	quant.	quant.	semi	semi	semi

* Be with windowless detector

Local Structure in Minerals

Conventional structural investigation techniques, such as diffraction methods, are not sensitive to short range order and provide information mostly about average structure. Mössbauer spectroscopy can provide information about local structure, i.e. bonding, valency state, and coordination number, for certain elements such as Fe, Au and Sn [21]. Analysis of the EELS spectra can also provide information about valency state; however, the technique is yet to find many usages [16]. In the recent past, X-ray absorption spectroscopy (XAS) has emerged as a powerful element specific probe which can be used to determine the local structure around a specific absorbing element, even when the element is at low concentration levels (~1000 ppm) [21,45]. XAS includes X-ray absorption near edge structure (XANES) and extended X-ray absorption fine structure (EXAFS), which give complementary information. The XANES part of XAS spectrum provides information about the coordination environment and the valence of a selected atom. Structural parameters such as the nature and number of nearest and next nearest atomic neighbours and interatomic distances can be derived from EXAFS. Furthermore, the technique can be used to probe most elements of the periodic table in most types of phases (e.g. crystalline and amorphous solids) and at structural sites ranging from those in the crystal and glasses to those at interfaces, such as mineral-water interface.

Surface Characterization

In most usages the "surface" and "bulk" are arbitrarily defined. Auger electron spectroscopy (AES or SAM), X-ray photoelectron spectroscopy (XPS or ESCA), low energy electron diffraction (LEED) and scanning probe microscopy, i.e. scanning tunnelling microscopy (STM) and atomic force microscopy (AFM), are generally regarded as surface characterization techniques. Bulk characterization techniques, IR and SIMS can also be used for surface characterization in internal reflectance spectroscopy (IRS) mode and static SIMS (SSIMS) mode, respectively. Table 5 presents a summary of commonly used techniques for elemental, chemical, microtopographic and crystallographic characterization of surfaces and interfaces in mineral processing.

Table 5 : Comparison of techniques available for the characterization of surface and interfaces

Criterion	AES/SAM	ESCA (XPS)	STM & AFM	LEED	FTIR-IRS	SIMS	
						DSIMS	SSIMS
Applicability (no depth profiling)							
Surface, 5-10 nm	yes	yes	yes	yes	no	no	yes
Near-surface ~ 1000 nm	yes	yes	no	no	yes	yes	no
Depth profile facility	yes	yes	no	no	no	yes	no
Depth resolution	< 10 nm	0.5-10 nm	atomic layer	~ 1 nm	~ 50 nm-1 μm	~ 13 nm	~ 1 nm
Lateral resolution	> 5 nm	100 μm - mm	atomic scale	~ mm	1 cm	~ μm	> 50 nm
Information							
Topography	yes	no,	yes	no	no	no	no
Chemical	some	yes	some	no	yes	some	yes
Elemental	yes	yes	no	no	no	yes	some
Quantitative	semi	yes	no	no	yes	semi	semi
Sensitivity, at%	10^{-2}-10^{-3}	10^{-2}-10^{-3}	single atom	mono-layer	submono-layer	10^{-5}-10^{-6}	10^{-5}-10^{-6}
Used to identify							
Interfaces	yes	yes	yes	no	yes	yes	no
Organics	no	yes	yes	yes	yes	no	some
Inorganics	yes	yes	yes	yes	some	yes	some
Environment	vacuum	vacuum	liquid, air etc	vacuum	liquid/air/ vacuum	vacuum	vacuum
In-situ (S) or ex-situ (Es)	Es	Es	S or Es	Es	S	Es	Es

In XPS, the kinetic energy of (photo-) electrons, ejected when the sample is bombarded with radiation, are measured and can be used to obtain directly the binding nergies of electrons in the core and valence orbitals. In AES, it is not the kinetic energy of these photo electrons that are measured but the kinetic energy of electrons (the Auger electrons) ejected when the ion resulting from photoemission undergoes relaxation (hence the final state is doubly ionized). For the purpose of this paper it is sufficient to state that limited depth of escape of both photoelectrons and Auger electrons makes them (XPS and AES) powerful techniques for surface analysis. Peaks associated with specific energy levels may be assigned to particular elements (and oxidation states) and variation in peak position related to changes in coordination and bonding. Peak intensities may also be used to estimate concentrations; combined with argon ion bombardment (etching) to remove surface layers, variations in the concentration of species with depth beneath the original surface may be measured (depth profiling) [15, 21]. Historically, LEED has been the primary tool for determining the surface crystallography [53, 54]. The analysis of LEED pattern, which originates from few monolayers, uses similar principle as in the analysis of conventional SAED pattern except that it assumes a two dimensional lattice. Surface imaging down to atomic level is possible with STM and AFM. The resolution in STM and AFM surpasses the resolution obtained in all other microscopic techniques (Fig. 1). These microscopes offer special advantages because they can be used under various environments, such as liquid, air, vacuum, and even high pressure. AFM can analyze both insulating and conducting surfaces, whereas STM is suitable only for conducting and semiconducting surfaces.

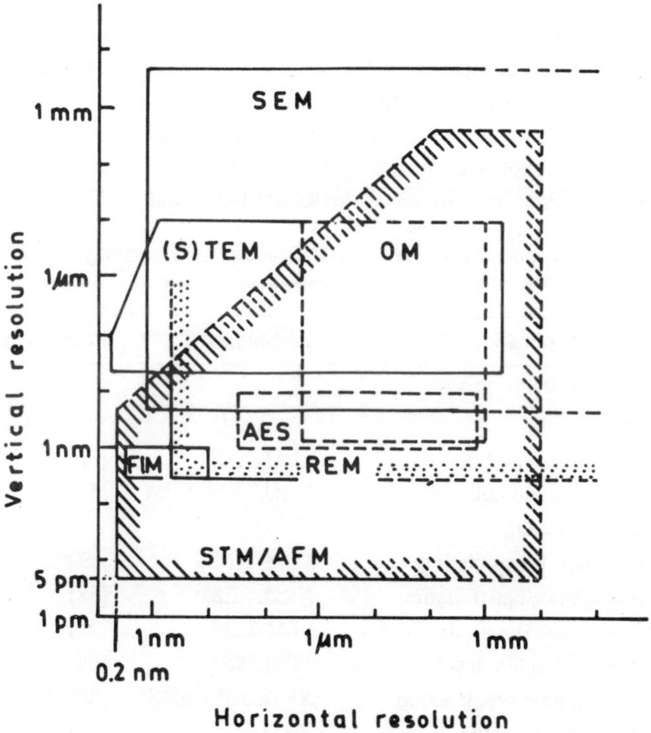

Fig. 1. Comparison of vertical and horizontal resolution ranges between the STM/AFM and other microscopes (after Ref. [51]) (Courtesy: JEOL Ltd., Tokyo, Japan)

In SSIMS, the goal is to remove a minimum amount of material necessary and obtain molecular mass spectral information from the surface. Mapping of the distribution of reagents on a mineral surface is also possible with the use of a time of flight (ToF) instrument [23, 37]. Internal reflectance spectroscopy with Fourier transform infrared interferometer, and not with the dispersion spectrometer as used in conventional IR instruments, (FTIR-IRS) has emerged as a sensitive IR technique for surface characterization in which the problem of interference from water spectra is minimized [47]. The most significant feature of this technique is that it is in-situ. The technique is suitable for the characterization of adsorbates on single mineral surfaces.

Apart from the techniques listed in Table 5, surface characterization is also possible by laser ionization mass spectroscopy (LIMS), EELS in reflection mode (REELS) and surface EXAFS (SEXAFS) [16, 44]. LIMS, in particular, offers quite attractive features for elemental analysis of surfaces, such as inert ionization source (unlike SIMS) suitable for nonconducting surfaces and variable powers of laser for probing different surface depths. The technique is expected to find many applications in future [44].

APPLICATIONS IN MINERALOGY AND MINERAL PROCESSING

Illustrative applications of the selected techniques (Table 2) in mineralogy and mineral processing, compiled on the basis of literature search for the last ten years, are presented in Table 6. The applications cover a wide range of materials and processes. Some representative examples are described in detail with a view to highlight the complexitiy of the problem involved in the characterization as well as potential and limitations of the techniques used.

Table 6 : Applications of selected techniques in mineralogy and mineral processing

Description of Application	Technique(s)	References
PHASE IDENTIFICATION		
A. General		
Deep sea manganese nodules	TEM, XRD	[8]
Ancient zinc slags and tailings	XRD,TEM, SEM	[56]
B. Mineral Processing		
Bauxite ultrafines, beneficiation	TEM, XPS	[57]
Lateritic nickel ores, beneficiation	XRD, SEM	[58]
Antimony minerals, flotation	SEM-EDS	[59]
Doped oxides of Mn, Fe, leaching	TEM, XRD	[60-64]
Strontium in uranium ore, leaching	XRD, SEM-EDS	[65]
Ancient zinc slags, leaching	XRD	[66]

(contd.)

Table 6 (*contd.*)

MORPHOLOGY AND ELEMENTAL ANALYSIS		
A. General		
Deep sea manganese nodules	SEM-EDS	[67, 68]
Mineral matter in coal	SEM-EDS	[69]
Sulfur in gold ores and coal	AEM	[70-72]
Gold in sulfides and silicates	AEM	[73, 74]
Gold in unoxidized sulfide ore	HRTEM	[75]
Anhydrite mineral and alloys	EELS	[33]
B. Mineral Processing		
Synthetic and real ores, liberation	SEM-EDS*	[76-82]
Ball wear mechanisms, grinding	SEM-EDS	[83]
Pt ores, electric pulse disintegration	SEM-EDS	[84]
Hematite concentrate, pelletization	SEM-EDS	[85]
Impurities in sulfides, flotation	SEM-EDS	[86
Au and Ag in leach residue, flotation	SEM-EDS, XRD	[5]
SMS slags, magnetic separation	SEM-EDS, XRD	[3, 4]
Diatomites slurry, magnetic separation	SEM-EDS	[88]
Sulfides, electrochemical leaching	SEM-EDS	[88, 90]
Sulfide minerals, bioleaching	SEM-EDS, AES	[91-93]
Silicates and oxides, leaching	HRTEM	[94, 95]
Gold ore, roasting and cyanidation	SEM-EDS	[96]
Activated carbon, gold extraction	SEM-EDS	[97]
TRACE ELEMENTS ANALYSIS		
Ag distribution in complex sulfide ores	SIMS	[98]
Gold in sulfide minerals and ores	SIMS, PIXE	[12, 99-102]
Pt group elements bearing ores	PIXE	[103, 104]
STRUCTURAL AND CHEMICAL CHARACTERIZATION		
Gold in pyrite and arsenopyrite	Mössbauer	[105-107]
Mn, Ti and Fe in minerals	EELS	[108, 109]
Oxyhydroxide of Mn and Fe	XAS	[110-114]
Ni, Co and Cu in oxides and silicates	XAS, HRTEM	[115-119]
Cr and Ti in oxides and silicates	XAS	[120, 121]
SURFACE CHARACTERIZATION		
A. General		
Apatite, calcite and silicates	XPS, AES	[122-124]
Sulfide minerals, surface oxidation	XPS, AES	[125-127]
Galena, surface oxidation	STM, XPS	[128]

(contd.)

Table 6 *(contd.)*

Galena, pyrite and hematite	STM, LEED, XPS	[129-133]
Hematite and albite	AFM, LEED	[134, 135]
Calcite, precipitation mechanisms	AFM, LEED	[136]
Alumina, goethite; cation adsorption	EXAFS	[13, 137, 138
B. Mineral Processing		
Pb, Zn, Cu and Fe sulfides, flotation	FTIR-IRS	[139-144]
Alumina, fluorite and calcite, flotation	FTIR-IRS	[144-149]
Cassiterite and cerussite, flotation	FTIR-IRS	[150-152]
Base metal sulfides, flotation	XPS, AES	[153-172]
Collectorless flotation of minerals	XPS	[172-175]
Sulfide minerals, flocculation	XPS	[176]
Galena, cassiterite, flotation	SIMS	[177, 178]
Sulfide minerals, flotation	LIMS, SIMS	[44, 179]
Scheelite, calcite and fluorite	IRS, XPS	[180]
Pyrite and pyrrhotite, flotation	FTIR-IRS, XPS	[181]
Hematite and apatite, flotation	FTIR-IRS, XPS	[182]
Cassiterite ores, flotation	SIMS, XPS	[183]
Noble metals, selective flotation	FTIR, XPS, SIMS	[184]
Chalcopyrite, electrochemical leaching	AES	[185]
Pyrite, role of silver ions in leaching	XPS, XRD**, SEM	[186]

* with advanced image processing system, e.g. QEM*SEM, MP-SEM-IPS etc.
** low angle X-ray diffraction

Applications in Mineralogy

Examples selected in this section concern deep sea manganese nodules and gold ores. Sea nodules represent ores characterized by fine grain size and complex assemblage of minerals. The example of gold ore is relevant for all other ores containing colloidal size inclusions and trace concentration of valuable elements. Determination of the local structure around the valuable elements is an important aspect in the characterization of ores such as gold ores and sea nodules. Applications concerning occurrence of gold in arsenopyrite and Ni, Co and Cu in oxides of manganese are presented in this context.

Minerals and microstructures in sea nodules

Recent characterization studies on nodules from Central Indian Ocean basin have thrown interesting light on the minerals and internal structure of these nodules [8, 67, 68]. Typical results of these studies are presented below.

The technique of conventional XRD has limited utility for the mineral identification in sea nodules due to poor crystallinity or fine grain size (100 nm to less than 10 nm) of minerals and presence of several minerals in small (less than about 5%) amount. TEM has been found to be very useful for the identification of minerals in nodules because of its ability to examine very small size particles of reasonable degree of crystallinity when operated in imaging mode coupled with SAED mode [8]. SAED pattern analysis revealed the presence of X-ray amorphous manganese minerals such as birnessite, todorokite, δ-MnO_2; iron minerals such as goethite, ferrihydrite, ferroxyhyte etc., and several silicate minerals [8]. Besides identification of minerals, TEM provides valuable information about structural disorder and crystallinity of minerals. Typical micrographs are presented in Fig. 2 to illustrate the structural disorder observed in manganese minerals. The crystallinity of iron minerals was so poor that it was not found possible to record a single crystal SAED pattern due to geometrical limitation of the microscope, i.e. the selected area was larger than the crystallite size.

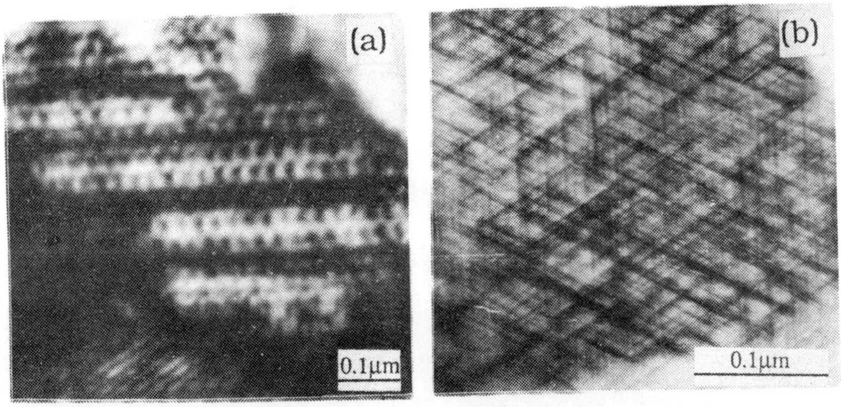

Fig. 2. TEM micrographs illustrating structural disorder in manganese minerals present in nodules from Central Indian Ocean basin : (a) intersecting stacking fault/Moire fringes in birnessite; and (b) trebble intergrowth in todorokite

Sea nodules are characterized by complex assemblage of minerals [67, 68]. SEM-BSE mode micrographs revealed specific Mn-Fe assemblages as columnar (or radial), dendritic and compact type [Fig. 3(a)-(c)]. Z-contrast BSE micrographs revealed change of chemical composition over an interval as small as 0.1-0.2 μm. The distinct looking regions in different morphology, as marked in Fig. 3, were characterized in terms of elemental composition, for major (Mn, Fe, Si) and minor (Ni, Cu, Co, Ti, Mg, Al, K, Ca etc) elements, using EDS analysis [68]. Large variation in composition was observed for each region due to higher spatial resolution of chemical analysis (1-2 μm). The nature of chemical heterogeneity was investigated using Q-mode factor analysis of the EDS data. It was found that four factors, based on Mn-Mg-Ni-Cu-Co- (factor 1), Si-Al-Fe-K-Mg-Ti- (factor 2), Fe-Co-Ti-Ca-Mn- (factor 3), and Mn-Al- (factor 4) assemblages, explained most of the variance in chemical composition [68]. On the basis of information available in literature on the composition of phases in nodules, the factors could be assigned, at least broadly, mineralogical significance. This enabled to correlate morphological features with chemical

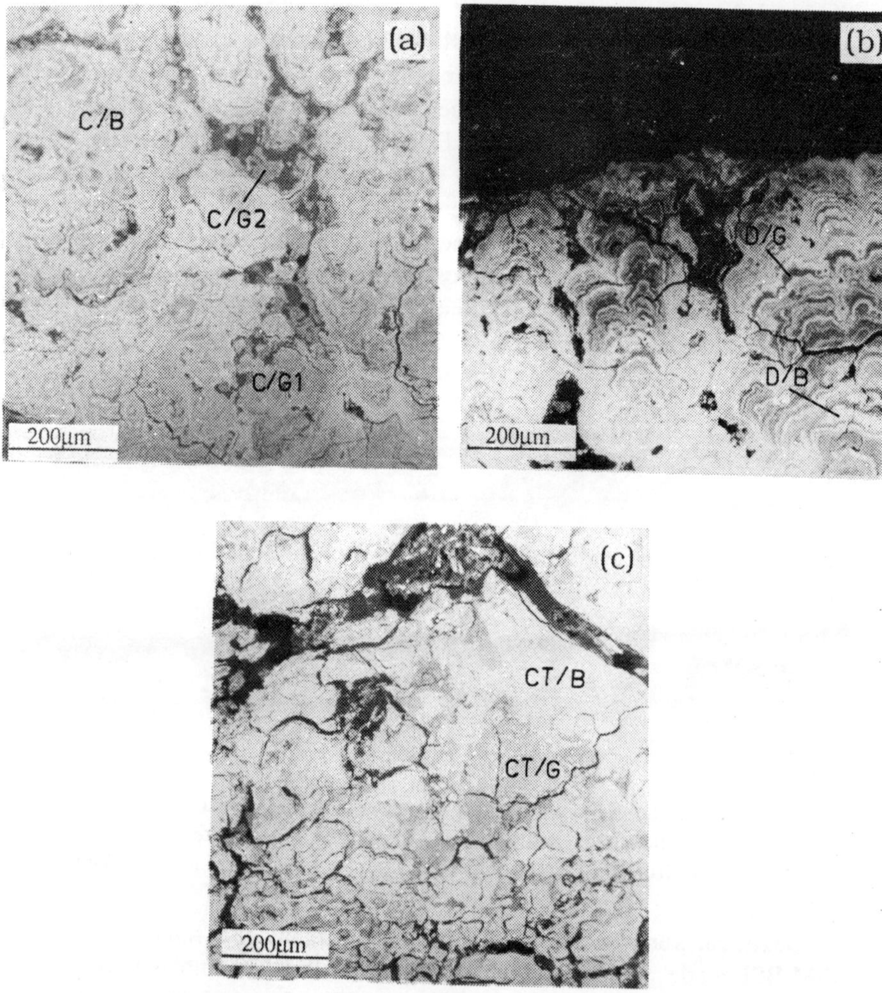

Fig. 3. SEM back-scattered electron micrographs of Mn-Fe assemblages observed in nodules from Central Indian Ocean basin : (a) columnar (or radial) pattern; (b) dendritic segregation; and (c) compact type of morphology (after Ref. [68])

composition (i.e. factor loadings for the region) as well as mineralogy [67, 68]. Figure 4(a) and 4(b) show typical line scan of elements in compact type of morphology and columnar or radial pattern, respectively. Line scan across typical regions (Fig. 4) corroborates the association of Ni and Cu with Mn as observed in factor 1. Co is associated with Mn in compact type of morphology (as in factor 1) and with Fe in columnar and radial pattern (as in factor 3).

Invisible gold in refractory ores

Until recently the process mineralogical studies were restricted to the determination of gold minerals that are visible under the optical (grains ≥ 1 μm) and

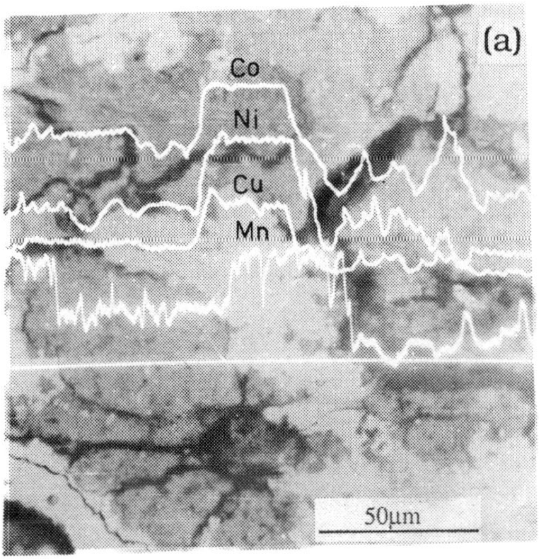

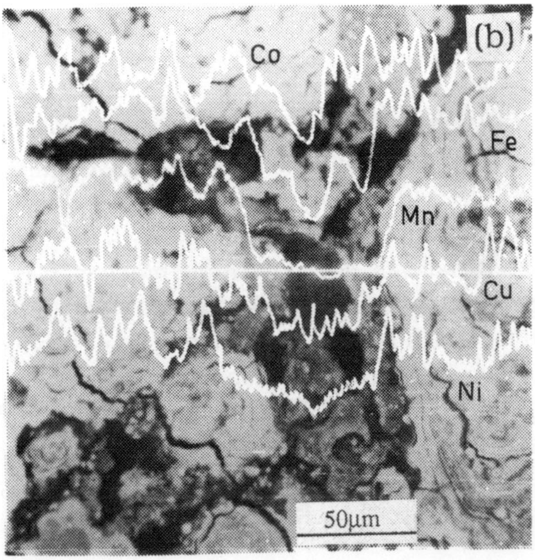

Fig. 4. Concentration profiles of different elements superimposed on typical regions in different morphology : (a) compact type of morphology; and (b) columnar (or radial) pattern (after Ref. [68])

scanning microscopes (grains ≥ 0.1 μm) [12]. The unaccounted submicroscopic or invisible gold was determined by difference. Invisible gold (a descriptive term [12]) includes both solid solution and colloidal size gold in the range of 0.1 μm to 10 nm. The characterization of invisible gold is important from the point of view of complete mineralogical balance of gold and selection of process for extraction. Excellent reviews on the characterization of gold ores, in particular invisible gold, have been provided by Chryssoulis and Cabri [12] and Chryssoulis *et al.* [98].

TEM-EDS has been utilized for the characterization of submicron gold present in sulphide ores [73,74]. Combined application of EPMA, SEM-EDS and TEM-EDS to a carlin type of ore established that invisible gold is composed of a series of ultramicro gold particles (500 nm to a few μm) encapsulated in pyrite, arsenopyrite and clay minerals [74]. HRTEM has been successfully used by Bakken *et al.* [75] to image colloidal size gold. Gold grains 5 nm in diameter were easily detected. Although suitable for the detection of colloidal size gold, TEM, irrespective of resolution, is not suitable for determination of gold mineralogical balance because of the extremely small area (less than 1 mm^2) examined.

Electron probe microanalysis is an important technique for determining the composition of gold minerals. However, it has very limited use for the determination of invisible gold because of its relatively high minimum detection limit (~200 ppm) [12]. SIMS, with a detection limit of 0.2-0.3 ppm and depth resolution of few atomic layer, has proved to be most successful for the quantification of invisible gold in ores [12, 97-101]. The minimum grain size that can be analyzed is about 25 μm. The depth profile facility in SIMS permit characterization of hidden submicroscopic inclusions as small as 20 nm across upto a maximum depth of 1-2 μm. Elemental image for different elements can also be obtained using SIMS. Figure 5 shows SIMS elemental images of As, S and Au across an arsenopyrite-löllingite grain. The image illustrates a general correlation between As and Au, suggesting a Au-As complex or that the substitution of gold into structure is facilitated by As. Conventional SIMS can not analyze insulating minerals due to charging problem. μ-PIXE (detection limit 3-6 ppm, depth resolution ~30 μm for minerals with density of quartz) has been successfully used for the determination of invisible gold in insulating minerals as well as sulphides [12, 102]. Cabri and Chryssoulis [100] reported that for the same arsenian pyrite grain the results of SIMS and PIXE analysis were quite different. This is attributed to the inhomogeneous distribution of gold and different shape and volume of regions analyzed in the two techniques; a thin disc of ~2400 μm^3 for SIMS and a parallelepiped of 5000 μm^3 for μ-PIXE.

Local structure and oxidation state of minor elements

The bulk characterization techniques, e.g. SEM-EDS, SIMS and PIXE provide valuable information about the association of minor elements with major elements or minerals. However, they do not provide any information about the local structure (or short range order) and chemical state of the minor elements. As illustrated in the examples presented below, techniques such as Mössbauer spectroscopy and XAS can provide valuable structural and chemical information and complement information obtained on average structure by X-ray and electron diffraction and microanalysis techniques.

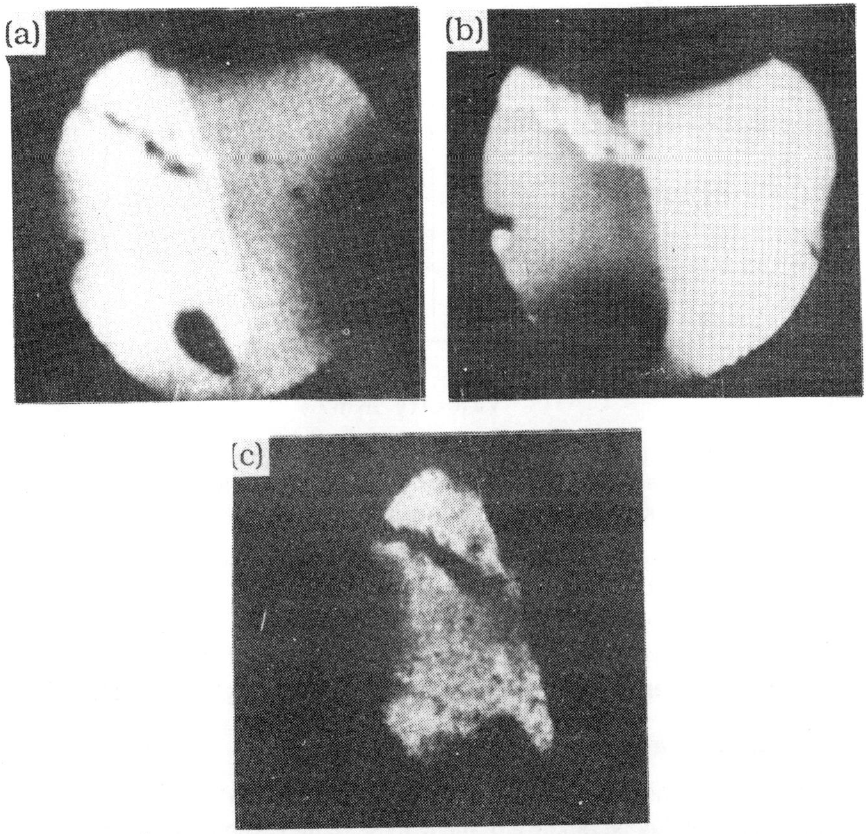

Fig. 5. Ion-microprobe images across arsenopyrite-löllingite grain boundary, for a sample from Mt. York, illustrating heterogeneous distribution of elements (A : ^{75}As, B : ^{32}S, C : ^{197}Au). Lighter areas are enriched in specified elements. Light areas in As image corresponds to löllingite grain. Diameter of image 150 μm (after Ref. [101]) (Courtesy : P. Neumayr, Karl-Franzens-University, Graz, Austria and L.J. Cabri, CANMET, Ottawa, Canada)

Gold in pyrite and arsenopyrite

Both pyrite and arsenopyrite are important host minerals of gold in naturally occurring sulphide ores. Au can occur as small discrete mineral or chemically combined with the sulphide. ^{197}Au Mössbauer spectroscopy is a promising technique for distinguishing between these two forms of gold [13, 21, 105-107]. The technique has a detection limit of 100-300 ppm Au [12]. Figure 6 illustrates the Mössbauer spectrum of Au in arsenopyrite. Oxidation state and chemical environment of Au can be determined using the isomer shift (the shift of peak on X-axis) and quadrupole spiliting (the spiliting of peak). Metallic Au with an isomer shift of −1.23 mm/sec (vertical line in the figure) can be easily distinguished from chemically combined gold which shows a doublet with an isomer shift (the mean for the two peaks) close to +4 mm/sec [13, 21 p.330]. Most of

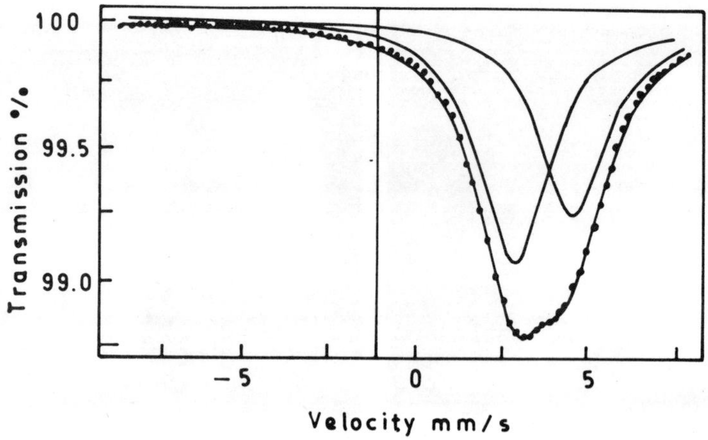

Fig. 6. Mössbauer spectrum of Au bearing arsenopyrite. The vertical line indicates the position of pure metallic gold absorption peak (after Ref. [106]) (Courtesy: Mineralogical Society of America, Washington, D.C., USA)

the arsenopyrite give Au doublet at +4 mm/sec and not the Au singlet, thus indicating the presence of Au in chemically combined form. Mössbauer spectroscopy studies on the roasted gold ore samples, containing various proportions of arsenopyrite and pyrite and 180-300 ppm gold (all in invisible form), indicated that the chemically combined gold converts into metallic gold during roasting [21].

Ni, Co and Cu in the oxides of manganese

An illustration of the type of structural information that XAS provides comes from the study of Co, Ni, and Cu in manganese oxides with two dimensional structure (e.g. asbolane, birnessite and lithiophorite) [110, 112, 114-118]. Figure 7 shows the EXAFS spectra of a lithiophorite (mixed layer mineral consisting of sheets of MnO_2 and $Al(OH)_3$) sample recorded at the Mn, Co, Ni and Cu K-edges. It comes out from the comparison of raw spectra that the local structure around Mn and Co, on the one hand, and around Ni, Cu, on the other hand, are quite similar. The detailed spectral analysis [112] led to the conclusion that Ni, Cu are diluted in $Al(OH)_3$ layer. Analysis of Co K-XANES and EXAFS spectra established presence of 6 fold coordinated trivalent cobalt in low spin configuration substituting for Mn atoms or building intimately mixed CoOOH layer. Other examples of XAS study of minor transition elements, i.e. Ni, Co and Cu in the oxides of Mn, are also available. These include Ni, Co in asbolane [115] and Ni, Co, Cu and Zn in birnessite [117, 118].

Applications In Mineral Processing

Development of a relationship between mineralogy and process response forms the crux of applications in mineral processing. Since it is difficult to develop this relationship for real ores, particularly complex ores, a number of applications concern pure minerals or their mixtures (synthetic ores). Applications in

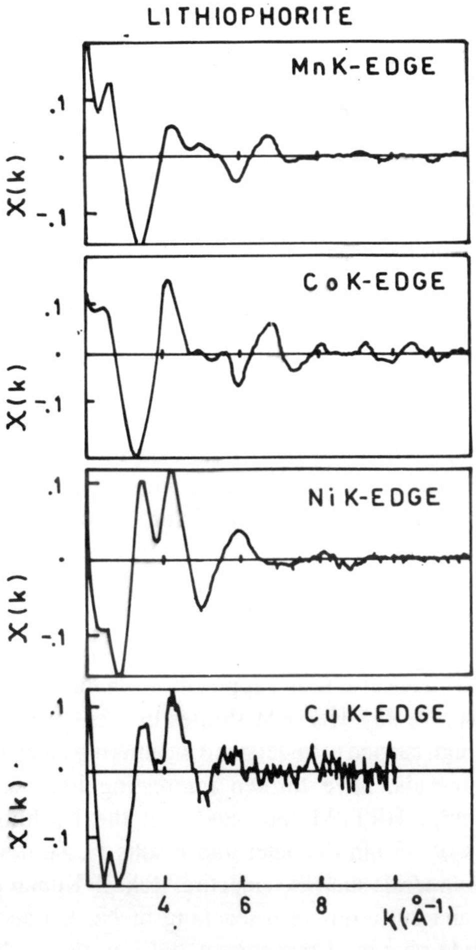

Fig. 7. EXAFS spectra of lithiophorite at Mn, Co, Ni and Cu K-edges (after Ref. [112]) (Courtesy: G. Calas, Laboratoire de Mineralogie-Crystallographie, Paris, France)

mineral processing are discussed in this section with emphasis on the leaching and surface characterization in flotation.

Leaching of minerals and ores

Leaching may be regarded as chemical beneficiation process. Generally, the process involves preferential dissolution or concentration of valuable element(s) in the solution phase leaving behind a solid product or leach residue. The characterization of leach residue is carried out to enhance understanding of leaching fundamentals with regards to a particular mineral and leaching medium [60-64, 94, 95] or to explain poor recovery in a leaching process [65, 66] (Table 6). An excellent review on the characterization of leach residues and other by-products in hydrometallurgical operations is presented by Chen and Dutrizac [9].

XRD and SEM-EDS have been widely used for the characterization of leach residues in terms of minerals, morphology and elemental associations. Page and Salinovich [65] used these techniques to detect the formation of uranium enriched strontianite in the leach residue formed during carbonate leaching of a strontium bearing uranium ore. The mineral strontianite, responsible for the loss of uranium during leaching, was found to form due to the reaction of U-poor celestite (present in the ore) with the carbonate rich leach liquor. Robinson [97] studied the roasting and cyanidation behaviour of South African sulphide gold ores, containing gold in pyrite and arsenopyrite minerals, and found that a maximum of 80-90% recovery of gold was possible. SEM-EDS revealed that gold was present as fine grains either at the end of pores in solid particles, or in the pore walls or in the solid matrix of particles. This indicated that the undissolved gold escaped contact with the cyanide solution owing to its encapsulation by other minerals. Pesic and Kim [91] studied the surface of pyrite, by SEM, after its interaction with T.Ferrooxidans bacteria. These workers [91] found that dense population of bacteria on the pyrite surface, after two days of interaction, hindered further leaching. Several other workers have used SEM to characterize the morphology of leached particles during electrochemical leaching of sulphide minerals [89, 90] and bacterial leaching of sulphide ores and mattes [92, 93].

TEM and HRTEM have also found applications in the characterization of leach residues [60-64, 94, 95]. HRTEM studies by Ceasy *et al.* [95] on the leached olivine, although carried to understand weathering mechanism of sheet structure silicate minerals, have thrown interesting light on the micro-mechanism of leaching. HRTEM indicated that the leaching selectively proceeds in channels of ~3 nm diameter and results in the development of topotactic secondary minerals such as semectite. Rakesh Kumar *et al.* [60-64] studied the sulfuric and sulfurous acid leaching of Ni, Co and Cu bearing synthetic oxyhydroxide phases of manganese and iron to develop an understanding of the leaching mechanisms. Two specific examples, based on these studies [60-64], will be presented here to illustrate the use of TEM for the detection of fine grain minerals and morphological features which are difficult to characterize by XRD and SEM, respectively. The first example relates to the leaching in sulphuric acid of synthetic birnessite (sheet structure manganese mineral) samples doped with Ni and Cu [62]. It was found that prolonged leaching of birnessite samples upto 6 days failed to leach significant fraction (20-30%) of Ni and Cu. XRD indicated that during the course of leaching birnessite undergoes phase transformation to another mineral. However, the identity of the mineral could not be established conclusively due to its fine grain size. TEM examination of original samples and leach residues revealed (Fig. 8) that the transformation product of birnessite [Fig. 8 (a), (b)] was essentially γ-MnO_2 (nsutite) [Fig. 8 (c), (d)]. It is known that nsutite has a compact structure as compared to the relatively open sheet structure of birnessite. Since diffusion of minor doped element is expected to be slow in host mineral having

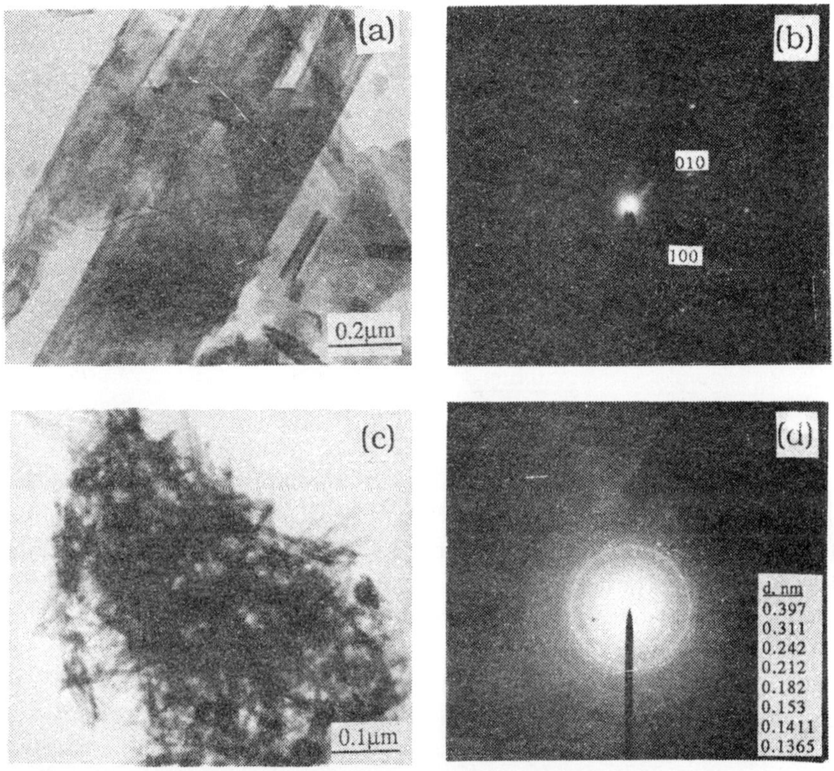

Fig. 8. TEM micrographs and SAED patterns : (a), (b) Ni bearing birnessite; (c), (d) Nsutite phase formed during sulphuric acid leaching of Ni-birnessite sample (after Ref. [62])

compact structure, poor leachability was explained in terms of transformation of birnessite into difficult to leach nsutite phase. The second example concerns the leaching in sulphuric and sulfurous acids of synthetic goethite samples having different physicochemical characteristics [61, 63]. In this case an attempt was made to correlate leachability of iron with surface area and other physical characteristics, e.g. microcrystallite size, structural disorder (measured using XRD [61]). It was found that the difference in the leachability of samples was not possible to explain only in terms of surface area of the samples, and both surface area and microcrystalline dimensions or structural disorder were required to explain the leaching results. TEM examination of the leach residues (Fig. 9) established the role of grain boundaries in leaching. It is evident from the micrograph [Fig. 9(a)] that particles are not uniformly leached and grain boundaries, which represent high energy regions, are preferentially attacked during early stage of leaching. Uniform attack takes place only at later time intervals [Fig. 9(b)].

Surface characterization, using techniques such as AES and XPS, has received little attention in the leaching systems relevant to metal extraction.

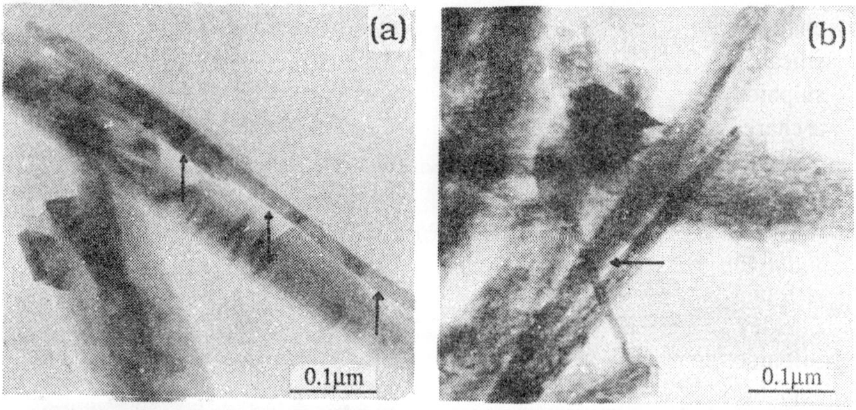

Fig. 9. TEM micrographs of goethite particles after leaching in sulphurous acid : (a) after about 2 min leaching; and (b) 20 min of leaching (arrows indicate possible regions of preferential attack) (after Ref. [63])

Hisky *et al.* [186] studied the interaction of Ag^+ ions and pyrite in 0.025 M H_2SO_4 and found through the analysis of reacted pyrite, using SEM, XPS and low angle XRD (suitable for the identification of surface species), that Ag^o and S^o were the primary species formed during the interaction. The surface Ag^o redissolved quickly in $FeCl_3$ and did not catalyse pyrite leaching in the ferric chloride medium. Warren *et al.* [185] studied the electrochemical reduction of chalcopyrite ($CuFeS_2$) in H_2SO_4 based electrolytes containing different concentration of Fe^{2+} and Cu^{2+}. XRD indicated chalcocite (Cu_2S) and djurleite ($Cu_{1.96}S$) as the only products of reduction. AES established, additionally, that a thin layer of sulphides containing different proportions of Cu and Fe also forms on the surface of pyrite.

Surface characterization in flotation

Chemically altered layers on the surface of minerals have long been known to interfere with the recovery and selectivity of mineral separation by flotation [16, 154]. Understanding of the surface of minerals, before and after reagent interaction, forms the core of a complete understanding of flotation chemistry. Surface characterization studies in flotation have concerned single minerals, mixture of minerals (or synthetic ores) and real ores. In this section we shall first take up the characterization of minerals before reagent addition, and then consider characterization of interaction of ions and collector with minerals.

Surface characterization of minerals and ores

XPS (information on composition and chemical state; spatial resolution ~ 150 μm) and SAM/AES (information on composition, spatial resolution ~ 1 μm) have found many applications for surface characterization of minerals in flotation and related studies. A number of studies have focused on the characterization of single sulphide minerals, e.g. galena [125, 155], bornite [126], chalcopyrite [156], pyrite [157], arsenopyrite [127], and pentlandite [128]; silicates and other minerals [122-124]. It is now well established for single sulphide minerals that hydroxides species, characterized by O 1s XPS signal at a charge corrected binding energy of 531.5 eV, form on the surface after exposure to ambient air or aqueous solution. Besides hydroxides, other surface species present on the surface may be sulphate and carbonate. Furthermore, in most cases a metal deficient sulphide layer, which gives rise to a broad S $2p_{3/2}$ signal shifted from the normal S^{2-} position (161.1 eV) to higher binding energy, is found beneath the hydroxide products [154]. The work, particularly of Buckley and Woods [155-157], has shown that this signal for metal deficient sulphide can systematically approach the binding energy of elemental S as the depletion of the metal increases; hence, a hydrophobic sulphide surface may be created beneath the hydroxide products. Smart [154] has demonstrated the utility of XPS, SAM/AES and SEM, as complementary techniques, in the surface characterization of base metal sulphide minerals and ores. These studies revealed that, in milled base metal sulphide ores, e.g. ore containing pyrite and chalcopyrite, iron may be present in the form of thin (~ 5-80 nm) layers, oxidized sulphide particles (~ 0.1-5 μm) attached to larger sulphide particles and colloidal, flocs precipitated from solution. XPS and AES have also been used to show that some complex sulphide ores contain a substantial proportion of iron sulphide minerals, as pyrite and chalcopyrite, which are hydrophobic as a result of a surface coating of a carbonaceous material [154]. The form of carbon can be graphitic, saturated hydrocarbons, or humic/fulvic residues (oxidized hydrocarbon) depending upon the degree of reduction experienced by the ore during its genesis.

The complete characterization of surface requires information not only about composition and chemical state of surface species but also on surface crystallography and microtopography. Laajalehto *et al.* [129] used STM and XPS to characterize the surface of galena oxidized in air. STM images of galena (100) crystal planes oxidized in air (0-270 min), showed systematic growth of oxidation products from regions < 0.6 nm to > 9 nm diameter. The combined results provided both structural and compositional evidence for the mechanism of reaction of galena surface to $PbCO_3.xPb(OH)_2$. A number of recent surface characterization studies, although not carried out directly in the context of flotation, have used STM/AFM, LEED, in conjunction with XPS and AES, for complete surface characterization [129-136]. STM, LEED and XPS have been used for the surface characterization of conducting minerals such as galena and pyrite [130, 131, 133]. AFM, LEED have found applications for surface characterization of nonconducting minerals such as albite, calcite etc [134-136].

Ion or collector interaction with minerals and ores

In-situ spectroscopic examination of interface between the minerals and aqueous solution is most desirable. FTIR-IRS has found many applications, in mid infrared region (4000-500 cm^{-1}), as an in-situ technique for the characterization of adsorbed surfactant molecules. The applications of this technique in flotation have been recently reviewed by Miller et. al. [144]. The short sampling depth in this method, usually a fraction of wavelength, minimizes the strong infrared absorbance of water, a condition which had limited the use of IR spectroscopy. Also, this technique is quite sensitive and very dilute collector concentrations, in the range of 10^{-3} to 10^{-7} M, have been measured [145]. Mixed mineral powders, such as an actual flotation concentrate from an ore, are almost impossible to investigate using this technique and all the studies have concerned single minerals. Three approaches have been used to study the flotation systems using IRS. The first two methods make use of an inert internal reflection element (IRE), e.g. Ge, ZnSe crystal, and involve pressing and dewatering a particulate system against the IRE (method 1) or coating (usually vacuum deposition) of the mineral on IRE which is then placed in contact with the solution (method 2). In method 3, reactive IRE is used which serves a dual purpose, i.e. internally reflect the light and as well as act as a substrate at which collector adsorption can be monitored. The various reported applications of method 1 are in the study of adsorption of ethylxanthate on sphalerite [139], marcasite [140], galena, chalcocite, chalcopyrite, pyrite [142-143] and cerussite [150-152]; dixanthogen and monothiocarbonate on marcasite [140]; dithiophosphate on galena [143]; and thionocarbamate on chalcocite, chalcopyrite and pyrite [141, 143]. Since reproducible contact between the mineral suspension and IRE is difficult to achieve, and consequently only qualitative measurements are possible with method 1. Also, collector contamination of inert IRE can not be ignored [144]. Method 2 has found limited applications, due to the difficulty in getting reproducible film of appropriate composition, crystal structure, thickness and coherency, e.g. cassiterite-styrene phosphonic acid system [150]. Methods 1 and 2 can provide information regarding the state of the adsorbed molecules, i.e. whether the collector is physically or chemically bonded at the surface. Method 3 allows direct calculation of the collector adsorption density, adsorption state and the orientation of adsorbed species [144-146, 149]. Flotation systems in which the application of method 3 has been demonstrated include fluorite-oleate, calcite-oleate, alumina-dodecylsulfate, sphalerite-xanthate and sylvite-n octylamine [144-149]. One major limitation in the use of method 3 has been that not all mineral crystals are IR transparent in the mid infrared region. Recently there have been attempts to use the near infrared region (15000-4000 cm^{-1}) in order to extend the utility of reactive IRE [146].

XPS has been successfully applied for studying the interaction of collectors and modifiers with mineral surfaces [16, 153, 154, 160-171]. Since the technique requires an ultra high vacuum, only ex-situ investigation of surface reactions in flotation is possible. Surface analysis can not be performed if the adsorbate is not stable under the high vacuum used. Sample preparation for XPS is quite critical and requires extra precautions in order to avoid surface contamination by extraneous factors [154]. As an illustration of the application of XPS, Figure 10 compares the Cu 2p signal from a ground chalcopyrite sample before and after addition of 5×10^{-5} M butyl xanthate. It is clear from the spectra that the high binding energy shoulder on the Cu $2p_{3/2}$ peak, due to charged hydroxide species, is removed by the addition of the xanthate. This indicates that xanthate, besides increasing the hydrophobicity of the sulphide surface, also counteracts the effect of hydroxides on the surface. The galena-xanthate system has been most extensively studied using XPS [160-164]. Laajalehto *et al.* [164] found that in the surface layer Pb : S : O : C ratio (1 : 4.2 : 2.75 : 6.73) was comparable to lead xanthate (1 : 4 : 2 : 6) suggesting its formation in the surface layer. Xanthate adsorption has also been studied for other sulphide minerals and metals [165-169]. XPS has been applied to correlate collectorless flotation of sulphide minerals with the formation of surface sulphur species [172-175]. The work of Buckley *et al.* [172] showed that under conditions where flotation commences, the surface species is generally a metal deficient sulphide rather than elemental sulphur. XPS can handle mixed mineral powders, synthetic as well as real ores [154]. Recently, Houot and Raveneau [175] studied the flotation of sphalerite in the presence of lead ions to explain the activation of sphalerite flotation in the presence of galena. XPS revealed that Pb^{2+}, produced in the solution due to dissolution of galena, may occupy on the surface of sphalerite 30% of cationic sites upto ten atomic layers and activate its flotation. Similarly, for the pyrite-chalcopyrite mixture it has been found, using XPS (and SAM), that copper can dissolve from chalcopyrite surface as Cu(II), adsorb on the pyrite surface via galvanic mechanism as Cu(I) and lead to the activation of pyrite into the concentrate via selective adsorption of xanthate [154].

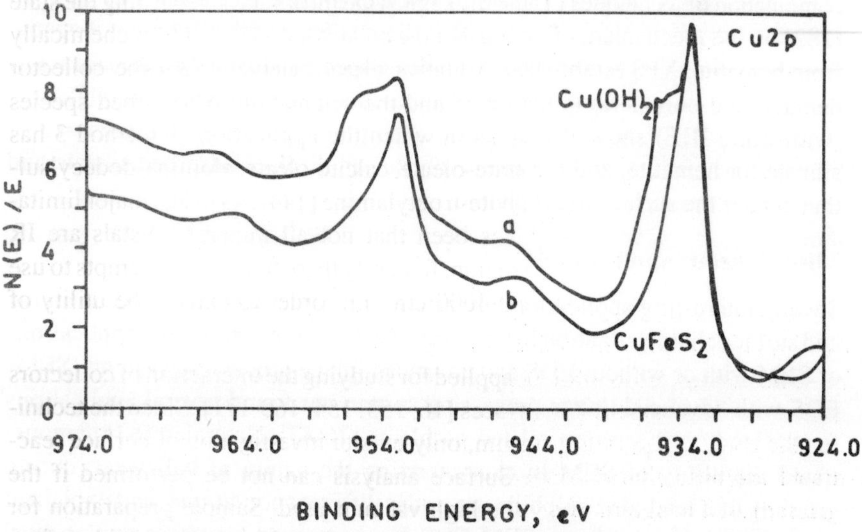

Fig. 10. XPS Cu 2p spectra from ground chalcopyrite surfaces derived from (a) distilled water (upper curve), and (b) pH 9 pulp with 5×10^{-5} M butyl xanthate addition (lower curve) (after Ref. [154])

Although quantification is possible in XPS, the sensitivity is much less compared to SIMS (Table 5). In the studies on the adsorption of monothiophosphates on noble metals, Nagraj *et al.* [184] demonstrated that very low adsorption of collector on gold could be clearly detected using SIMS but not by other techniques such as FTIR, XPS etc. The potential of SIMS and SIMS imaging have also been demonstrated recently in the study of interaction of natural galena crystallite surfaces with dialkyldithiophosphinate [178]. SIMS used in the imaging mode allowed the mapping of the distribution of dithiophosphinate and showed preferential non-uniform adsorption on the crystallite surfaces, possibly in the oxygen rich regions.

Recently, Chryssoulis *et al.* [44, 179] has reported applications of LIMS in the flotation of sulphide minerals, e.g. Brunswick Zn-Pb-Cu-Ag complex sulphide ore, Sundbury Ni-Cu-PGE sulphide ore. Several advantages have been claimed for this technique : a neutral primary beam which allows examination of insulating particle without any charging problem; flexibility in terms of available laser wavelength, permitting the analysis of adsorbed organics; the power of primary laser is adjustable, giving control over sampling depth; and a small beam size, permitting the analysis of individual mineral grains as little as 5-10 μm across. It may be added that insulating particles are difficult to analyze by XPS as well as SIMS. Also spatial resolution in the case of XPS is only about 1 mm and for SIMS it is greater than 10 μm. So in these respects LIMS offers distinct advantages.

In view of the fact that each of the surface characterization techniques is associated with certain limitations, a number of researchers have used the combination of techniques (Table 6). Typical example is recent study by Qi *et al.* [182] on the mechanism of action of sodium silicate in the flotation of apatite from hematite. XPS established that silicate species adsorbed on the surface of hematite are polymeric in nature. Combined application XPS and FTIR-IRS (with ZnSe IRE) showed that oleate adsorption reduces in the presence of silicate for hematite, and for apatite oleate can displace adsorbed silicate and thus render the surface hydrophobic.

Miscellaneous applications

Several interesting applications are also found in the domain outside leaching and surface characterization in flotation. Most of these concern the applications of SEM with or without EDS and advanced image processing systems. SEM-EDS with advanced image processing system has found several applications for the study of liberation related problems [76-82]. Natarajan [83] demonstrated the utility of SEM in characterizing the nature of ball wear, during grinding of a lead-zinc sulphide ore under different conditions, and explained the ball wear mechanisms. SEM-EDS has been used for phase and/or morphological characterization of electric pulse disintegrated precious metal ores [84], self fluxed pellets of hematite concentrate [85], and products in the HGMS (high gradient magnetic separator) separation of impurities in a slurry of diatomites from different Bulgarian deposits [88]. Fujita and Iwasaki [3, 4] investigated occurrence of phosphorous in steel making slags, by XRD and SEM-EDS, for the separation of phosphorous bearing minerals by HGMS and recycling of the slag.

CONCLUDING REMARKS

In this paper we have illustrated the potential of selected microscopic, diffraction and spectroscopic techniques in solving complex mineralogical problems related with minerals, microstructure, trace element analysis, local structure determination and surface characterization. The techniques considered have their own merits. However, these techniques are not a substitute for conventional methods of characterization, e.g. wet chemical analysis, XRF, optical microscopy etc. The techniques described are only complementary tools. A single characterization technique is generally not sufficient and a judicious combination of techniques, depending upon the nature of problem, is required to solve a specific problem at hand. For example, a combination of optical microscopy, AEM, SEM-EDS, PIXE, SIMS, and Mössbauer spectroscopy, together with leaching test and bulk chemical analysis, may be necessary for the complete characterization of gold ores. Besides, depending upon the requirement it may be necessary to couple these techniques with other measurements, e.g. contact angle and electrokinetic measurements in the study of flotation systems.

ACKNOWLEDGEMENT

The authors are grateful to Professor P. Ramchandra Rao, Director, National Metallurgical Laboratory (NML), Jamshedpur for his kind support during the preparation of this article. We would like to express our thanks to the many authors of cited papers who have provided valuable literature for the preparation of this paper. In particular, we would like to acknowledge Profs. L.J. Cabri (CANMET, Ottawa, Canada), P. Neumayr (Department of Mineralogy-Crystallography and Petrology, Karl-Franzens-University, Graz, Austria), G. Calas (Laboratoire de Mineralogie-Crystallographie, Paris, France), R.St.C. Smart (School of Chemical Technology, University of South Australia, The Levels), M.F. Hochella (Department of Geological Sciences, Virginia Polytechnic Institute and State University, USA) and J.J. Keller (Department of Metallurgical Engineering, South Dakota School of Mines and Technology, USA). The transmission electron micrographs presented in this paper were very kindly provided by Mr. Samar Das (Scientist, Electron Microscopy Laboratory, NML, Jamshedpur).

REFERENCES

1. Abishev, D.N. : In Forssberg, K.S.E. (Ed.) Development in Mineral Processing 10B. Amsterdam : Elsevier 1988, p 1087.
2. Biswas, A.K. : Biswas, A.K. (Ed.) : Frontiers in Applied Chemistry. New Delhi : Narosa (Springer-Verlag) 1989, pp 11-52.
3. Fujita, T., Iwasaki, I. : In Carson, D.J.T., Vassiliou, A.H. (Eds.) : Process Mineralogy VIII. Warrendale PA : The Minerals Metals and Materials Society 1988, pp 293-308.
4. Fregeau-Wu, E., Pignolet-Brandom, S., Iwasaki, I. : In Petruk, W., Hagni, R.D., Pignolet-Brandom, S., Hausen, D.M. (Eds.) : Process Mineralogy IX. Warrendale PA : The Minerals Metals and Materials Society 1989, pp 429-439.
5. Dutrizac, J.E., Chen, T.T. : Trans. Instn. Min. Metall. (Section C : Mineral Process. Extr. Metall.) 97 (1988) 180-190.
6. Schapiro, N., William, J.M., Park, W.C. : In Hausen, D.M., Park, W.C. (Eds.) : Process Mineralogy. Warrendale PA : The Metallurgical Society of AIME 1981, pp 25-30.
7. Jones, M.P. : Applied mineralogy - a quantitative approach. London : Graham and Trotman Ltd. 1987.
8. Rakesh Kumar, Das, S., Ray, R.K., Biswas, A.K. : Trans. Ind. Inst. Met. 44 (1991) 119-130.
9. Chen, T.T., Dutrizac, J.E. : In Petruk, W., Hagni, R.D., Pignolet-Brandom, S., Hausen, D.M. (Eds.) : Process Mineralogy IX. Warrendale PA : The Minerals Metals and Materials Society 1990, pp 289-303.
10. Rakesh Kumar : Characterization and leaching studies on deep sea nodules and synthetic analogs, Ph.D Thesis. Indian Institute of Technology, Kanpur 1988.
11. Kuhnel, R.A., Roorda, H.J., Steensma, J.J.S. : Bull. B.G.R.M. 11(3) (1978) 91-106.
12. Chryssoulis, S.L., Cabri, L.J. : Trans. Instn. Min. Metall. (Section C : Mineral Process. Extr. Metall.) 99 (1990), 1-10.
13. Bancroft, G.M., Hyland, M.M. : In Hochella, M.F. Jr., White, A.F. (Eds.) : Mineral Water Interface Geochemistry; Reviews in Mineralogy, Vol. 23. Washignton : Mineralogical Society of America 1990, pp 552-555.
14. Chander, S., Gebhardt, J.E. : In Sastry, K.V.S., Fuerstenau, M.C. (Eds.) : Challenges in Mineral Processing. Littleton CO : Society of Mineral Engineers Inc. 1989.
15. Hochella, M.F. Jr. : In Hochella, M.F. Jr., White, A.F. (Eds.) Mineral Water Interface Geochemistry; Reviews in Mineralogy, Vol. 23. Washington : Mineralogical Society of America 1990, pp 87-131.
16. Marabini, A.M., Contini, G., Cozza, C. : Inter. J. Mineral Process. 38 (1993) 1-20.
17. Prosser, A.P. : In Proceedings of Ninth Commonwealth Mining and Metallurgical Congress, London 5-24 May 1969 (Mineral Processing and Extractive Metallurgy). London : Institution of Mining and Metallurgy 1970, pp 59-79.
18. Hagni, R.D. : In Hagni, R.D. (Ed.) : Process Mineralogy II. Warrendale PA : The Metallurgical Society of AIME, 1982, pp 29-38.

19. Chan, R.W., Haasen, E.J., Kramer, E.J. (Eds.) : Materials science and technology – characterization of materials, Vol. 2A. Weinheim (FRG) : VCH 1992.
20. Chan, R.W., Haasen, E.J., Kramer, E.J. (Eds.) : Materials science and technology – characterization of materials, Vol. 2B. Weinheim (FRG) : VCH 1994.
21. Hawthorne, F.C. (Ed.) : Spectroscopic methods in mineralogy and geology, Reviews in mineralogy, Vol. 18. Washington DC : The Mineralogical Society of America 1988.
22. Brundle, C.R., Evans, C.A., Wilson, S.L. (Eds.) : Encyclopedia of materials characterization : surfaces, interfaces, thin films. Stoneham MA : Butterworth-Heinmann 1992.
23. Cromia, R.D. : Adv. Mater. Process. 142(6) (1992) 16-23.
24. Bish, D.L., Post, J.E. (Eds.) : Modern powder diffraction, Reviews in mineralogy, Vol. 20. Washington DC : The Mineralogical Society of America 1989.
25. Zussman, J. : Mineral. Mag. 51 (1987) 129-138.
26. Champness, P.E. : Mineral. Mag. 51 (1987) 33-48.
27. Drits, V.A. : Electron diffraction and high resolution electron microscopy of mineral structures. Berlin : Springer-Verlag 1987.
28. Thomas, G. : Images of Materials. New York : Oxford University Press 1991, pp 143-173.
29. Lloyd, G.E. : Mineral. Mag. 51 (1987) 3-19.
30. Lloyd, G.E., Schmidt, N.H., Mainprice, D., Prior, D.J. : Mineral. Mag. 55 (1991) 331-345.
31. Brindell, J.B. : Adv. Mater. Process. 143(3) (1993) 20-27 (see also the references cited in the paper).
32. Heinrich, K.F.J., Newbury, D.E. (Eds.) : Electron probe quantitation. New York : Plenum Press 1991, 400 p.
33. Champness, P.E. : In Polt, P. et. al. (Eds.) Microanalysis Techniques in Geosciences. London : Chapman and Hall 1994 (in press).
34. Goldstein, J.I., William, D.B. : Microbeam Analysis 1 (1992) 29-53.
35. Wang, Z.L., Cowley, J.M. : Surface Sci. 193 (1988) 501-512.
36. Hunt, J.A., William, D.B. : Ultramicroscopy 38 (1991) 47-73.
37. Reed, S.J.B. : Mineral. Mag. 53 (1989) 3-24.
38. Metson, J. : In Perry, D.L. (Ed.) : Instrumental Surface Analysis of Geologic Materials. New York : VCH Publisher 1990, pp 311-352.
39. Elphick, S.C., Graham, C.M., Walker, F.D.L., Holness, M.B. : Mineral. Mag. 55 (1991) 347-356.
40. Williams, D.B., Soni, K.K., Tseng, M.W., Chabala, J.M., Setti, L. : J. Microscoy 169(2) (1993) 163-172.
41. Cabri, L.J. : Nucl. Instrum. Methods Phys. Res. B30(3) (1987) 459-465.
42. Sie, S.H., Ryan, C.G., Cousens, D.R., Griffin, W.L. : Nucl. Instrum. Methods Phys. Res. B40-B41(1) (1988) 690-697.
43. Beusen, J.M., Surkyn, P., Gijbels, R., Adams, F. : Spectrochim. Acta. 38B (1983) 843-851.

44. Chryssoulis, S.L., Stowe, K.G., Reich, F. : Trans. Instn. Min. Metall. (Section C : Mineral Process. Extr. Metall.) 101 (1992) 1-6.
45. Teo, B.K. : EXAFS : Basic principles and data analysis. Berlin : Springer-Verlag 1986.
46. Sperline, R.P., Muralidharan, S., Freiser, H. : Langmuir 3(2) (1987) 198-202.
47. Miller, J.D., Kellar, J.J., Cross, W.M. : In Chander, S., Klimpel, R.R. (Eds.) : Advances in Coal and Mineral Processing using Flotation. Littleton : SME Inc. 1990, pp 33-46.
48. Perry, D.L., Taylor, J.A., and Wagner, C.D. : In Perry, D.L. (Ed.) : Instrumental Surface Analysis of Geologic Materials. New York : VCH Publisher 1990, 45-86.
49. Bining, G., Quate, C.F., Gerber, Ch. : Phys. Rev. 56 (1986) 930.
50. Albrecht, T.R., Quate, C.F. : J. Vacuum Sci. Technol. A6 (1988) 271-274.
51. Iwatsuki, M., Kitamura, S. : JEOL News 28E(1) (1990) 24-28.
52. Ogletree, F., Salmeron, M. : Scanning tunneling microscopy and atomic structure of solid surfaces. Berkley : Lawrence Berkley Laboratory June 1991, pp 1-123.
53. Clarke, L.M. : Surface crystallography, an introduction to low energy electron diffraction. New York : John Wiley and Sons 1985, 329 p.
54. Van Hove, M.A., Weinberg, W.H., Chan, C.M. : Low energy electron diffraction : experiment theory and surface structure determination. New York : Springer Verlag 1986, 603 p.
55. Product Spotlight in Adv. Mater. Processes 143(3) (1993) 31-32.
56. Gangopadhyay, A., Ramachandran, T.R., Biswas, A.K. : Trans. Ind. Inst. Met. 37 (1984) 234-238.
57. Kuys, K., Ralston, J., Smart, R.St.C. : Mineral Eng. 3(5) (1990) 421-435.
58. Sen, R., Yarar, B., Spottiswood, D.J., Schowengerdt, F.D. : In Park, W.C., Hausen, D.M., Hagni, R.D. (Eds.) : Applied Mineralogy. Warrendale PA : The Metallurgical Society of AIME 1985, pp 441-466.
59. Lager, T., Forssberg, K.S.Ẹ. : Scand. J. Metall. 18(5) (1989) 251-257.
60. Rakesh Kumar, Ray, R.K., Biswas, A.K. : In Koch, M., Taylor, J.C. (Eds.) : Productivity and Technology in Metallurgical Industries. Warrendale PA : The Mineral Metals and Materials Society 1989, pp 561-586.
61. Rakesh Kumar, Ray, R.K., Biswas, A.K. : Hydrometallurgy 25 (1990) 61-83.
62. Rakesh Kumar, Das, S., Mohanty, O.N., Biswas, A.K. : Characterization and leaching studies on Ni, Cu bearing oxyhydroxides of tetravalent manganese. Report of Investigation 13490140, Jamshedpur : National Metallurgical Laboratory, July 1992, 52 pp.
63. Rakesh Kumar, Ray, R.K., Biswas, A.K. : Hydrometallurgy 32 (1993) 39-59.
64. Rakesh Kumar : Materials Transactions JIM 35(1) (1994) 27-34.
65. Page, M.L., Salinovich, T. : In Park, W.C., Hausen, D.M., Hagni, R.D. (Eds.) : Applied Mineralogy. Warrendale PA : The Metallurgical Society of AIME 1985, pp 583-597.
66. Rakesh Kumar, Biswas, A.K. : Hydrometallurgy 15 (1986) 267-280.
67. Rakesh Kumar, Ray, R.K., Biswas, A.K. : Trans. Ind. Inst. Met. 45 (1992) 287-301.

68. Rakesh Kumar, Das, S.K., Ray, R.K., Biswas, A.K. : Trans. Ind. Inst. Met. 47 (1994) 273-285.
69. Ramesh, R., Somasundran, P. : Coal Preparation 8(1/2) (1990) 1-16.
70. Tseng, B.H., Wert, C.A., Hausen, D.M., Hill, D.L. : Vassiliou, A.H., Hausen, D.M., Carson, D.J.T. (Eds) : Process Mineralogy VII. Warrendale PA : The Mineral Metals and Materials Society 1987, 67-80.
71. Wert, C.A., Hsieh, K.C., Buckentin, M., Tseng, B.H. : Scanning Microscopy 2(1) (1988) 83-96.
72. Ge, W., Wert, C.A. : In Orr, W.L., White, C.M., (Eds.) : Geochemistry of Sulfur in Fossil Fuel; ACS Symposium Series 429. Washington DC : The American Chemical Society 1990, pp 316-325.
73. Gao, A., Chen, X., He, K. : J. Central South Institute of Mining and Metallurgy (China) 22(6) (1991) 666-669 (In Chinese, Metal Abstract : 41-0164, July 1992).
74. Ye, X., Sun, Z., Chen, T., Ying, J., Zheng, N. : J. Chinese Electron Microscopy Society 11(2) (1992) 125-130 (Metal Abstract : 41-0257, November 1992)..
75. Bakken, B.M., Hochella, M.F.Jr., Marshall, A.F., Turner, A.M. : Econ. Geol. 84 (1989) 171-179.
76. Mckee, D.J., In Park, W.C., Hausen, D.M., Hagni, R.D. (Eds.) : Applied Mineralogy. Warrendale PA : The Metallurgical Society of AIME 1985, pp 103-116.
77. Jeulin, D. : Scanning Microscopy (Suppl.2) (1988) 165-183.
78. Choi, W.Z., Adel, G.T., Yoon, R.H. : Int. J. Mineral Process. 22(1-4) (1988) 59-73.
79. Reid, K.J., Pignolet-Brandom, S. : In Carson, D.J.T., Vassiliou, A.H. (Eds.) : Process Mineralogy VIII. Warrendale PA : The Metallurgical Society of AIME 1988, pp 369-377.
80. Gong, W.Q. : Int. J. Mineral Process. 37(1, 2) (1993) 89.
81. Austin, L.G., Sutherland, D.N., Gottlieb, P. : Mineral Eng. 6(5) (1993) 491-507.
82. Petruk, W. : 1994 (in this monograph).
83. Natarajan, K.A. : Int. J. Mineral Process. 34(1, 2) (1992) 161-175.
84. Rudashevsky, N.S., Lupal, S.D., Shuloyakov, A.D., Burakov, B.E. : In ICAM' 91 : International Congress on Applied Mineralogy, Vol. 2, Pretoria, South Africa, 2-4 September, 1991. Pretoria : CSIR, pp 47/1-9.
85. Jena, B.C., Panigrahy, S.C., Rigaud, M. : In Proceedings of W.O. Philbrook memorial symposium, Toronto, Ontario, Canada, 17-20 April, 1988. Warrendale PA : Iron and Steel Society, Inc. 1988, pp 67-75.
86. Malvik, T., Pilstrom, G. : In Forssberg, K.S.E. (Ed.) : Flotation of Sulfide Minerals. Amsterdam : Elsevier 1985, pp 221-237.
87. Bronsnaham, D.R., Jones, J.A. : In Vassiliou, A.H., Hausen, D.M., Carson, D.J.T. (Eds.) : Process Mineralogy VII. Warrendale PA : The Metallurgical Society of AIME 1987, pp 526-537.
88. Videnov, N., Shoumkov, S., Dimitrov, Z., Mogilski, L., Brakalov, L. : Int. J. Mineral Process. 39(3,4) (1993) 291-298.
89. Rao, M.K.Y., Natarajan, K.A. : Key Eng. Mater. 20-28(1) (1988) 403-416.
90. Dutrizac, J.E. : Hydrometallurgy 35(3) (1994) 275-292.

91. Pesic, B., Kim, I. : In EPD Congress' 90, Anaheim, California, USA, 19-22 February 1990. Warrendale PA : The Minerals Metals and Materials Society 1990, pp 133-160.

92. Ballester, A. : In Davies, G.A. (Ed) : Separation Processes in Hydrometallurgy. West Sussex (UK) : Ellis Horwood Ltd. 1987, pp 99-110.

93. Ballester, A., Blazquez, M.L., Gonzalez, F. : Canad, Met. Q. 28(4) (1989) 361-365.

94. Casey, W.H., Bunker, B., In Hochella, M.F. Jr., White, A.F. (Eds.) Mineral Water Interface Geochemistry; Reviews in Mineralogy, Vol. 23. Washignton : Mineralogical Society of America 1990, pp 397-426.

95. Casey, W.H., Banfield, J.F., Westrich, H.R., McLaughlin, L. : Chem.Geol., 105 (1993) 1-15.

96. Robinson, J.J. : Extraction of gold from eastern transvaal concentrate by roasting and cyanidation, Report No. M257D. South Africa : MINTEK 1986, pp 1-49.

97. Jones, M.H., Macrae, C.M., Sparrow, G.J., Woodcock, J.T. : AustIMM Bull. Proc. 293(6) (1988) 121-126.

98. Chryssoulis, S.L., Chauvin, W.J., Surges, L.J. : Canad. Met. Q. 25(3) (1986) 233-239.

99. Chryssoulis, S.L., Cabri, L.J., Salter, R.S. : In Salter, R.S., Wyslouzil, D.M., McDonald, G.W. (Eds.) : Proceedings of International Symposium on Gold Metallurgy, Vol. 2 Winnipeg, Canada, 23-26 August, 1987. New York : Pergamon Press 1987, pp 235-244.

100. Cabri, L.J., Chryssoulis, S.L. : Applied Geochemistry 6 (1991) 225-230.

101. Neumayr, P., Cabri, L.J., Groves, D.I., Mickucki, E.J., Jackman, J.A. : The Canadian Mineralogist 31 (1993) 711-725.

102. Ren, C.G., Zhou, S.J., Hu, W.M., Huang, F.Y., Tang, J.Y., Yang, F.J., Wang, K.R., Zhou, Y.Q. : Chinese Science Bulletin 37(5) (1992) 375-378 (In Chinese, Metal Abstract 23-0580, August, 1992).

103. Cabri, L.J., Laflamme, J.H.G. : In Park, W.C., Hausen, D.M., Hagni, R.D. (Eds.) : Applied Mineralogy. Warrendale PA : The Metallurgical Society of AIME 1985, pp 911-922.

104. Czamanske, G.K., Kunilov, V.E., Zientek, M.L., Cabri, L.J., Likhachev, A.P., Calk, L.C., Oscarson, R.L. : The Canadian Mineralogist 30 (1990) 249-287.

105. Wagner, F.E., Marion, Ph., Regnard, J.R. : In Gold 100, Vol. 2, Extractive Metallurgy of Gold. Johannesburg : SAMM 1986, pp 435-443.

106. Cathelineau, M., Boiron, M.C., Holliger, P., Marion, P. : In Bicentennial Gold 88. Melbourne : Morphet Press 1988, pp 235-240.

107. Swash, P.M. : J. S. Afr. Inst. Min. Metall. 88(5) (1988) 173-180.

108. Otten, M.T., Miner, B., Rask, J.H., Busek, P.R. : Ultramicroscopy 18 (1985) 285-290.

109. Rask, J.H., Miner, B.A., Busek, P.R. : Ultramicroscopy 21 (1987) 321-326.

110. Brown, G.E. Jr., Calas, G., Waychunas, G.A., Petiau, J. : In Hawthorne, F.C. (Ed.) : Spectroscopic Methods in Mineralogy and Geology, Reviews in Mineralogy, Vol. 18. Washington DC : The Mineralogical Society of America 1988.

111. Manceau, A., Combes, J.M. : Phys. Chem. Minerals 15 (1988) 283-295.

112. Manceau, A., Combes, J.M., Calas, G. : J. de Chimie Physique 86(7/8) (1989) 1533-1545.
113. Combes, J.M., Manceau, A., Calas, G., Bottero, J.Y. : Geochim. Cosmochim. Acta 53 (1989) 583-594.
114. Combes, J.M., Manceau, A., Calas, G. : Geochim. Cosmochim Acta 51 (1990) 1083-1091.
115. Manceau, A., Llorca, S., Calas, G. : Geochim. Cosmochim Acta 51 (1987) 105-103.
116. Manceau, A., Buseck, P.R., Miser, D., Rask, J., Nahon, D. : American Mineralogist 75 (1990) 400-404.
117. Stouff, P., Boulegue, J. : American Mineralogist 73 (1988) 1162-1169.
118. Manceau, A. : American Mineralogist 74 (1989) 1386-1389.
119. Manceau, A., Clas, G. : Clay Minerals 21 (1986) 341-360.
120. Calas, G., Manceau, A., Navikoff, A., Boukil, H. : Bull. Minéral. 107 (1984) 755-766.
121. Waychunas, G.A. : American Mineralogist 72 (1987) 89-101.
122. Hochella, M.F. Jr., Brown, G.E. Jr. : Geochim. Cosmochim. Acta 52 (1988) 1641-1648.
123. Somasundaran, P., Kunjapppu, J.T. : Miner. Metall. Proc. 5 (1988) 68-79.
124. Stipp, S., Hochella, M.F. Jr., Parks, G.A., Leckie, J.O. : Geochim. Cosmochim. Acta 52 (1992) 1941-1954.
125. Tossel, J.A., Vaughan, D.J., : The Canadian Mineralogist 25(3) (1987) 381-392.
126. Vaughan, D.J., Tossel, J.A., Stanley, C.J. : Mineral. Mag. 51 (1987) 285-293.
127. Richards, S., Vaughan, D.J. : Mineral. Mag. 53 (1989) 213-222.
128. Richards, S., Vaughan, D.J. : Mineral. Mag. 53 (1989) 223-229.
129. Laajalehto, K., Smart, R.St.C., Ralston, J., Suoninen, E. : Applied Surface Science 64 (1993) 29-39.
130. Hochella, M.F. Jr., Eggleston, C.M., Elings, V.B., Parks, G.A., Brown, G.E. Jr., Wu, C.M., Kjoller, K. : American Mineralogist 74 (1989) 1233-1246.
131. Eggleston, C.M., Hochella, M.F. Jr. : Geochim. Cosmochim Acta 54 (1990) 1511-1517.
132. Eggleston, C.M., Hochella, M.F. Jr. : American Mineralogist 77 (1992) 221-224 and 911-922.
133. Elggeston, C.M., Hochella, M.F. Jr. : American Mineralogist 78 (1993) 877-883.
134. Hochella, M.F. Jr., Eggleston, C.M., Elings, V.B., Thompson, M.S. : American Mineralogist 75 (1990) 723-730.
135. Johnsson, P.A., Eggleston, C.M., Hochella, M.F. Jr. : American Mineralogist 76 (1991) 1442-1445.
136. Dove, P.M., Hochella, M.F. Jr. : Geochim. Cosmochim. Acta 57 (1993) 1442-1445.
137. Hayes, K.F., Leckie, J.O. : J. Colloid Interface Science 115 (1987) 564-572.
138. Chisholm, C.J., Hayes, K.F., Roe, A.L., Park, G.A., Hodgson, K.O., Leckie, J.O. : In situ X-ray absorption study of Pb(II) and Co(II) sorption complexes at the γ-Al_2O_3/ water interface Report No 88/01. Stanford Synchrotron Radiation Laboratory 1988.

139. Mielczarski, J. : Int. J. Mineral Proc. 11 (1986) 179-194.

140. Mielczarski, J. : Colloids Surfaces 17 (1986) 251-271.

141. Leppinen, J.O., Basilio, C.I., Yoon, R.H. : Colloids Surfaces 32 (1988) 113-125.

142. Leppinen, J.O., Basilio, C.I., Yoon, R.H. : In Richardson, P.E., Woods, R. (Eds.) : Electrochemistry in Minerals and Metals Processing. Pennington NJ : The Electrochemical Society 1988, pp 49-59.

143. Leppinen, J.O., Basilio, C.I., Yoon, R.H. : Int. J. Mineral Process. 26 (1989) 259-264.

144. Miller, J.D., Keller, J.J., Cross, W.M. : In Chander, S. (Ed.) : Advances in Coal and Mineral Processing using Flotation. Littleton CO : SME 1990, pp 33-44.

145. Kellar, J.J., Cross, W.M., Miller, J.D. : Applied Spectroscopy 43(8) (1989) 1456-1459.

146. Kellar, J.J., Cross, W.M., Miller, J.D. : Applied Spectroscopy 43(8) (1990) 1508-1512.

147. Keller, J.J., Young, C.A., Kuntson, K., Miller, J.D. : J. Coll. Interface Sci. 144(2) (1991) 381-389.

148. Keller, J.J., Young. C.A., Miller, J.D. : Int. J. Mineral Proc. 35 (1992) 239-251.

149. Keller, J.J., Cross, W.M., Yalamanchili, M.R., Young, C.A., Miller, J.D. : Miner. Metall. Proc. 10 (1993) 75-80.

150. Kuys, K., Roberts, N.K. : Colloids Surfaces 24 (1987) 1-17.

151. Popov, S.R., Vuvcinic, D.R. : Int. J. Mineral Process. 34(4) (1992) 307.

152. Popov, S.R., Vuvcinic, D.R. : Int. J. Mineral Process. 35(1, 2) (1992) 85-100.

153. Cecile, J.L. : In Forssberg, K.S.E. (Ed.) : Flotation of Sulfide Minerals. Amsterdam : Elsevier 1985, pp 61-80.

154. Smart, R.St.C. : Mineral Eng. 8(7-11) (1991) 891-901.

155. Buckley, A.N., Woods, R. : Appl. Surf. Sci. 17 (1984) 401-414.

156. Buckley, A.N., Woods, R. : Aust. J Chem. 37 (1984) 2403.

157. Buckley, A.N., Woods, R. : Appl. Surf. Sci. 27 (1987) 437-452.

158. Buckley, A.N., Woods, R., Wonterlood, H.J. : Int. J. Mineral Proc. 26 (1989) 29-49.

159. Mielczarski, J., Suoninen, E. : Colloids Surf. 33 (1988) 231-237.

160. Page, P.W., Hazell, L.B. : Int. J. Mineral Proc. 25 (1989) 87-100.

161. Buckley, A.N., Woods, R. : Int. J. Mineral Proc., 28 (1990) 301-311.

162. Buckley, A.N., Woods, R., : Colloids Surfaces 53 (1991) 33-45.

163. Laajalehto, K., Nowak, P., Pomianowski, A., Suoninen, E. : Colloids Surfaces 57 (1991) 319-333.

164. Laajalehto, K., Nowak, P., Pomianowski, A., Suoninen, E. : Int. J. Mineral Proc. 37 (1993) 123-147.

165. Johansson, L.S., Juhanoja, J., Laajalehto, K., Suoninen, E. : Surf. Interface Anal. 9 (1986) 501-505.

166. Laajalehto, K., Johansson, L.S., Mielczarski, J., Anderson, S., Suoninen, E. : In Forssberg, K.S.E. (Ed.) : XVI Int. Miner. Prcess. Congr. Amsterdam : Elsevier 1988, pp 691-702.

167. Mielczarski, J., Suoninen, E., Johansson, L.S., Laajalehto, K. : Int. J. Mineral Proc. 26 (1989) 181-191.

168. Cases, J.M., de Donato, P., Kongolo, M., Michot, L.J. : In 3rd Internatonal symposium on beneficiation and agglomeration, Bhubaneswar, Orissa, India, 16-18 Jan. 1991. Bhubaneswar : Indian Institute of Metals 1991, pp 121-131.

169. Cases, J.M., Kongolo, M., de Donato, P., Michot, L.J., Erre, R. : Int. J. Mineral Process. 38(3, 4) (1993) 267-299.

170. Wang, X.H., Jiang, C.L., Xuan, D. : In Forssberg, K.S.E. (Ed.) : Electrochemistry and Mineral Processing III, St. Louis, Missouri, USA, 17-22 May, 1992. Pennington NJ : The Electrochemical Society Inc. 1992, pp 235-258.

171. Houot, R., Raveneau, P. : Int. J. Mineral Process. 35 (1992) 253-271.

172. Buckley, A.N., Hamilton, I.C., Woods, R. : In Forssberg, K.S.E. (Ed.) : Flotation of Sulfide Minerals. Amsterdam : Elsevier 1985, pp 41-60.

173. Garbassi, F., Marabini, A.M. : J. Chem. Soc. Faraday Trans. 82 (1986) 2043-2055.

174. Wang, D., Hu, Y., Li, B., Sun, S. : Nonferrous Metals (China) 43(3) (1991) 34-39 (In Chinese, Metal Abstract 41-0079, March1992).

175. Zachwiega, J.B., McCarron, J.J., Walker, G.W. Buckley, A.N. : J. Colloid Interf. Sci. 132 (1989) 462-468.

176. Acar, S., Somasundaran, P. : Miner. Metall. Proc. 7 (1990) 94-99.

177. Houchin, M.R. : In International Symposium on Beneficiation and Agglomeration, Bhubaneswar, Orissa, India, 17-19 Dec. 1986. Bhubaneswar : The Indian Institute of Metals 1986, pp 65-71.

178. Brinen, J.S., Greenhouse, S., Nagaraj, D.R., Lee, J. : Int. J. Mineral Proc. 38 (1993) 93-109.

179. Chryssoulis, S., Kim, J.Y., Stowe, K. : LIMS Study of Variables Affecting Sphalerite Flotation, Paper no 28. Presented in Canadian Mineral Processor Meeting, Ottawa, January 20, 1994.

180. Jiang, X., Lu, Y. : Nonferrous Met. (China) 4 (1987) 52-88 (In Chinese, Metal Abstract 410122, 1988).

181. Prestidge, C.A., Ralston, J., Smart, R.St.C. : Int. J. Mineral Proc. 38 (1993) 205-233.

182. Qi, G.W., Klauber, C., Warren, C.J. : Int. J. Mineral Process. 39(3, 4) (1993) 251-273.

183. Senior, G.D., Poling, G.W., Frost, D.C. : Int J. Mineral Proc. 27(3-4) (1989) 221-242.

184. Nagaraj, D.R., Brinen, J.S., Farinato, R.S., Lee, J. : Langmuir 8 (1992) 1943-1949.

185. Warren, G.W., Sohn, H.J., Wadsworth, M.E., Wang, T.G. : Hydrometallurgy 14(2) (1985) 133-149.

186. Hisky, J.B., Phule, P.P., Pritzker, M.D. : Metall. Trans. B 18B(4) (1987) 641-647.

Recent Developments in Process Mineralogy as Applied to Liberation Problems in Mineral Beneficiation

WILLIAM PETRUK
CANMET, Department of Natural Resources Canada,
555 Booth Street., Ottawa, Ontario, Canada, K1A OG1

ABSTRACT

Mineral liberation is an important characteristic determined by process mineralogy techniques when evaluating ores and mill products with respect to mineral beneficiation. The liberation analysis includes measuring grain sizes and textures to predict approximate grinds for liberating the minerals, measuring mineral liberations and mineral associations, and calculating recoveries of liberated and unliberated grains. Grain sizes and textures are measured in polished sections of unbroken ores by optical microscopy, scanning electron microscopy and image analysis; mineral liberations and mineral associations are measured in polished sections of mill products by SEM based image analysis systems using either linear intercept analysis or areal fraction analysis; and recoveries of liberated and unliberated grains are calculated by materials balancing techniques. The liberation data obtained by analysing ground material in a polished section is different than true liberation due to a slicing effect. A satisfactory mathematical correction has been developed for liberation measurements when linear analysis is used, but not for liberation measurements by areal fraction analysis. The values measured by areal fraction analysis, however, are closer to the true values than those measured by linear analysis, and can be used for routine analysis without correction. Qualitative mineral liberation measurements have been made for many years, by point counting using an optical microscope, and are still widely used.

INTRODUCTION

The continuing trend of processing low grade ores and of increasing demands for higher grade concentrates sustains the search for improving metallurgical efficiencies. An important step is to achieve optimum liberation of the minerals so that they can be concentrated. Applied mineralogy plays a major role by predicting the proportion of a mineral that can be liberated at a particular grind, measuring the percentage of mineral that has been liberated, and determining the proportions of liberated and partly liberated minerals that have been recovered by mineral beneficiation. The mineralogical analysis involves analysing unbroken ores to determine textures and size distributions of the minerals, analysing ground materials to determine liberations, and performing mineral-materials balances to determine recoveries. The texture analysis is performed with an optical microscope, scanning electron microscope, and an image analysis system, and the size distributions and mineral liberations are measured with an image analysis system [1, 2]. In some instances, where the mineralogist did not have access to an image analysis system, acceptable results were obtained by using techniques developed by image analysis, and making a limited number of measurements with an optical microscope [2, 3]

ANALYSING UNBROKEN ORES

Liberation Models for Predicting Grind by Analysing Unbroken Ores

Mineral processing engineers generally determine the grind for liberating minerals by a series of grinding and beneficiation tests, but for some ores the "grind" is difficult to establish. In contrast, approximate grinds for liberating minerals can be predicted by determining the size distributions of minerals in an unbroken ore by image analysis. Theoretical liberation models can be used to calculate the liberation that would be obtained in either the total sample [4] or in each sieved fraction of the ground ore [5-9]. The calculated liberation values are approximate and frequently do not correspond to measured liberations; probably because of assumptions in the liberation models, and because liberation is dependent upon the bonding between minerals and the grinding method.

Most theoretical liberation models (except Petruk, 1986 [4]) assume random breakage. However, the minerals in some ores are weakly bonded to their neighbours and break preferentially along grain boundaries, particularly during the early stages of grinding [10]. The minerals in other ores are strongly bonded and break both across (random breakage) and along grain boundaries (preferential breakage). Furthermore, the bond strength between minerals is different for each mineral pair and is stronger in recrystallized (metamorphosed) ores than in unmetamorphosed ores. Weakly bonded grains tend to separate readily during grinding and to produce liberated grains that are nearly the same size as the grains in unbroken ore. Well bonded grains break randomly, and liberation is not achieved until the grains are broken into smaller sizes than the grain sizes in unbroken ores. Laslett *et al.* [11] recommended that random breakage be rejected as a basis for predicting liberation.

Some metamorphosed base metal ores contain large recrystallized pyrite grains which are difficult to break during grinding. In contrast, unmetamorphosed ores do not contain recrystallized grains and less grinding is needed to achieve the same degree of liberation. In particular, it took 60 minutes to grind the Faro base metal ore, which contains large recrystallized pyrite grains, to

80% minus 53 μm rod mill discharge (RMD), and 23 minutes to grind the relatively unmetamorphosed Brunswick Mining and Smelting base metal ore to the same degree of fineness from the same sized RMD in the same 2 kg laboratory ball mill using steel balls [12]. Similarly, it was observed that coarse-grained magnetite in an iron ore from the Labrador Trough is recrystallized and is strongly bonded to quartz, but coarse-grained hematite is not recrystallized and is weakly bonded to quartz [13]. The coarse-grained hematite is readily liberated and easily concentrated into a high grade product. In contrast, the coarse-grained magnetite is not fully liberated at the coarse grind, and a high grade magnetite concentrate cannot be produced without regrinding.

A liberation model that uses preferential breakage along grain boundaries is not yet available. Nevertheless, a liberation model that can handle preferential breakage was proposed by Weedon [14]. Weedon's model involves shattering a specific weight of unbroken ore with a controlled force, such as a pendulum, and determining the mineral liberation in the resulting material. It is likely that because of the many uncertainties with respect to preferential and random breakage, Weedon's model presents the most realistic approach to predicting the actual grind for liberating the mineral of interest.

Texture and Size Analysis Related to Predicting Grind

Textures of minerals are studied and analyzed in unbroken ores to ascertain whether the grains are strongly bonded and would break randomly during grinding, or weakly bonded and would break preferentially along grain boundaries. Size distributions of the minerals are used in all liberation models, except in the Weedon's model [14], to predict the grind that would be required to liberate the minerals.

Texture analysis is performed somewhat differently for ores that are composed of a heterogeneous mixture of minerals, than for ores where the mineral of interest occurs as inclusions in a mass or matrix of another mineral. In both cases the results are used to predict how finely the ore must be ground to release the mineral of interest from the ore. The main measurements in texture analysis are grain size distributions of all the mineral grains in the ore and grain size distributions of the minerals of interest. The bond strength between the minerals of interest and its neighbour is classified as low, medium or high.

Analysing a sample composed of a heterogeneous mixture of minerals by image analysis involves:

(1) obtaining images of each mineral and of each particle;
(2) measuring size distributions of the mineral of interest and of the particles;
(3) evaluating whether the grains of the mineral of interest are weakly or strongly bonded to a neighbour. Grains that are weakly bonded have a relatively straight grain boundary, whereas strongly bonded grains have a sinuous boundary with many re-entrants. The irregularity of the grain

boundary can be determined qualitatively with an optical microscope, or quantitatively with an image analysis system.

Texture analysis of an ore where the mineral of interest occurs as inclusions in a matrix poses a problem for determining the size distributions of the matrix mineral grains, since polished sections of the matrix mineral do not display grain boundaries of the matrix mineral. The problem can be handled by image analysis if the grains are poorly bonded. For such grains, incipient fractures and pits are present along grain boundaries on the surface of the polished section, and can be used as a guide for an image analysis program that delineates the grains in the matrix mineral (Fig. 1). Size analysis can be performed on the delineated grains to predict the grind for separating the grains and exposing or releasing the inclusions.

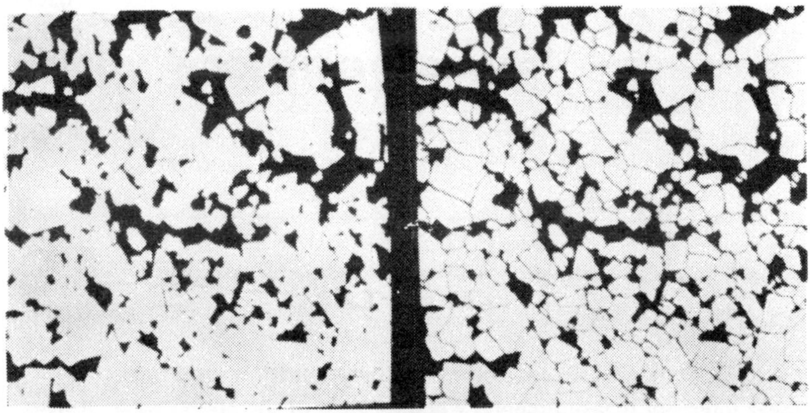

Fig. 1. Binary image of massive pyrite (white area, left diagram); after grain boundary reconstruction (right diagram)

ANALYSES OF GROUND MATERIALS

Measuring Mineral Liberations

Mineral liberations are measured by analysing crushed and ground materials (i.e. mill products). The procedure involves analysing polished sections of the ground material and measuring the proportion of mineral that is liberated, partly liberated, and unliberated. The standard procedure is to subdivide the mineral as occurring in particles that contain 0 to 10, 10 to 20, 20 to 30, 30 to 40, 40 to 50, 50 to 60, 60 to 70, 70 to 80, 80 to 90, 90 to 99.9 and + 99.9 area % mineral (% linear intercept if the liberation is performed by linear analysis). The data are obtained by measuring either the areal fractions or linear intercepts of the mineral with respect to the host particle. The results are commonly plotted as cumulative liberation yield on the basis of the areal fraction of the grain in particle category (Fig. 2).

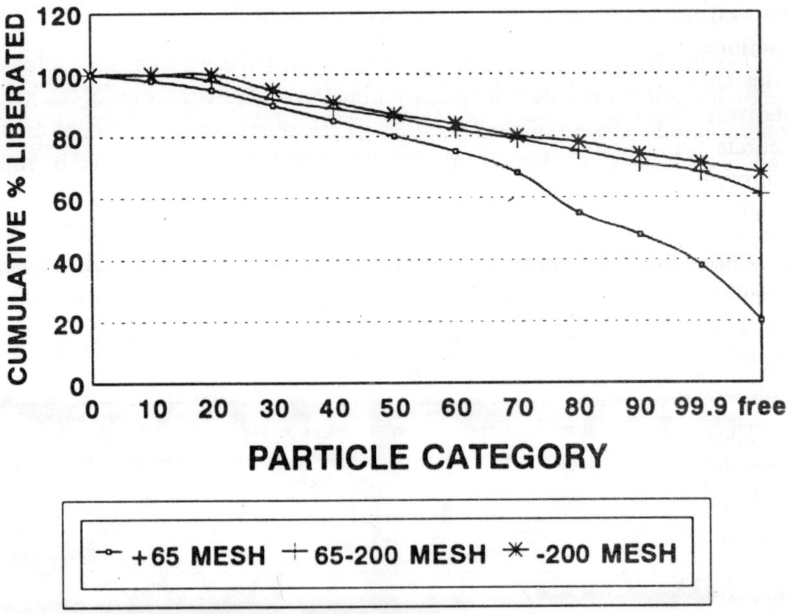

Fig. 2 Cumulative liberation yield for mineral in +65 mesh, 65 to 200 mesh and – 200 mesh fractions

Most of the theoretical studies on liberation analysis have been done with linear intercept measurements, whereas most of the analysis performed with respect to commercial operations are done by measuring areal fractions. Many of the small image analyzers on the market do not have the capability of measuring the full liberation spectrum by either linear analysis or areal fraction analysis. They can only measure the proportion of mineral that is totally liberated, and unliberated.

The measured data give apparent liberations, and present a bias towards amount liberated due to slicing effect that is produced by making polished sections. The polishing can slice a particle and make an unliberated grain appear like a free grain, but can not make a free grain appear unliberated.

King and Schneider [15], King [16] and King and Stirling [17] have shown that, for **linear analysis**, the apparent liberation for particles in every category (i.e. 0 to 10%, 10 to 20%, etc) is different from the true liberation, and it is necessary to make liberation corrections before using the data. King and Schneider [15] have developed and tested a mathematical correction for liberation data obtained by linear analysis.

An acceptable method for correcting measured liberation by **areal fraction** analysis has not yet been established. The author has determined by mineral-materials balancing techniques of mill products, using uncorrected data for liberated and unliberated grains, that the **error of liberation analysis by areal fractions** is relatively small. He used the same technique to test error of analysis using an empirical correction [18]. The empirical correction introduced a relatively large, unknown error of analysis. He concluded that it is more accurate to use uncorrected liberation analysis data when the liberation is determined by measuring areal fractions, than to use a correction that has not been fully tested.

Two unexpected anomalies commonly occur when the liberation is measured from a polished section of an ore; (1) the measured percentage of totally liberated grains is frequently anomalously low, and (2) the measured liberation for the minerals in finest sieved fraction, is lower than for the minerals in the coarser-grained sieved fractions.

The anomalously low liberation for totally liberated grains is observed for minerals that have irregular grain boundaries. The anomaly is observed only for measurements of totally liberated grains and is not observed for particles containing 90 to 99.9%, 80 to 90%, 70 to 80%, etc of the mineral of interest. It is interpreted that the anomaly is due to small amounts of minute inclusions of other minerals in particles that contain more than 95% of the mineral of interest. The problem is obviously related to ore textures, and is observed when essentially liberated grains retain a trace of the neighbouring mineral as an impurity.

The anomaly of measuring a low liberation in the fine-grained sieved fraction is also due to the texture of re-entrants of a neighbouring mineral. The finest-grained sieved fraction contains only small particles, many of which represent edges that have broken off larger grains. These edges commonly contain remnants of re-entrants with other minerals. These small re-entrants drastically reduce the percentage of liberated mineral when only small particles are measured. In contrast, the re-entrants on larger particles, which are observed in the coarser grained fractions, make a very small (insignificant) contribution to the degree of liberation of the particles.

Mineral Liberation and Optimum Recovery

From the point of view of mineral beneficiation the following liberation characteristics should be determined:

(1) percentage of mineral that is totally liberated (+99.9%),

(2) percentage unliberated but in types of particles that can be recovered by mineral beneficiation (> 70% mineral in particles),

(3) percentage unliberated but can be liberated by regrinding (> 20% and < 70% mineral in the particle),

(4) percentage unliberated and in too small grains to be liberated or recovered (< 20% mineral in the particle).

It has been observed that concentrates recover both totally liberated grains and particles that contain a high percentage of the mineral of interest. Therefore, the type of unliberated grain (i.e. particle with a high percentage of the mineral of interest) that is recovered by mineral beneficiation is related to the concentrate grade. In a typical study Petruk [19] observed that concentrates having grade equal to about 93% of the theoretical grade for the concentrated mineral contained particles with more than 90% of the concentrated mineral; concentrates grading about 83% of the theoretical grade, contained particles with more than 70% of the mineral of interest; and concentrates grading about 65% of the theoretical grade, contained particles with more than 50% of the mineral of interest. Therefore, it has been proposed that the term "liberated" be used to describe the type of particle that is recovered in a particular type of concentrate [19].

Recoverable particles are also size dependent. For example, spirals recover "liberated" iron oxides in the size range of about 100 μm to 600 μm in diameter; flotation cells recover "liberated" sphalerite between about 5 μm and 100 μm in diameter and "liberated" chalcopyrite between about 3 μm and 75 μm. An ore must therefore be ground to produce the "liberated" grains within a specific size range. Consequently **the optimum recovery** is defined as *the percentage of mineral that, in a properly ground product, is in "liberated" grains with respect to the required concentrate grade, and in the appropriate size range for the concentrating process.*

The types of particles recovered in concentrates can be determined by materials balance by analysing a suite of samples, including feed, concentrate, and tailings from a circuit, and calculating the recovery of each type of particle in the concentrate. Results for recovery of chalcopyrite in the copper concentrate from Trout Lake ore are given in Figure 3. They show that 90 to 100 percent of the free chalcopyrite is recovered in 4.7 μm to 75 μm particles; and in 3.3 μm to 37.5 μm particles containing 60 to 90 vol % chalcopyrite. The recoveries drop drastically for particles larger than 4.7 μm and containing less than 60 vol % chalcopyrite. Hence, for Trout Lake ore, the term "liberated" chalcopyrite applies to particles containing more than 60 vol % chalcopyrite. It is noteworthy that the minimum size for high recovery is 4.7 μm for free chalcopyrite particles.

Mineral Associations

The minerals with which unliberated grains are associated may have a bearing on selecting a concentration process. Therefore a technique of determining mineral associations needs to be used. At least four techniques have been used by the author:

(1) a binary minerals technique,
(2) dominant mineral in particle,
(3) average composition of average particle, and
(4) proportion of mineral that is in contact with another mineral.

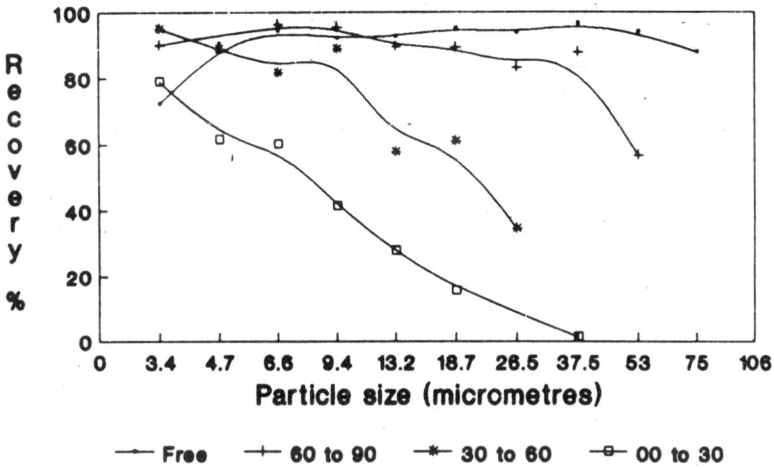

Fig. 3. Recoveries of chalcopyrite in Trout Lake first cleaner with respect to particle sizes for chalcopyrite in different types of particles

The binary mineral technique is used only in optical microscopy for manual grain counting. The approach is too imprecise to be used by image analysis because most particles that contain the unliberated mineral also contain inclusions of two or more other minerals [20]. The optical microscopist makes the judgement that the major other mineral is the other component of the so-called "binary particle" and counts it accordingly.

The dominant mineral technique is a variation of the binary technique, but is adapted to image analysis. The associated dominant mineral in the particle is defined as the mineral which, excluding the mineral of interest, accounts for more than 50% of the remaining minerals in the particle. The analysis involves establishing a series of classes, with a different dominant associated mineral for each class [20].

The "average composition of an average particle" technique is performed by producing images of particles from the following arbitrarily selected categories: 90 to 99.9%, 80 to 90%, 10 to 79.9%, and 0 to 9.9% of mineral of interest (or a variant of this classification) and performing a modal analysis on the images that contain each category of particles. The results give an average composition for each category of particle, and show the mineral associations for average particles from each category, i.e. for particles containing 90 to 99.9% of the mineral of interest, 80 to 89.9% of mineral of interest, etc. [20].

The proportion of mineral that is in contact with another mineral is determined by measuring the interface lengths between the minerals, and calculating the relative percentages of the interface lengths between the different minerals.

MATERIALS BALANCE

All the data produced by image analysis relates to quantities of minerals and in this respect is equivalent to assays. If the data were obtained for a suite of samples from a flowsheet, they can be used to calculate materials balances for each mineral category. Recoveries and weight units can be calculated for each mineral, for liberated grains of each mineral, for partly liberated grains, for unliberated grains, for particles or grains at different sizes, for mineral associations, etc.

Chemical assays are normally used to calculate the flow of material by MATBAL (computer program [21]), since the chemical assays generally represent the most reliable data. The materials flows are fixed and a BILMAT (computer program [22]) or MATBAL calculation is used to calculate recoveries of each mineral, of each category (degree of liberation) of the mineral, of sizes of each category of mineral, and of each association of mineral. The BILMAT calculation is preferable because it balances daughter data in relationship to parent data. For example, the mineral liberation and mineral association balances will relate to the mineral quantities balances, and the balance for sizes of liberated minerals will relate to the mineral liberations balance.

BEHAVIOUR OF AN ORE IN CIRCUIT OR CONCENTRATOR

When it is desirable to determine the behaviour of an ore in a circuit, a complete suite of samples needs to be analyzed. Representative samples of feed, concentrate, and tailings are taken from each cell over a period of time. The samples are analyzed chemically, and materials balances are performed to determine and fix the flow of material through each sample point. The samples are then analyzed with an image analyzer to determine mineral quantities, percentages of mineral as free grains, percentages of mineral in unliberated grains but in recoverable types of particles, percentages of mineral in unliberated particles that can be liberated by regrinding, and percentages of mineral that is too small to be liberated by regrinding. The size distributions of the mineral in each category of particle, as well as the size distributions of particles, are determined. These data are processed by a materials balancing program, either MATBAL [21] or BILMAT [22], and the proportion of mineral in different types of particles at each point in a flow-sheet are calculated [12, 23].

The recoveries of liberated, partly liberated, unliberated, and unrecoverable particles are determined according to grain sizes, and the data are presented with respect to the proportion of each category of particle and to the proportion of particle category in ore.

REFERENCES

1. Petruk, W. : In Petruk, W., Rule, A.R. (Eds.) : Process Mineralogy XII. Warrendale PA : The Minerals Metals and Materials Society 1994, pp 173-188.
2. Gaspar, O. : In Petruk, W., Rule, A.R. (Eds.) : Process Mineralogy XII. Warrendale PA : The Minerals Metals and Materials Society 1994, pp 241-258.
3. Bonifazi, G., Gorga, R : In Petruk, W., Rule, A.R. (Eds.) : Process Mineralogy XII. Warrendale PA : The Minerals Metals and Materials Society 1994, pp 259-272.
4. Petruk, W. : In Hagni, R.D. (Ed.) : Process Mineralogy VI. Warrendale PA : The Minerals Metals and Materials Society 1986, pp 393-403.
5. King, R.P. : Int. J. Mineral Process. 6 (1979) 207-220.
6. Finch, J.A., Petruk, W. : Int. J. Mineral Process. 12 (1984) 305-311,
7. Barbery, G., Leroux, D. : In Hagni, R.D. (Ed.) : Process Mineralogy VI. Warrendale PA : The Minerals Metals and Materials Society 1986, pp 415-427.
8. Klimpel, R.R. : In Petruk, W. (Ed.) : Process Mineralogy III. New York NY : SME/AIME 1984, pp 65-81.
9. Barbery, G. : Liberation measurement. Holland, Quebec Canada : G. Escher Heirn/Cordon Art-Baarn 1992.
10. Petruk, W. : In Carson, D.J.T., Vassiliou, A.H. (Eds.) : Process Mineralogy VIII. Warrendale PA : The Minerals Metals and Materials Society 1988, pp 181-193.
11. Laslett, G.M., Sutherland, D.N., Gottlieb, P., Allen, N.R. : Powder Tech. 60 (1990) 83-97.
12. Petruk, W. : In Proceedings of Conference on Evaluation and Optimization of Metallurgical Performance. New York NY : AIME/SME 1991, pp 181-191.
13. Petruk, W., Chung, G., Doyle D. : Evaluation of Spiral Performance for Processing Iron Ore from the Carol Lake Deposit by the Iron Ore Company of Canada, Paper no 19. In Proceedings of Canadian Mineral Processors 25th Annual Meeting. Ottawa : Can. Inst. Mining and Metallurgy 1993.
14. Weedon, D.M. : A model of mineral liberation in tumbling mills, Ph.D Thesis. Julium Kruttschnitt Mineral Research Centre, University of Queensland, Brisbane Australia 1992.
15. King, R.P., Schneider, C.L. : XVIII International Mineral Processing Congress (Australian Institute of Mining and Metallurgy) 1 (1993) 203-211.
16. King, R.P. : In Petruk, W., Rule, A.R. (Eds.) : Process Mineralogy XII. Warrendale PA : The Minerals Metals and Materials Society 1994, pp 189-200.
17. King, R.P., Stirling, P.A. : In Petruk, W., Rule, A.R. (Eds.) : Process Mineralogy XII. Warrendale PA : The Minerals Metals and Materials Society 1994, pp 201-212.
18. Petruk, W. : Trans. Instn. Min. and Metall. (Section C : Mineral Process. Extr. Metall.) 87 (1978) 272-277.
19. Petruk, W. : In Petruk, W., Hagni, R.D., Pignolet-Brandom, S. and Hausen, D.M. (Eds.) : Process Mineralogy IX. Warrendale PA. : The Minerals Metals and Materials Society 1990, pp 31-36.

20. Lastra. T., Petruk, W. : Determining Association of Unliberated Minerals in Ground Particles by Image Analysis. In Proceedings of Canadian Mineral Processors 26th Annual Meeting. Ottawa : Can. Inst. Mining and Metall. 1994.
21. Laguitton, D. : The SPOC manual, CANMET Report SP 95-1/3.2E, CANMET, Department of Energy, Mines and Resources, Ottawa, Canada 1985.
22. Hodouin, D., Flament, F. : BILMAT - Computer Program for Materials Balance Data Adjustment. In Laguitton, D. (Ed.) : SPOC Manual 3.1, CANMET, Department of Energy, Mines and Resources, Ottawa, Canada 1985.
23. Petruk, W. : An Image Analysis and Materials Balancing Procedure for Evaluating Ores and Mill Products to Obtain Optimum Recoveries, Paper no 19. In Proceedings of Canadian Mineral Processors 23rd Annual Meeting. Ottawa : Can. Inst. Mining and Metall. 1991.

Grinding Aids: A Review of their Use, Effects and Mechanisms

P. SOMASUNDARAN and S. SHROTRI
Henry Krumb School of Mines, Columbia University, New York, N.Y. 10027.

ABSTRACT

With the decrease in quality of ores as well as available energy resources there is a dire need for improving the energy efficiency of unit operations like comminution which is used extensively in a wide range of industries. One way to improve the efficiency of grinding mills and drilling operations is by the use of chemical additives. Comminution operations have been found to be influenced significantly by both the physical and the chemical nature of the grinding environment. Grinding aids, particularly some organic compounds can improve grindability by modifying the surface chemistry of the mineral and rheology of the pulp. Use of such aids can also be detrimental depending on the system and operating conditions. The past work on the effects and mechanisms of grinding aids is critically reviewed in this paper. The grinding aids have been proposed by one school to improve the comminution efficiency by assisting the propagation of cracks. Two major mechanisms have been put forward to explain this effect: Rehbinder's theory based on the adsorption induced surface energy changes and Westwood's mechanism based on the adsorption-induced mobility of near surface dislocations. According to the other school, the grinding aids act mainly by modifying the rheological characteristics of the mill charge. These mechanisms are discussed in the light of available literature data and key research opportunities are identified.

INTRODUCTION

Efficiency and Energy Consumption /Utilization Issue in Ore Grinding

Comminution is one of the oldest and most important industrial processes. It finds application in the processing of ores which involves size reduction, generation of large surface area and / or liberation of valuable minerals from their matrices. In addition, comminution is also an important unit operation in the construction industry, ceramic processing, and in various branches of paint, pharmaceutical and the chemical industries [2]. With the development of advanced technologies to produce new materials the potential field of application of the grinding process has substantially enlarged. At the same time the constraints have become more stringent in terms of energy consumption and purity specifications. However, the energy efficiency of comminution remains very low [1] and it is still the single most energy-intensive and least efficient unit operation in mineral processing [5]. Energy consumed in ore grinding in various mills as compiled by Hartely *et al.* [3] is shown in Table 1. The high energy consumed, ~70% of the whole beneficiation process, in ore grinding in various mills becomes evident upon examining Table 2 (reproduced from 1934 article [4]). Corresponding figures for the present operations can be expected to much higher. With the diminishing quality of ores and increase in the degree of fineness of values in the ore matrix, the need for fine and ultra fine grinding is steadily increasing. Consequently, the grinding costs as a proportion of total beneficiation cost are also increasing.

Table 1: Energy consumed in grinding at various mills (after Ref. [3])

Concentrate Produced	Plant	Energy (K Wh/t)
Cu	Calumet & Hecla, Michigan	8.5
Cu	Anaconda Co,. Montana	7.5
Cu	P.D. Lavendar Pit, Arizona	7.0
Cu	Mulfulira, N. Rhodesia	8.2
Cu	Silver Bell	9.1
Cu	Galena (Idaho)	11.2
Cu	Copper Cities	7.0
Cu-Fe	Cyprus Mines, Cyprus	10.9
Cu-Fe	Kerreti, Finland	10.7
Cu-Fe	Boliden, Sweden	13.3
Cu-Fe	San Manuel, Arizona	11.2
Cu-Ni	Kotalahti, Finland	10.5
Cu-Ni	Inco-Copper Cliff, Ontario	7.0
Cu-Ni	Inco-Creighton, Ontario	17.2
Pb-Zn, Cu-Zn, Zn	Tennessee Copper	6.2
Pb-Zn, Cu-Zn, Zn	St. Joseph Indian Creek	8.8
Pb-Zn, Cu-Zn, Zn	A.S. & R Buchans	15.8
Pb-Zn, Cu-Zn, Zn	A.S. & R Page	20.8
Pb-Zn, Cu-Zn, Zn	Idarado	9.8
Pb-Zn, Cu-Zn, Zn	Gilman	5.9
Pb, Zn	St. Joseph Lead, Montana	7.5
Pb, Zn	Broken Hill, Australia	4.1
Pb, Zn	New Broken Hill, Australia	6.3
Pb, Zn	New Broken Hill, Australia	7.0
Pb, Zn	Broken Hill, Australia	7.3
Pb, Zn	Soc. Algerienne, Morocco	7.9
Pb, Zn	Bunker Hill, Idaho	14.7
Mo	Climax Colorado	4.9
Au	Carlin Gold Mining Co.	7
Au	Benguet Consolidated	10.18
Au	Coretz Gold Mines	7
Au	Homestake Mining Co.	9.5
Au	Rosacio Domainicana, Domanican Rep	4.5
Au	Hogan- Suyoc Mines, Inc.	19.5

Table 2: Energy consumed in grinding in different plants compared with the total energy needed for the whole beneficiation (after Ref. [4])

Plant	Power consumed in grinding (% plant total)
A Copper ores	
Hayden	30
Harmony	50
United Verde	60
Copper Queen	35
B Lead and zinc ores	
Chief consolidated	38
Tybo	65
Hughesville	36
Morning	46
C Gold ores	
Porcupine	44

The basic process of comminution is the particle breakage. In the mill, particles are always stressed by contact forces, whether by contact with grinding media directly and or by contact with other particles in a bed of materials. Specific particle loading in different kinds of machine, such as jaw crushers, ball mills and impact mills, differ in number and direction of the contact forces and the deformation velocity [6]. Figure 1 is a schematic of the particle breakage process highlighting the interdependence of various phenomena [6]. In practice, however, the energy consumption is not only for particle breakage, but accounted for by other phenomena also. The flow of energy and utilization is illustrated in Figure 2 [7-9]. It can be seen that within the particles energy is expended for elastic deformation, plastic deformation, lattice rearrangements and mechano-chemical reactions, and new surface energy. External to the particles the energy is expended for friction between the particles and the grinding media as well as between particles and particles, sound energy, kinetic energy of the products, and deformation and wear of the grinding media [10]. The figures and the additional data given in Table 3 and Table 4 indicate that only a small fraction of the total energy input is utilized for the production of the new surface (fracture). A great proportion of the energy input (more than 75%) is lost as heat, probably due to friction, non-productive collisions, and elastic and plastic deformation [10]. However, as shown by Piret [11], the energy consumed for the plastic and elastic deformation might be necessary activation energy for subsequent fracture processes. If this is taken into account then the grinding efficiency might be considered to be 20-50% [9]. A better understanding of the various grinding sub-processes and their dependence on the environment in the mill is needed in order to identify suitable strategies for improving the grinding efficiency.

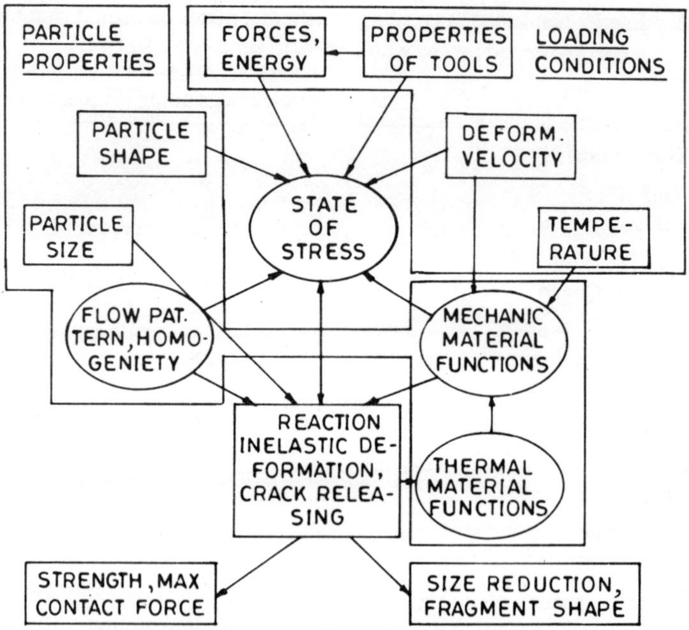

Fig. 1. Scheme of particle breakage (after Ref. [6])

Table 3: Experimental data of surface and crushing energies (after Ref. [9])

Material	Surface energy (erg/cm^2)	Crushing energy (erg/cm^2)
Quartz	920	117000
Glass	1210	8200
Halite	276	26100
Barite	1250	71600
Calcite	1100	32400

1 erg = 10^{-7} J

Table 4: Distribution of energy in ball mill (after Ref. [34])

Energy distribution	Energy consumption (%)
Bolt friction	4.3
Gear losses	8.0
Heat losses from drum	6.4
Heat absorbed by air circulation	31.0
Theoretical energy for size reduction	0.6
	97.6

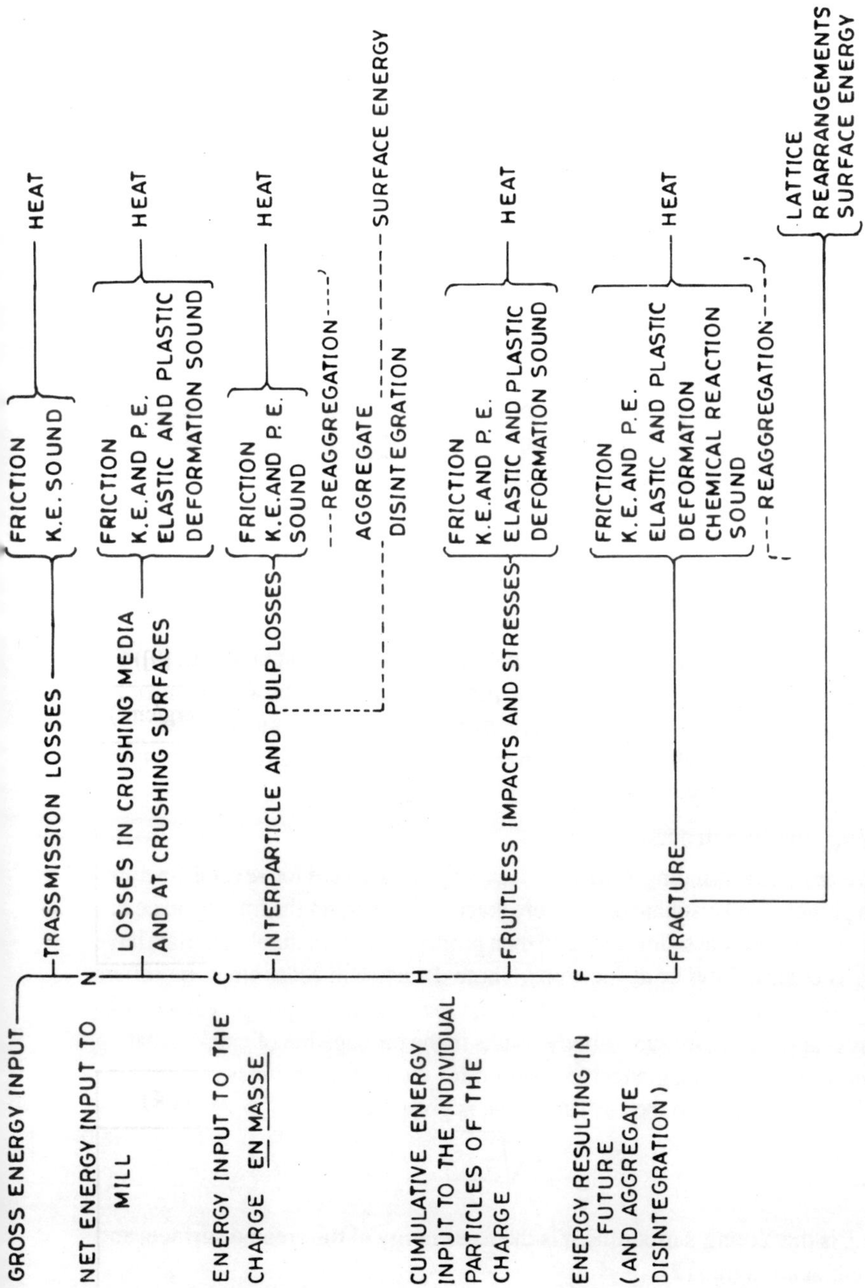

Fig. 2. Energy utilization in modern tumbling mill: gross energy > N > C > H > F.
(* residual kinetic energy after possible secondary fracture has occurred.) (after Ref. [7-9])

Grinding Component Processes

Grinding operations involve two major sub-processes, namely pulp flow and stress application. These processes include specifically transporting the material to and from the grinding zone, and subjecting the material to the grinding actions leading to initiation and propagation of cracks. Any parameter affecting any or all of these processes will affect the grinding efficiency [10].

Particle breakage can be considered to be a probabilistic event. Breakage can be achieved if the particle is captured in the grinding zone leading to a fruitful breaking action. Probability of breakage, P (overall breakage), is thus the product of probabilities for the above two basic sub-processes [3] :

P(overall breakage) = P(capture) * P(breakage upon capture)

It is expected that probability of particle capture in the grinding zone would depend, among other things, on the fluidity and the particle transport in the mill and the flocculation of particles. The probability of breakage, provided there is a fruitful capture, is directly related to mechanical properties of the particle such as strength, and toughness.

Pulp flow process

Pulp fluidity is an important parameter governing the transport of material inside a mill, which is affected, among other, by the state of aggregation or dispersion of the fine particles inside the mill. This is determined essentially by inter particle and particle-grinding media interactions that take place in the mill during grinding. The fluidity of the pulp can determine how well particles are transported to the regions where grinding action is most severe [3]. Modification of the pulp flow thus provides a potential means for improving the grinding efficiency.

Stress application process

The stresses in the grinding zone are essentially responsible for several fractures in each particles and resultant particle breakage. However, for the fracture to occur, cracks must be present or initiated and then propagated. All natural materials have defects as cracks, flaws or dislocations. These defects will act as stress modifiers leading to the strength values lower than the theoretically expected values [10]. The stress application on such a body results in the propagation of cracks creating new surfaces. The mathematical expression for the minimum stress required for the fracture, according to the Griffith's law, is given by

$$\sigma = \sqrt{\frac{4E\gamma}{L}} \tag{1}$$

where E is the Young's modulus, γ is the free energy of the created surface, and L is the crack length [12].

In real cases, however, some energy is also expended for the plastic deformation of material in the surface and the sub surface regions. Hence, energy consumed for plastic deformation in the vicinity of the crack should also be included in the Griffith's formulation [13, 14]. The radius of the crack tip is also an important parameter that might affect the strength of the material. The sharper the crack tip, the higher the localized stress concentration and hence there is reduction in the strength of the material. There also exists a possibility of interaction between the surrounding environment and the crack surface, leading to changes in the crack length and/or the radius of the crack tip, which in turn affects the strength of particles.

Grinding efficiency is very much dependent on the factors that affect the above component processes, but unfortunately, the present understanding of such processes and their interactions with the mill environment is limited and mostly intuitive or speculative at best [15].

We now consider some of the commonly used grinding aids and reported effects of the physico-chemical environment on mechanical properties and grindability of the materials and the mechanisms proposed for such observations.

EFFECT OF THE PHYSICO-CHEMICAL ENVIRONMENT ON MECHANICAL PROPERTIES AND GRINDABILITY OF MATERIALS

Rehbinder [33] were probably the first to investigate the effect of liquid media with and without additives on the failure of solids in comminution processes. They found that liquids, particularly water, played an active part in the process of failure and that this effect could be modified by suitable additives. This initiated various researchers to look at the effect of various additives on mechanical properties of solids and on comminution process, such as drilling and grinding.

GRINDING AIDS

As mentioned earlier, in the past, grinding has been considered as a physical process controlled only by the mechanical conditions of the grinding system. It was only after Rehbinder's [33] findings on the effect of chemicals in enhancing the mechanical failure of the material were published, that researchers began to study the comminution process in the presence of chemical additives. Such additives, generally some organic compounds, are termed as grinding aids. Lowrison [34], has compiled a list of such additives used in the grinding investigations together with the changes claimed (see table 5). The data suggests as much as a 20-fold increase in the grinding rate due to use of chemical additives.

Grinding aids based on their effect on environment of grinding

Dry grinding systems

A lot of effort has been directed in developing grinding aids for dry grinding systems. Most of such work is oriented towards the cement industry, where huge amount of clinkers are dry ground in ball mills. Cement clinker is particularly difficult to grind and fine grinding of this material is currently a major problem of the cement industry. Similar work has been done on the effect

Table 5: Additives used in comminution (after Ref. [34])

Additive	% added	Material ground	Wet or dry	Grinding rate factor*
Water	0.06	Ceramic enamels	D	–
	0.06	Marble	D	1.6
	0.04	Cement clinker	D	1.3
Alcohols and Phenols				
Methanol	–	Quartz		–
Isopentanol	–	Quartz		1.29
		Iron Powder		20.0
sec-Octanol	–	Quartz		1.4
Series normal alcohols	–	Soda lime glass	D	20.0
Glycerol	–	Iron powder		0.5
Phenols and polyphenols	0.01	Cement		–
	0.25			–
	–	Gypsum		–
Ketones				
Acetone	0.2	Cement clinker	D	1.37
Amines				
Triethanolamine	–	–		
Flotigan	0.02	Quartzite		2.2
	0.02	Limestone		1.7
Sulphonic acids				
Arylalkyl sulphonic acid (RDA)	0.06	Graphite		–
		Cement		1.3
Fatty acids				
Oleic acid	0.03	Limestone		1.1
		Zinc blende		–
Butyric acid	–	Quartz		1.27
Stearic acid	1.0	Pumice		–
	–	Limestone		–
Sodium oleate	0.1	Quartz		2.0
	0.1	Limestone		2.0
Vinsol resin (Calcium stearate)	0.05	Limestone		–
	0.1			
Sodium stearate	–	Dolomite		–
	0.15	Cement clinker		1.2
Aluminium stearate	–	Cement		–
Carpilic acid	0.5	Chrome ore-magnesite		1.2
Marine oil	0.5	Chrome ore-magnesite		1.1
Beef tallow	–			–

(contd.)

Table 5 (*contd.*)

Additive	% added	Material ground	Wet or dry	Grinding rate factor*
Wool Grease	5.0	Limestone		–
		Gypsum		–
Other carboxylic acids				
Napthenic acid	0.1	Cement clinker	D	1.33
Sodium napthenate	1.0	Quartzite	W	1.40
Sodium sulphonapthenate	1.0	Quartzite	W	1.80
Hydrocarbons				
n-alkanes, various	–	Soda lime glass	D	10.0
Esters				
Amylactetate	–	Quartz		1.23
Others				
Carbon black	0.08	Cement		1.3
	0.32	Cement		–
Sodium silicate	1.16	Clay slip		–
Sodium hydroxide	–	Magnesite		–
	0.008	Limestone		1.5
Sodium carbonate	0.02	Limestone		2.0
Sodium chloride	0.08	Quartzite		1.2
Carbon dioxide	–	Magnesite		–
	–	Dolomite		–
	0.03	Quartzite	W	1.55
Aluminium Chloride	–	Carbon black		–
		Graphite		–
		Talc		–
Aluminium carbonate	–	Mica		–
		Vermicullite		–
Haerdwood pitch	–	Pumice		–
Sodium polymetaphosphate (Calgon)	0.01- 0.04	Lead-zinc ore		1.3
Kaolin	10.0	Sulphur	D	–
Thalium chloride	0.02	Quartzite	W	1.65

* Grinding rate factor = (new surface produced with additive)/(new surface produced without additive)

of grinding aids on the fine grinding of limestone, quartz and Portland cement clinker, by Sohoni *et al.* [35]. They used seven grinding aids, namely triethanolamine, mono and diethylene glycols, oleic acid, sodium oleate, sulphite waste liquor and dodecylbenzene sulphonic acid. They concluded that while the effect on limestone grinding became appreciable with increasing fineness of the product, quartz grinding was not affected significantly, a mixed effect was observed for cement clinkers. Another work on dry grinding of calcite with the addition of fourteen liquids was conducted by Parmasivam and Vedaraman [36]. They found that the additive action was influenced by the nature of the additive, the size distribution of the mill hold up and the adsorption of additive molecules on particle surfaces.

Grinding in water

Wet grinding in general is more efficient than dry grinding [37-40]. This has been attributed mainly to the chemical reactions between broken surface bonds and water molecules [40]. Other than chemical parameters, investigators have also considered physical factors such as reduced cushioning in water of coarse particles, as well as effects of viscosity and the specific gravity [1, 15, 23-31, 41-52].

Grinding in organic liquids

There are several reports in the literature showing that grinding in organic media is more efficient than in water [54, 17, 43-48]. For example, work done by Kiesskalt [43] shows that with organic grinding media, there was a twelve fold increase in the surface area. Similar work done by Engelhardt [45, 46] using alcohol in quartz grinding showed less energy to be needed to produce the same surface area than that required for grinding in water. It was suggested [44, 47] that the vapors of the organic liquids reduced the forces of adhesion between the particles leading to reduced agglomeration and enhanced material flow which in turn improved grinding. Savage *et al.* [17] studied the grinding of silicon carbide in organic liquids. They obtained a 100% increase in the grinding rate in benzene in comparison with that of in the water alone. The effect of ethanol was found to be considerably lower than benzene.

Effect of surfactant addition

Several workers in the past have reported that surface active agents can influence the grinding efficiency. von Szantho [53] found, Flotigam P (C12-C14) amine from coconut oil to produce as much as a 100 % increase in the surface area of the quartz and a 75% increase in the surface area of limestone at reagent concentrations in the range of 0.005 to 0.02% (see Table 6). However, at additive concentration above 0.02% the effect was reduced. Table 6 shows other similar results. The observed decrease in surface area above a certain concentration of the additives could be due to experimental artifacts introduced by the possible aggregation of the ground product.

Similar work by Gilbert and Hughes [54] involved the study of effect of additions of Armac T (a mixture of saturated and unsaturated C16 and C18 amine acetate) on the ball milling of quartz. They found a 4 to 12% decrease in the amount of minus 200 mesh fraction produced in the pH range of 2 to 5, at a concentration of 0.01%. An increase in the surfactant concentration from 0.01% to 0.1% decreased the grinding efficiency further at all pH values; however, there was no explanation offered. Somasundaran and Lin [1] analyzed

Table 6: Effect of surface active agents on grinding of quartzite and limestone (after Ref. [54])

Feed	Additive	Additive conc. (%)	% increase in new surface
Quartzite	Flotigam P	0	0
		0.005	+ 103.2
		0.01	+ 114.7
		0.03	+ 123.4
		0.05	+ 71.4
		0.1	+ 65.8
Limestone	Oleic acid	0	0
		0.008	0
		0.0017	+ 7.3
		0.003	+ 10.6
		0.006	– 17.2
		0.06	– 61.3
		1.33	– 99.9
Limestone	Sodium oleate	0	0
		0.04	– 35.7
		0.08	– 1.8
		0.5	– 84.5
		1.0	– 92.4
Limestone	Flotigam P	0	0
		0.005	+ 29.2
		0.01	+ 56.4
		0.02	+ 74.6
		0.05	+ 9.6

these results and suggested that adsorption of surfactant Armac T (cationic) on the quartz surface (negative above pH 2) causes a decrease in the charge of the particles thereby leading to an increase in the flocculation of fines. In this regard, El-Shall *et al.* [55] have emphasized the importance of proper dispersion of the ground product

Another similar work reported by Carter *et al.* [56], on the influence of surfactant addition on the mechanical properties and comminution of the wet milled calcite. They concluded that near surface mechanical properties of calcite were affected by the external environment. The surfactants which increase the extent of cracking are most likely to be beneficial for enhanced rate of comminution.

Other reports of the effect of surfactant addition include that of Ryncarz and Laskowski [51], on the effect of dodecylammonium chloride on grinding of quartz. The effect of the additive was found to be dependent on its concentration and the solution pH. They also correlated their grinding results with the zeta potential of the quartz in amine solutions, and concluded that the grindability was minimum under conditions of point of zero charge, where the aggregation tendency is maximum. The possibility of surface contamination, in this particular case, by the aluminium species from the porcelain mill used, was, however, ignored by the authors.

El-Shall *et al.* [50] obtained finer products in the neutral and alkaline pH range, on adding dodecylammonium chloride during quartz grinding in a stainless steel ball mill. Detrimental effects were obtained in the acidic solutions, as shown in Fig. 3. Authors correlated their results with the effect of amine on the interfacial properties such as, zeta potential and flocculation characteristics, under chemical conditions simulated to take into the account of release of iron from the balls and mill during grinding. This is an important consideration since as shown in Figs. 4 and 5, ferric ions can even reverse the charge of quartz particles and consequently affect its flocculation properties.

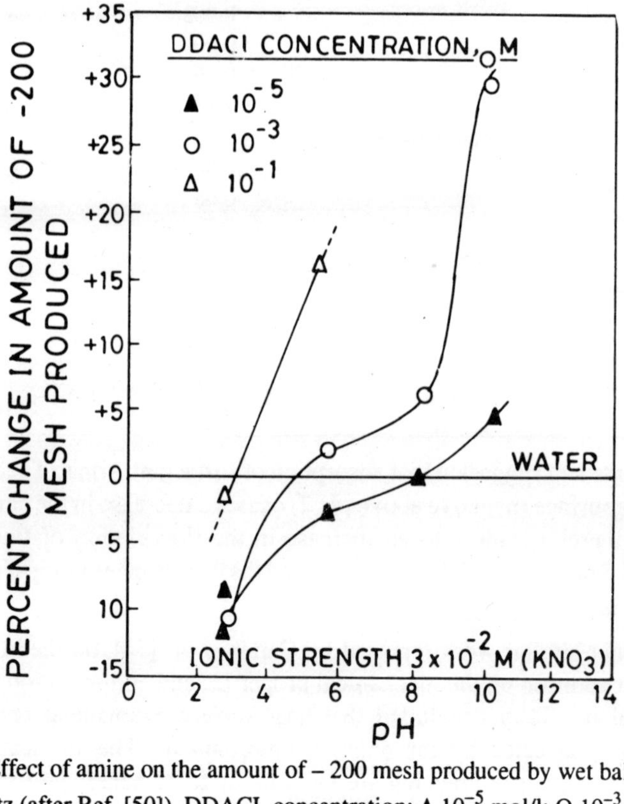

Fig. 3. Effect of amine on the amount of – 200 mesh produced by wet ball milling of quartz (after Ref. [50]). DDACL concentration: ▲ 10^{-5} mol/l; O 10^{-3} mol/l; Δ, 10^{-1} mol/l. Ionic strength: 3×10^{-2} mol/l (KNO_3)

Effect of addition of inorganic electrolytes to grinding environments

Several workers have reported the effect of inorganic electrolytes on grinding of minerals [53, 57, 58]. Although the reported results are often contradictory to each other, grinding has been found in general to be more efficient in the

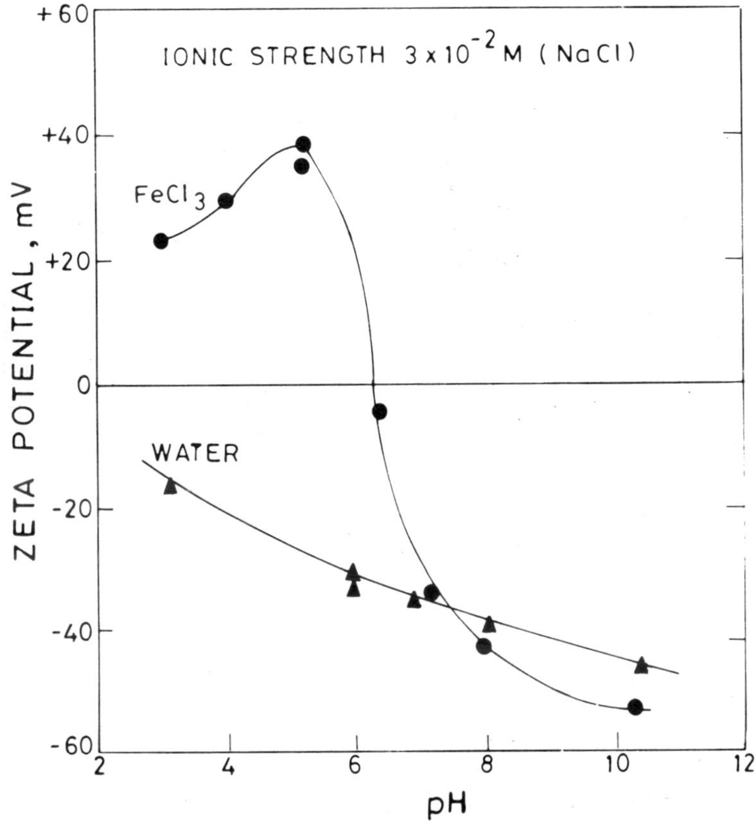

Fig. 4. Effect of 10^{-4} mol/l $FeCl_3$ on zeta potential of quartz as a function of pH (after Ref. [10]). ▲, water, •, $FeCl_3$. Ionic strength: 3×10^{-2} mol/l (NaCl)

presence of electrolytes than in its absence [59]. Kukolev and Melnishenko [60] studied the effect of caustic soda and soap on the ball milling of the magnesite ($MgCO_3$) and dolomite ($MgCO_3.CaCO_3$). They found caustic soda to be beneficial for the grinding of magnesite, but with no effect on the grinding of dolomite; a reverse effect was observed with soap. This behaviour was attributed to the difference in the adsorption of sodium hydroxide on the two minerals. However, Somasundaran and Lin [1] have noted that there is no reason for sodium hydroxide not to have the same adsorption capacity on the above two minerals.

Similar work has been done by Frangiskos and Smith [61] and some of it was repeated by Ghosh and coworkers [62], using sodium hydroxide and sodium carbonate in the grinding of limestone. Both results indicate beneficial effects except that Ghosh et al. report a maxima at 0.04% of additive dosage. The presence of such a maximum has also been reported by other workers [53, 63, 64], who obtained poor grinding upon increasing the additive concentration above certain levels.

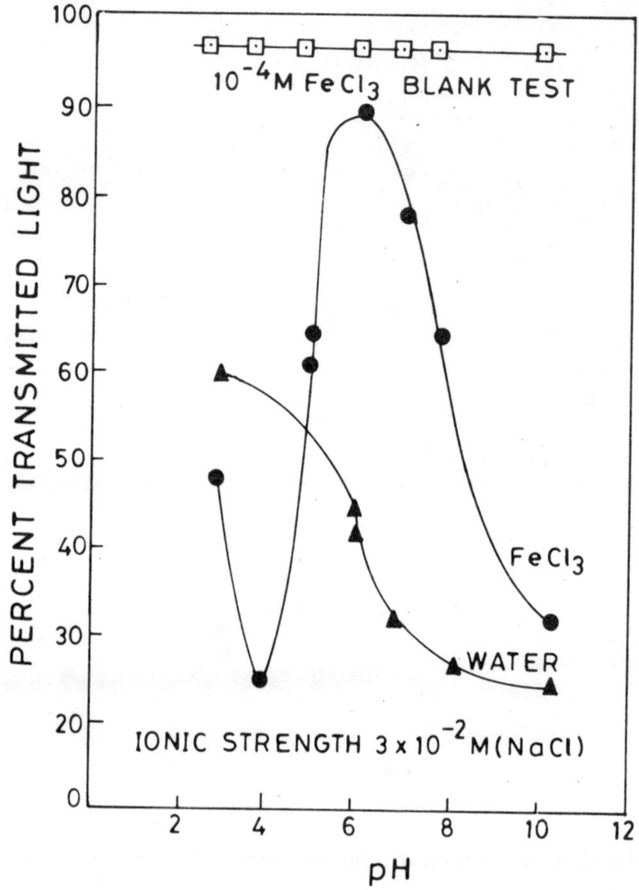

Fig. 5. Effect of 10^{-4} mol/l $FeCl_3$ on turbidity of quartz slimes on a function of pH (after Ref. [50]). ⊡, 10^{-4} mol/l; $FeCl_3$ blank test; ▲, water, •, $FeCl_3$. Ionic strength 3×10^{-2} mol/l (NaCl)

Savage et al. [17] studied the effect of sodium chloride and various salts of multivalent ions on the grinding of the silicon carbide, using water or ethanol as the grinding fluid. They classified the salts into flocculants (MgC12, CaC12, and FeC12, and FeC13) and dispersants (NaCl and Na4P2O7). The authors concluded that in both water and ethanol, dispersants improved the grinding rate significantly.

The effect of multivalent ions has been studied also by various workers. For example, Frangiskos and Smith [61] have reported as much as a 50% and 15% increase in the surface area of the ground quartz due to the addition of 2.0 mol/l $AlCl_3$ and $CuSO_4$, respectively. They also found iron-stained quartz crystals to grind faster than the clear ones, suggesting that ferric ions might have affected the fracture of quartz. Again it is stressed here that there was no indication as to whether pH or ionic strength was controlled.

Effect of the physical properties of the grinding environment

As mentioned earlier, one of the important component processes in the grinding mill is the transport of the material through it. The material transport in the mill depends primarily on pulp rheology. The rheology of the mineral suspension is highly complex and is difficult to predict, as there is no single parameter which can solely explain it [16]. Physical properties of pulp, such as pulp fluidity, chemical environment, particle size and shape, temperature of the environment, state of dispersion or flocculation of the pulp, density of the solids and density of the medium influence the rheology and hence the material transport [10,16]. Thus, such properties can be expected to have an effect on the hydrodynamic behaviour of particles as well as on the grinding media such as balls or rods, and consequently on the grinding performance [1].

It has been reported [17-22] that pulp viscosity can influence the grinding of ores. Schweyer [18] found grinding in pebble mills to be dependent on the viscosity of the medium up to a certain grinding duration, and then to be independent of it. Similarly Kawatra and Eisele [16], concluded the efficiency of the mill proper to be maximized when the percent solid is held just below the level which will produce a yield value in the output slurry and the viscosity reduced as far as possible. Another work by, Hockings *et al.* [19] studied the effect of pulp viscosity (using corn syrup to control the viscosity) on the rate of production of quartzite fines in a rod mill. They found an increase in pulp viscosity from 1 to 1000 cP to cause as much as a 50% decrease in the grinding rate when the grinding was conducted for 1 min. The observed effect was of a smaller magnitude for 2, 3, and 4 minutes of grinding (refer Fig 6). Additional

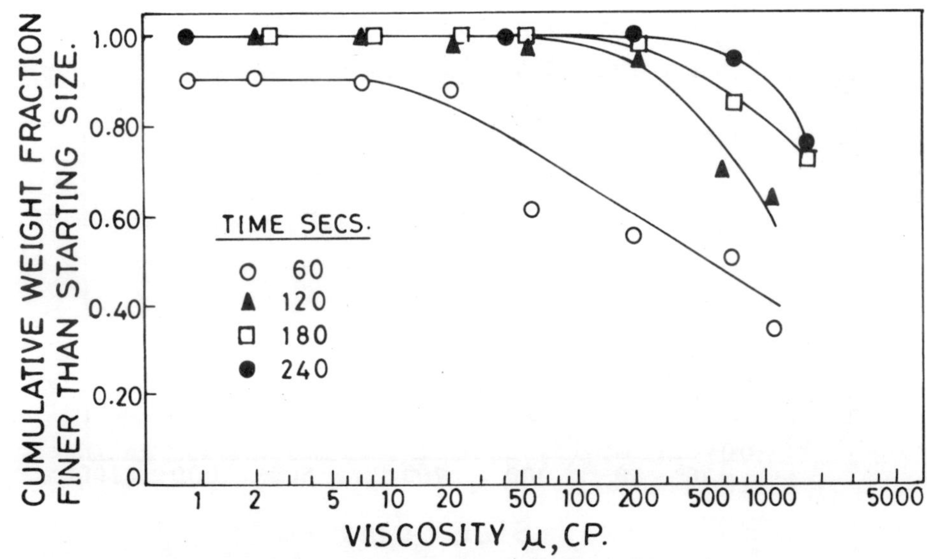

Fig. 6. Cumulative percentage finer than the starting size vs. viscosity for rod mill grinding of 10 × 14 mesh quartzite in corn syrup. water mixture for various grinding times; after Hockings *et al.* (after Ref. [19])

work is also done in this area by Savage *et al.* [17], who used liquids of different viscosity for grinding silicon carbide. The results indicated the grinding rate to be faster in the liquids of lower viscosity.

It is to be noted that the fluidity modifiers and dispersants have however been found to produce both an increase and decrease in the grinding efficiency [3, 21, 23, 24-31]. For example, Hartley, *et al.* [3] have obtained a marked decrease in the grinding efficiency of the molybdenum ore when they added sodium tripolyphosphate to their grinding circuits. Similarly Fuerstenau *et al.* [32] found that in wet grinding there is a critical pulp viscosity above which the grinding media "centrifuge" because the balls cannot be detached from the mill wall during the time of rotation. They also estimated the critical viscosity by analyzing the forces acting on the grinding media and found certain additives to be able to reduce the pulp viscosity to a value less than the critical viscosity providing conditions where the balls did not centrifuge so that the mills drew more power generating more fines. On the other hand, Hanna and Gamal [21] found that sodium silicate enhanced the grinding of limestone.

Klimpel and co-workers [24-31] have done a lot of work in identifying possible grinding aid chemicals. They have conducted several industrial and laboratory scale tests using fluidity modifiers. Some of the results are shown in Figs. 7 and 8. Figure 7 suggests an increase in the specific rate of breakage due to the grinding aid addition. Similarly, Fig. 8, dealing with the industrial

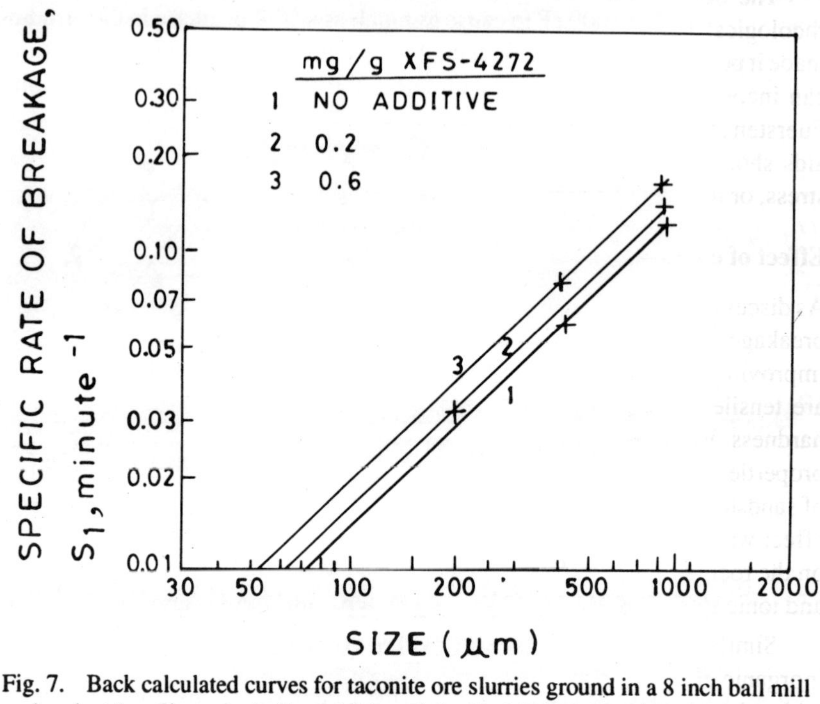

Fig. 7. Back calculated curves for taconite ore slurries ground in a 8 inch ball mill showing the effect of grinding additives (after Ref. [28]). All slurry densities at 84 wt. %.
+, experimental (one size fraction method): –, computer back calculation

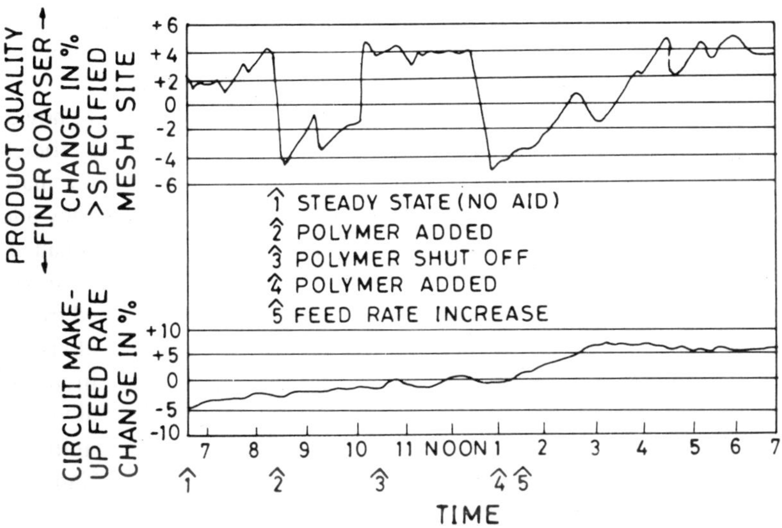

Fig. 8. Typical response curve for industrial-scale mill (after Ref. [28])

data shows that with the help of grinding aid, finer products could be obtained at approximately constant feed rate.

The behaviour of these reagents has been analyzed by correlating the rheological data and laboratory grinding results [30-31]. Such correlations have made it possible to find optimum slurry conditions when the chemical additives can increase rates of breakage. In this regard, Kawatra and Eisele [17] and Fuerstenau *et al.* [32] have concluded that chemicals that can work as grinding aids should maintain pseudo plastic behaviour in the slurry without a yield stress, or reduce the yield stress in a dense pseudo plastic slurry.

Effect of environment on mechanical properties of materials

As discussed earlier, one of the important grinding sub-processes is the particle breakage. There have been several studies indicating the role of additives in improving its efficiency. The key material properties for the particle breakage are tensile strength of the material, compressive strength, and the material hardness. Various workers have found the effect of additives on these material properties. Boozer *et al.* [65], found a decrease in the ultimate tensile strength of sandstones when the pores were saturated by oleic acid or oleylamine. This effect was attributed by the authors to the adsorption of surfactant molecules on the rock grains. However, no adsorption test itself was performed, also pH and ionic strength were not apparently monitored properly.

Similar work done by Wang [66], deals with the effect of organic and inorganic electrolytes on the compressive strength of sandstone's. They used dodecylammonium chloride, aluminium chloride and sodium carbonate as grinding aids and the results showed the effect to be pH dependent. Again, it was attributed to the adsorption of such additives on the solid surface, resulting in a reduction of the surface energy.

The decrease in the compressive strength of some rocks due to the presence of water has also been observed [67], see Table 7.

Table 7: Effect of water on the compressive strength of rocks (after Ref. [67]).

Type of rock	Compressive Strength (MPa)		
	Dry	Saturated	Saturated/dry
Basalt	168.5	84.79	50.3
Basalt with sandstone	64.71	28.39	44.0
Diabase	137.86	173.10	125.6
Granite	114.55	94.97	83.0
Gniess	155.48	106.72	67.7
Limestone	85.18	47.98	112.3
Quartzite	256.52	205.61	80.2

Vutkuri [68] has found decrease in the tensile strength of quartzite in the presence of aluminium chloride solution as compared to water, however the data showed a statistical variance of 30% which made conclusions difficult. Daryl *et al.* [69] working on a similar system, fused quartz thread and aluminium chloride, failed to find any effect on the tensile strength of the material.

Other studies have focused on the enhancement of fracture propagation with the addition of grinding aids. Carter *et al.* [56], have studied mechanical properties of wet milled calcite with the addition of surfactants. They have concluded that hardness and fracture of calcite can be affected by the external environment and that surfactants which increase the extent of cracking are particularly beneficial.

Similar work on the effect of environment on mechanical properties has been conducted by other investigators also [59, 70-79]. They have observed a decrease in fracture strength of glass like materials due to different environments. Some investigators [80, 81] have also considered the role of stress corrosion cracking.

In addition to the strength, material hardness is also affected by the surrounding environment [33, 82-85]. In this context, Rehbinder [33] classified hardness reducers into two groups. The first group contained inorganic salts such as NaCl, NaOH, Na_2CO_3, $MgCl_2$, $CaCl_2$ and $AlCl_3$, and second group contained organic chemicals with the polar molecules. Rehbinder found that with the first group of additives, the effect increases rapidly with the level of addition, reaching a maximum at a given level for each reagent. For the organic compounds, the effect depended mainly on the molecular weight of the compound used. For example, in a homologous series of organic acids, hardness reduction is observed with an increase in molecular weight [10].

Westwood and co-workers have done a comprehensive study on the effect of different environments on the hardness of MgO and CaF_2 [84, 85]. They found the hardness to increase or decrease depending on the surrounding environment. The authors ascribed the observed effects as adsorption-induced changes in the flow properties at the near surface region as controlled by the movement of surface dislocations.

Westwood *et al.* [86, 87] in another investigation correlated the observed effects with the surface charge properties of the materials in different environments.

Another study by Shchukin and Yushchenko used a molecular dynamic (MD) simulation of strain and failure of crystal, [88]. They looked at the environment sensitive mechanical behaviour of the materials on an atomic scale using MD simulation. The results indicated that adsorption of active atoms often caused brittle cracking. This was attributed to a sharp decrease in the surface energy of the solid in contact with the environment. Besides interacting with the cracked walls, the environment around atoms create a force compatible with the interatomic bond strength, which promotes crack propagation.

GRINDING AIDS: REPORTED MECHANISMS, AND DISCUSSIONS

On the basis of the above discussions, one can conclude that comminution process can, in general, be enhanced with the addition of grinding aids under selected conditions. Indeed, in order to fully develop and utilize such effects, it is necessary to establish the mechanisms involved in such effects. However, before considering the various mechanisms proposed for their action, it is important to keep certain aspects in perspective. Firstly, most of the past work done, lacked a strict control on the important experimental parameters like pH and ionic strength. Also, the size analysis is possibly questionable due to the formation of aggregate after contacting with surfactants and lack of proper dispersion. Also, due to the presence of surfactants, there can be foaming and this can lead to entrapment of the particles, thus rendering a reduction in grinding efficiency. It is hence necessary to be cautious of experimental artifacts, problems and misleading observations.

Focusing on the mechanisms proposed, let us first consider what is typically called 'Rehbinder's effect' [89]. It is based on the reduction of the surface free energy of solids induced by the adsorption of surface active agents. As fracture of the material involves new surface formation, the amount of energy required should be proportional to the surface free energy of the created surfaces [10]. If the surface free energy could be somehow reduced, then we can expect a decrease in the energy requirement for breakage. On this basis, adsorption of surface active agents during grinding should be expected to improve the efficiency of the process. However, as discussed earlier, the actual amount of energy needed to create a new surface, represents only a tiny fraction (~1%), of the total energy of the comminution system. In this regard, Moothedath *et al.*[90] have suggested that, according to this theory a reduction in the surface energy and a consequent reduction in the critical stress of crack propagation is expected in the presence of the grinding aids. However as pointed out by Somasundaran and Lin [1] and also by Moothedath *et al.* [90] the crack propagation velocities in the normal grinding practices are very high and the grinding aids may not be able to diffuse down to the crack tips at such high velocities.

The second mechanism, based on the adsorption-induced mobility of near surface dislocations was proposed by the Westwood *et al.* [84-87], who studied the effect of various additives on the hardness of different crystalline and non crystalline materials. They correlated hardness and drilling rates with near surface dislocation mobility and interpreted 'Rehbinder's effect' as a result of changes in the electronic states near the surface, and point and line defects caused by the adsorption of the additive on the surface of the solid. Such changes can influence specific interactions between dislocations and point defects which control the dislocation mobility and hence the hardness [10]. However, this model can not be used to explain the effect of grinding environment exclusively, because important parameters such as rheological properties of slurry are not considered.

Another altogether different mechanism, is based mainly on the role of reagents in dispersion [17, 23-31, 32, 38, 90] and, hence the flow of particles in the mill during comminution process. As mentioned earlier, the underlying principle is that, in the presence of grinding aids the flowability is improved, thus increasing the possibility of the particles reaching the grinding zone. This in turn increases the probability of capture of the particles between the grinding media.

Other researchers have proposed chemisorption or formation of additive complexes between the additive molecules and crack surface [1], are also the stress corrosion cracking [81] as responsible mechanisms.

CONCLUSIONS

It is clear from the above review that chemical additives can influence grinding for a number of reasons:

- modification of flow of pulp in the grinding mill;
- influence on the reagglomeration of the freshly produced fines; and modification of the frothing characteristics of the pulp during grinding;
- influence of interaction between mineral particles and balls and mill wall and among particles themselves (due to changes in frictional characteristics);

influence on the strength of material due to: (1) effect on crack initiation and crack extension energy, and (2) retardation of rejoining or sealing of the freshly created cracks.

Grinding is an integral process composed of several simultaneous sub-processes, viz. breakage, attrition, classification and transport of the material in the mill. Understanding of mechanisms associated with the grinding aids would essentially depend on identification of the effects on all the above mentioned factors and interactions between them.

REFERENCES

1. Somasundaran, P. and Lin, I.J. : I and EC Processes Des. Dev. 11 (1972) 321.
2. Larrison, G C : Crushing and grinding. London : Butterworth 1974.
3. Harteley, J.N, Prisbery, K.A. and Wick, O.J. : E/MJ October (1978) 105.
4. Gow, A.M., Guggenhiem, M. and Coghill, W.H. : Review of fine grinding in ore concentrators, I.C. 6757. Washington DC : US Department of Interior, Bureau of Mines 1934.
5. National Materials Advisory Board : Comminution and energy consumption. NMAB364,1981 Congress.
6. Schonert, Klauss : In Somasundaran, P. (Ed.) : Advances in mineral processing. Society of Mining Engineers, Inc. 1986, pp 19.
7. Harris, C. C. : Trans. Inst. Mining and Metallurgy 75 (1966) C50.
8. Rumpf, H. : Powder Technol. 7(1973) 145.
9. Orr, C. Jr. : In Particulate technology. New York : Macmillan 1966, pp. 44-86.
10. Shall, H.L. : D.E.Sc. Thesis, Columbia University, New York 1980.
11. Piret, E.L. : Chem. Eng. Progress 49 (1953) 56.
12. Griffith, A.A. : Phil. Trans. Roy. Soc. London, Ser. A 221 (1920-1921) 163.
13. Irwin, G.R. : Proc. ASM Symposium, Chicago, 1947, p. 147.
14. Orowan, E. : Rept Progr. Phys. 12 (1949) 185.
15. Somasundaran, P. : In Onoda, G. Jr. and Hench, L. (Eds.) : Ceramic Processing Before Firing. John Willey 1978, pp 105-123.
16. Kawatra, S.K. and Eisele, T.C. : In Forssberg, K.S.E. (Ed.) : Proceedings of the XVI International Mineral Processing Congress. Amsterdam : Elsevier 1988, pp 195.
17. Savage, K.I., Austin, L.G. and . Sun, S.C : Trans. AIME 255 (1974) 89.
18. Schweyer, H.E. : Ind. Eng. Chem., 34 (1942) 1060.
19. Hockings, W.A., Volin, M.E. and Mular, A.L. : Trans. AIME, 232 (1965) 59.
20. Clarke, B. and Kitchner, J.A. : British Chem. Eng. 13 (1968) 991.
21. Hanna, K. M. and Gamal, A.E. : Powder Technol. 17 (1977) 19.
22. El-Shall, H. and Somasundaran, P. : Powder Technol. 38 (1984) 267-273.
23. Revnivtsev, V.I. : In Forssberg, K.S.E. (Ed.) : Proceedings of the XVI International Mineral Processing Congress. Amsterdam : Elsevier 1988, pp 93.
24. Klimpel, R. and Manfroy, W. : Development of chemical grinding aids and their effect on selection-for-breakage and breakage distribution parameters in the wet-

grinding of ores. Paper presented at XII Int.Min.Proc.Cong. Sao Paulo, Brazil, 1977.

25. Klimpel, R.R. and Manfroy, W. : In Ramani, S. (Ed.) : In Proceedings of 14th International Symposium on Application of Computers in the Mineral Industry. New York : AIME, 1977, pp. 197-206.
26. Klimpel, R. R. and Manfroy, W : Ind. Eng. Chem. Process Des. Def. 17 (1978) 518
27. Klimpel, R.R. and Samuels, R. : In Proceedings of Annual Meeting of Canadian Mineral Processors. Ottawa : Can. Inst. Mining Metal. 1979.
28. Klimpel, R.R. : In Somasundaran, P. (Ed.) : Proceedings of Symposium on Fine Particles Processing, Las Vegas, 1979, New York : AIME 1980, p. 1129.
29. Katzer, M., Klimpel, R. and Sewel, J. : Mining Engineering 33 (1981) 1471.
30. Klimpel, R.R. and Austin, L.G. : Powder Technol. 31 (1982) 239.
31. Klimpel, R.R. : Powder Technol. 32 (1982) 267.
32. Fuerstenau, D.W., Venkataraman, K.S. and Velamakani, B.V. : Int. J. of Mineral Processing 15 (4) (1985) 251-268.
33. Rehbinder, P.A. : Physik 72 (1931) 191.
34. Lowrison, G.C. : Crushing and grinding. Clevland, OH : CRC Press Inc. 1979, p. 60.
35. Sohoni, S., Sridhar, R. and Mandal, G. : Powder Technol. 67 (1991) 277-286.
36. Parmasivam, R. and Vedaraman, R. : Powder Technol. 70 (1992) 43-50.
37. Bond, F.C : Mining Congress J. January (1975), 38-40.
38. Rose, H.E. and Sullivan, R.M.E : Ball, tube, and rod mills. New York : Chem. Pub. Co. 1958, p 30.
39. Somasundaran, P. and Fuerstanau, D.W. : Trans AIME 226 (1963) 132.
40. Lin, I.J. and Metzmager, A. : Trans AIME 241 (1968) 412.
41. Kapur, P.C., Muler, A.L. and Fuerstenau, D.W. : Can J. Chem. Eng. 43 (1965)119.
42. Oberson, H.J. and Brown, J.H. : Trans. AIME 220 (1961) 110.
43. Kiesskalt, S.Z. : Ver Deut. Ing. 91 (1949) 313.
44. Locher, F.W. and Seebach, H.M.V. : I and EC Process Des. Dev. 11 (1972) 190.
45. Engelhardt, W.V. : Nachr. Akad. Wiss. Gottingen, Math. Phys.Kl., No. 2 (1942).
46. Engelhardt, W.V. : Naturwisseenschaften 33 (1946) 195.
47. Fischer W. : Zem Kalk Gops 20 (1967) 138.
48. Schneider, H. : Zem Kalk Gops 22 (1969) 193.
49. Malati, M.A., Orphy, M.K and Yousef, A. A. : Powder Technol. 2 (1968) 21.
50. El-Shall, H., Gorken, A. and Somasundaran, P. : Effect of chemical additives on wet grinding of iron ore minerals. Paper presented at the XIII Int. Minl. Proc. Cong., Warsaw, June 1979.
51. Ryncarz, A. and Laskowski, J. : Powder Technol. 18 (1977) 179.
52. Fuerstenau, M.C., Eligillani, D.A. and Miller, J.D. : Trans. AIME 247 (1970) 11.
53. von Szantho, E. : Z. Fur Erzberglass unf Metallhuttenwesen, 2 (1949) 353.
54. Gilbert, L.A. and Hughes, T.H. : Symposium Zerkleinern 1, Verlag Chem. Duseldorf, Germany,1962, pp. 170-193

55. El-Shall, H., Vidarage, S. and Somasundaran, P. : International Journal of Mineral Processing 6 (1979) 105.
56. Carter, G.M., Henshall, J.L. and Wakeman, R.J. : Powder Technol. 65 (1991) 403-410.
57. Brown, J.H. : M.I.T. Progress Report: N.Y.O., 7172 (1955).
58. Stanzyk, M.H. and Feld, I.L. : U.S. Bureau of Mines Report 7168 (1968).
59. Joffe, A., Kripitschewa, M.W. and Lewitsky, M.A. : Z. Physik 22 (1924) 286.
60. Kukolev, G.V. and Melnishenko, L.G. : Fireproof Mater. (Moscow) 13 (1948) 447.
61. Frangiskos, A.Z. and Smith, H.G. : Trans. Miner. Dressing Congr., Stockholm, Sweden 1957, pp. 67 -84.
62. Ghosh, S.K., Harris, C.C. and Jowett A. : Nature 188 (1960) 1182.
63. Mallikarjunan, R., Pai, K.M. and Halasyamani, P. : Trans. Indian Institute of Metals 18 (1965) 79.
64. Halasyamani, P., Venkatachalam, S. and Mallikarjunan, R. : Ind. Eng. Chem. Process Des. Develop. 7 (1968) 79.
65. Boozer, G.D., Hiller, K.H and Serdengecti, S : In Fairhust, C. (Ed.): Rock mechanics. New York : Pergamon Press,1963, pp 579-625.
66. Wang, F.D. : Ph.D. Thesis, University of Illinois 1966.
67. Ruiz, M.D. : Proc. 1st Cong. Int. Soc. Rock 1 (1966) 115.
68. Vutukuri, V.S. : Trans. AIME. 252 (1972) 407.
69. Daryl, R.T., Tweeton, R.D., Engelmann, W.H., Savanick, G.A. and Johnson, D.I. : U.S.B.M., Rep. of Invest. No. 8186(1976).
70. Berdinnikov, V.P. ; Physik, Z. Sowjetunion 4 (1933) 397.
71. Gregg, S.J. : Surface chemistry of solids. New York: Reinhold 1951, p. 297, Ceram. Abstr. (Sept. 1952) 173e.
72. Benedicks, C. and Ruben, G. : Jernkontorets Ann., 129 (1945) 37, (Ceramic Abstract, February 1946, 32).
73. Baker, T.C. and Preston, F.W. : J. Appl. Phys. 17 (1946) 162.
74. Petch, N.J. : Phil. Mag. (8), 1(4) (1956) 331.
75. Moorthy, V.K. and Tooley, F.V. : J. Am. Ceram. Soc. 39 (1956) 215.
76. Johnston, T.L. and Parker, E.R. : Sixteenth ONR Tech. Report. Ser. 127, Issue No. 16, University of California, Mineral Research Laboratory, 1957.
77. Levengood, W.C. and Johnston, W.H. : J. Chem. Phys., 26 (1957) 1184 (Ceramic Abstract, October 1957, 2311).
78. Sherby, O.D. : ONR Tech. Rept., ONRL (1958) 34.
79. Hammond, M.L. and Ravitz, S.F. : J. Am. Ceram. Soc. 46 (1963) 329.
80. Weyl, W.A. : J. Am. Ceram. Soc. 32 (1949) 367.
81. Uhlig, H.H. : In Liebowtiz, H. (Ed.), Fracture: an advanced treatise. New York : Academic Press 1971, pp. 646-675.
82. Lichtman, V.I., Rehbinder, P.A. and Karpenko, G.V. : Effects of a surface active medium on the deformation of metals, H.M. Stationary Office, London, England, 1958.

83. Lichtman, V.I., Shchuken, E.D. and Rehbinder, P.A.: Physico-chemical mechanics of metals. Moscow, USSR : Academy of Sciences 1962.
84. Westwood, A.R.C., Goldheim, D.L. and Lye, R.G. : Phil. Mag. 16 (1967) 505.
85. Westwood, A.R.C. and Goldheim, D.L. : J. Appl. Phys. 39 (1968) 3401.
86. MacMillan, N.H., Huntington, R.D. and Westwood, A.R.C. : 7th Rept. to Office of Naval Research, Contract No. 0014-70-C-0300, July (1973) 15.
87. Westwood, A.R.C., MacMillan, N.H. and Kalyoncu, R.S. : Trans. AIME 256 (1974) 106.
88. Shchukin, E.D. and Yushchenko, V.S. : J. Materials Sc. 16 (1981) 313.
89. Rehbinder, P.A., Schreiner, L.A. and Zhiyach, K.F. : Hardness reducers in drilling- A physico-chemical method of facilitating the mechanical destruction of rocks during drilling. Akad Nauk, Moscow, USSR (1974), Trans. by Council for Scientific and Industrial Research (CSIR), Melbourne, Australia, 1948, p. 163.
90. Moothedath, S.K. and Ahluwalia, S.C. : Powder Technol. 71 (1992) 229-237.

An Appraisal of the Linear First-order Kinetic Model Based Ball Mill Design Correlations

V.K. GUPTA
Research & Development Centre,
Steel Authority of India Ltd., Ranchi 834 002

ABSTRACT

This paper reviews various ball mill design correlations that have been proposed to describe the variation of the grinding rate parameter with several important design and operating variables. It has been shown, using available experimental data on several mill-material systems, that the correlations lack sufficient generality and precision from the design point of view. These deficiencies have been attributed to two main factors : (i) the use of erroneous experimental and calculation procedures for determination of the S values, and (ii) the lack of a broad data base in respect of the number and types of materials, particle-size range, and combinations of the important variables studied. The current S parameter estimation technique is based on two basic assumptions : (i) first-order disappearance of the coarsest size class particles in the batch mode of grinding, and (ii) particle-size-distribution independent grinding rates for particles of all sizes, i.e. linear breakage kinetics. The validity of these two basic assumptions in general, and their significance in the context of model parameter estimation for precise design work have been discussed. Several examples, based on published experimental data and the author's own work, are included to conclude that the present technique of rate parameter estimation cannot be used for carrying out precise design work. The basic considerations for the development of a new methodology have been outlined.

INTRODUCTION

Over the period of last three decades, a number of researchers [1-3] have worked towards developing a procedure for carrying out ball mill scale-up and design based on the following linear kinetic model of the batch grinding operation [4]:

$$\frac{dM_1(t)}{dt} = -S_1 M_1(t) \qquad (1a)$$

$$\frac{dM_i(t)}{dt} = -S_i M_i(t) + \sum_{j=1}^{i-1} B_{i,j} S_j M_j(t), \; i > 1 \qquad (1b)$$

where M_i is the mass fraction of the particulate charge in the i-th discrete size interval defined by the openings of the two successive screens in a standard series (such as the Tyler "$\sqrt{2}$ screen aperture ratio" series); t is grinding time; S_i, the grinding rate parameter, denotes the fractional rate of loss of material from size interval i due to breakage [4, 5]; and $B_{i,j}$, the breakage distribution parameter, denotes the fraction of material leaving a coarser size interval j at any time that reports to size interval i [4, 5]. So far, a major part of the activity in this area has been the development of empirical correlations for describing the effects of particle load [6-11], ball load [10, 11], mill speed [1, 17], ball size [1, 3, 17, 20], mill diameter [2, 3, 14, 16, 18] and mill power [2, 16, 23-25] on the rate parameters, S_i, in the dry mode of operation. The scope of this paper is, therefore, restricted to only the dry operation rate parameter correlations.

A careful survey of the literature shows that most of the proposed correlations are based on S values determined for only one selected size class of a single mineral [6, 8-11, 13, 14, 16, 20, 23]. Not surprisingly, due to the lack of a broad data base, there have been several controversies in the past [3, 6-11, 18, 19, 21, 22, 24, 26], which remain unsettled even to date. This paper highlights some of the important points pertaining to the rate parameter correlations :

AN OVERVIEW OF VARIOUS CORRELATIONS

Particle Load Correlations

Many correlations have been proposed for describing the effect of particle load on the rate parameters [6-11]. Some of the more cited ones can be listed as :

$$S_i \;\alpha\; (a_i + b_i\, W^{-0.5}), \qquad \text{for all } W \qquad \text{(Corr. 1)}$$

$$S \;\alpha\; (a + b f_c^{-0.55}), \qquad 0.05 < f_c < 0.3 \qquad \text{(Corr. 2)}$$

$$S \;\alpha\; (a\, e^{-10.25\, f_c/J} + e^{-2 f_c/J}), \qquad f_c < 0.8\, J \qquad \text{(Corr. 3)}$$

$$S \;\alpha\; e^{-3 f_c/J}, \qquad 0.2\, J < f_c < 0.6\, J \qquad \text{(Corr. 4)}$$

where W is particle load; f_c and J denote the formal fractional filling of the mill volume by particles and balls, respectively (based on a particle-bed or ball-bed porosity of 0.4 [1]); and a, a_i, b and b_i are constants.

Correlation 1 was developed by Le Houillier *et al.* [8] based on the S values obtained by them for 14/20, 28/35 and 35/48 mesh size fractions of quartz [mill diameter D: 29 cm; mill speed (fraction of critical speed) ϕ_c: 0.7; J: 0.36]. This correlation was also verified for the S values reported by Mika *et al.* [6] for 7/9 mesh dolomite (D: 25 cm; J: 0.5; ϕ_c: 0.7) and the S values reported by Austin *et al.* [9] for 14/16 mesh anthracite (D: 19.5 cm, J: 0.2, ϕ_c: 0.65). However, in these two cases, a second set of constants a_i and b_i were used for the particle loads corresponding to $U > 0.8$, where U denotes formal fractional filling of the interstices of the ball-bed at rest, and is given by

$$U = f_c / 0.4\,J \qquad (2)$$

It should be noted that Corr. 1 is the only correlation that suggests nonuniform effect of particle load on grinding rate of different size fractions.

Correlation 2 was proposed by Austin *et al.* [9] based on von Seebach's data on 1410/1250 μm cement (*D*: 15 cm; *J*: 0.5) [9] and Austin *et al.*'s data on 14/16 mesh anthracite (*D*: 19.5 cm; *J*: 0.2) [9]. It was stated that the values of the constants *a* and *b* were independent of *J* over the range $0.2 < J < 0.5$.

Shoji *et al.* [10] carried out a detailed investigation in which the ball load was varied over the range $0.16 < J < 0.6$. A 19.5 cm mill was used for grinding 20/30 mesh quartz at 70% critical speed. Correlation 3 was proposed based on the data obtained in this study. It can be seen that unlike Corr. 2 this correlation is not independent of *J*, the ball load. Later, Shoji *et al.* [11] carried out experiments in a 61 cm diameter mill also. Based on the additional data acquired by them, correlation 4 was proposed [11]. It was stated that the variation of *S* with particle load predicted by this correlation over a somewhat limited range of *U* is similar to that of Corr. 3. Neither of these two correlations was verified for data reported by other investigators.

In their recently published book, Austin, Klimpel and Luckie [1] have selected Corr. 4 of Shoji *et al.* [11] over other correlations. Using eq. (2), this correlation can be rewritten as follows

$$S \;\alpha\; e^{-1.20\,U}, \qquad 0.5 < U < 1.5 \qquad \text{(Corr. 4)}$$

This form provides a better basis for comparing the effect of particle load at different levels of ball load. From the data presented in Table 1 it can be seen that the predictions of change in grinding rate with particle load, based on Corr. 4, vary considerably from case to case. This explains why so many different correlations have been proposed.

Table 1: Particle load effect: A comparison of correlation 4 predictions with the experimental results.

Material	Particle size (mesh)	ϕ_c	J	U_1	U_2	$S(U_1)/S(U_2)$ Exptl.	$S(U_1)/S(U_2)$ Corr. 4
Quartz [12]	14/20	0.7	0.32	0.38	1.54	5.2	4.0
			0.42	0.38	1.54	6.0	4.0
			0.32	0.38	1.15	2.9	2.5
			0.53	0.38	1.15	3.1	2.5
			0.21	0.77	1.54	2.1	2.5
	20/28	0.7	0.42	0.47	1.40	3.5	3.0
Cement [12]	28/35	0.7	0.42	0.62	1.40	2.9	2.5
Dolomite [6]	7/9	0.6	0.43	0.49	1.30	2.1	2.6
				0.49	0.81	1.3	1.5
Quartz [8]	14/20	0.7	0.36	0.44	1.32	3.2	2.9
	28/35	0.7	0.36	0.74	1.32	2.4	2.0

Ball Load Correlations

Based on the estimates of the S values obtained for 20/30 mesh quartz ground in a 19.5 cm diameter mill (ϕ_c : 0.7, lifter size : 2 mm), Shoji *et al.* [10] proposed the following correlation

$$S \; \alpha \; \frac{(2.8\, e^{-10.25 f_c J J} + e^{-2 f_c J J})}{(1 + 10\, J^5)}, \; J < 0.6, \; f_c < 0.8\, J \qquad \text{(Corr 5)}$$

In a later study, Shoji *et al.* [11] replaced the 2 mm lifters by more effective bigger lifters. Also, an additional larger mill of 61 cm diameter was used. The results for both the mills were fitted by the following empirical function

$$S \; \alpha \; \frac{(e^{-3 f_c J J})}{(1 + 6.6\, J^{2.3})}. \; 0.2 < J < 0.5, \; 0.2\, J < f_c < 0.6\, J$$

or

$$S \; \alpha \; \frac{(e^{-1.2\, U})}{(1 + 6.6\, J^{2.3})}, \; 0.5 < U < 1.5 \qquad \text{(Corr 6)}$$

The data given in Table 2 provides a comparison of Corr. 6 predictions with the experimental results reported by several researchers [12-14].

Table 2: Ball load effect:: A comparison of correlation 6 predictions with the experimental results

Material	Particle size (mesh)	ϕ_c	U	J_1	J_2	$S(J_1)/S(J_2)$	
						Exptl.	Corr. 6
Quartz [12]	14/20	0.7	0.77	0.21	0.53	1.53	2.01
				0.42	0.53	1.24	1.30
				0.21	0.42	1.24	1.55
			1.16	0.21	0.42	1.51	1.55
				0.32	0.42	1.39	1.27
Dolomite [13]	7/9	0.6	0.81	0.17	0.43	1.38	1.75
				0.30	0.43	1.31	1.38
Limestone [14]	8/10	0.6	0.81	0.26	0.43	1.35	1.50

Mill Diameter Correlations

All the researchers [1-3, 14, 16] have used a common functional form for describing the variation of grinding rate, S, with mill diameter, D:

$$S \,\alpha\, D^{\Gamma} \qquad \text{(Corr. 7)}$$

where Γ is a constant. Austin [15], Malghan and Fuerstenau [16], Herbst and Fuerstenau [2] and Austin *et al.* [1, 17] have tried to show, using S values determined from grinding data on a single size fraction of a single material, that the mill diameter exponent Γ has a general value of about 0.55. This value of Γ is said to be in agreement with the Bond's empirical correlation which states that for mills of same length/diameter ratio, the mill capacity increases as $D^{3.5}$ [1, 2]. However, in our investigation, which involved a detailed study of the grinding behaviour of six different minerals and ores (limestone, quartz, pyrochlore, a vanadium ore, a soft cement clinker and a hard cement clinker) in several laboratory and pilot-scale mills (0.29 m $< D <$ 0.76 m), the Γ value was found to be as low as 0.25 and as high as 1.35 [3, 18]. In view of the types of materials covered in the investigation [3, 18], it is not unreasonable to expect that for many more materials of general interest the Γ value can be significantly different from the proposed general value of 0.55.

It can easily be shown that for mills of same length/ diameter ratio, the mill capacity varies as $D^{3+\Gamma}$ [19]. Table 3 gives some data on the effect of error in Γ value on the mill capacity predictions based on this relationship [19]. By way of example, if a 0.6 m diameter pilot-scale mill is used to predict the capacity of a 3.6 m diameter industrial mill, for a 10% accuracy in the mill capacity prediction the error in the Γ value to be used in calculations should not be more than 0.05. Clearly, a universal Γ value is not acceptable for design work, and the Γ value for each material must be precisely determined.

Table 3: Effect of error in Γ on the mill capacity prediction

Absolute error in Γ	% error in the predicted mill capacity				
	Mill diameter scale-up factor (D/D*)				
	2	4	6	8	12
0.05	3.5	7.2	9.4	11.0	13.2
0.10	7.2	14.9	19.6	23.1	28.2
0.20	14.9	32.0	43.1	51.6	64.4

Ball Diameter Correlations

Gupta *et al.* [3] obtained relevant batch grinding data on ball diameter effect for quartz, limestone, a soft cement clinker and a hard cement clinker. Three mills of 0.29, 0.41 and 0.61 m diameter were used, and the diameter of the balls used ranged from 1.27 to 3.81 cm. Based on these data, the following correlation has been proposed by Gupta *et al.* [3]:

$$S \propto 1/(d - d^{\circ})^{\beta} \qquad \text{(Corr. 8)}$$

where d is ball diameter and d° and β are material characteristic constants. Values of the constant β have been reported to range from 0.19 to 0.55 and that of d° from 0.67 to 0.89. In their recent publications [1, 17], Austin *et al.* have mentioned the following correlation

$$S \propto 1/d^{\theta} \qquad \text{(Corr. 9)}$$

where θ is a constant. The value of θ for milling in general is in doubt. A θ value of 1.0 has been reported for quartz in a 0.6 m diameter mill and θ = 0 has been reported for a 0.2 m diameter mill [1, 17]. This correlation is actually based on very limited experimental data of Smaila [20].

It should be noted that the effect of ball size on *S* parameters is strongly material dependent. For example, if the quartz correlation is used for cement clinker, the predicted value of the ratio "*S* (2 cm balls)/*S* (4 cm balls)" will be 50% higher than the actual value.

Mill Speed Correlation

Based on the assumption that the *S* parameters vary with mill speed in the same way as the net mill power, Austin *et al.* [1, 17] have proposed the following correlation

$$S_i \propto \frac{\phi_c - 0.1}{1 + \exp(15.7(\phi_c - 0.94))} \qquad \text{(Corr. 10)}$$

It should be pointed out that this correlation has not been verified by actual determination of the *S* values at different mill speeds. Using some available published data [13, 14, 23] and our own data [21, 22] on variation of grinding rate with mill speed, a comparison of observed change in *S* values with that predicted by Corr. 10 was carried out for several mill-material systems. The data presented in Table 4 very clearly show that the basic assumption made by Austin *et al.* [1, 17] for obtaining this correlation is not valid in most cases. Mill speed effect on grinding rate is actually not uniform for all materials [21, 22]. Also, in general, as particle size decreases, mill speed effect on grinding rate becomes progressively less significant [21, 22].

Specific Power Correlation

Based on limited data obtained by grinding 7/9 mesh dolomite in a 0.25 m diameter ball mill, Herbst and Fuerstenau [23] proposed the following general correlation:

$$S \propto P/W \qquad \text{(Corr. 11)}$$

where *P* is net power input to the mill, *W* is particle load, and (*P*/*W*) denotes specific net power input to the mill. This correlation has been claimed to be

Table 4: Mill speed effect: A comparison of correlation 10 predictions with the experimental results

Material	Particle size (mesh)	J	U	ϕ_c^1	ϕ_c^2	$S(\phi_c^1)/S(\phi_c^2)$	
						Exptl.	Corr 10.
Dolomite [16, 23]	7/9	0.43	0.8	0.60	0.53	1.29	1.16
				0.70	0.53	1.57	1.37
Limestone [14]	8/10	0.43	0.8	0.60	0.50	1.41	1.25
				0.70	0.50	1.41	1.47
Clinker I [21]	14/20	0.42	1.0	0.70	0.55	1.31	1.31
	100/150					1.00	1.31
Clinker II [21]	28/35					1.18	1.31
	100/150					1.00	1.31
Limestone [21]	14/20					1.06	1.31
	20/28					1.00	1.31
Quartz [21]	10/14					1.63	1.31
	20/28					1.35	1.31
	100/150					1.22	1.31

independent of ball load, particle load, mill diameter and mill speed - with the only exception of the variable ball size [13, 14, 16, 23, 25].

Actual power measurements have shown that for small laboratory mills power varies as $D^{2.5}$ and for the industrial mills it varies as $D^{2.3}$ [1,14]. Therefore, according to Corr. 11, for small mills grinding rate should vary as $D^{0.5}$ and for the industrial mills as $D^{0.3}$ – uniformly for all materials. However, as has been discussed earlier in this paper, the effect of mill diameter has been observed to be considerably different for different materials. Therefore, it should be concluded that Corr. 11 cannot be used for mill diameter scale-up as a general correlation.

In the previous section, it was shown that, at constant particle load, S is not proportional to net power input to the mill. Therefore, it follows that Corr. 11 cannot correctly account for the mill speed effect also. In fact, it can be seen from the data of Herbst and Fuerstenau [23] in Tables 4 and 5 that, upto about 70% critical speed, grinding rate increases considerably faster than the net mill power input.

Table 5: Variation of grinding rate with specific power input for 7 × 9 mesh dolomite (after Ref. [23])

J	U	ϕ_c	S (hr^{-1})	P/W (kW/ton)	SW/P $(kW\ hr/ton)^{-1}$
0.43	0.81	0.53	28.6	15.9	1.80
		0.60	36.9	18.6	1.98
		0.70	45.0	20.7	2.18
		0.90	39.0	20.5	1.90

BASIC ASSUMPTIONS MADE IN CORRELATION DEVELOPMENT

In the previous section, it was demonstrated, using published data from various sources, that the correlations proposed by different investigators lack general applicability. Besides the fact that most of the correlations were proposed based on very limited experimental data, it is likely that the experimental and calculation procedures currently used for determination of the S values have certain important deficiencies. These procedures are based on two basic assumptions: (i) first-order disappearance of the coarsest size-class particles in the batch mode of grinding [1, 4-6] and (ii) particle size distribution independent grinding rates for particles of all sizes, i.e. linear breakage kinetics [1, 4, 6]. The validity of these two basic assumptions in general, and their significance in the context of model parameter estimation for precise design work is the subject of this section.

First-Order Disappearance Kinetics

The technique commonly used for a direct determination of the S values is based on an assumption that the disappearance kinetics for the top size in the mill feed are first-order, as represented by eq. (1a) with a time independent S_1 value [1, 6, 13]. Equation (1a) gives

$$\ln M_1(t) = \ln M_1(0) - S_1 t \tag{3}$$

and

$$S_1 = \frac{\ln M_1(0) - \ln M_1(t^*)}{t^*} \tag{4}$$

where t^* is the duration of the batch grinding experiment. Generally, the product obtained from the first test is ground sequentially several times, and, in accordance with eq. (3), the S value is obtained by fitting a straight line to the $\ln M_1(t)$ versus t plot [1, 14, 16, 20].

Figure 1 shows four first-order decay plots for 20/30 mesh quartz ground under different operating conditions. These plots, reproduced here from the thesis of Smaila [20], are typical of numerous such plots found in the literature [1, 6, 8, 10, 14]. It has been claimed, based on such graphical representations of the experimental data, that very frequently the first-order hypothesis is an excellent approximation of the grinding kinetics [1]. However, this is not confirmed by actual calculation of the rate values for different intervals of grinding time, as shown in Fig. 1. It can be seen that, in general, the grinding rate varies quite significantly from one time interval to the other. The average deviation from the "best fit" *S* value reported by Smaila [20] (given in the lower left hand corner of the figure, for each case) is about 15% for the data set 1, 20% for set 2, 7% for set 3 and 40% for set 4.

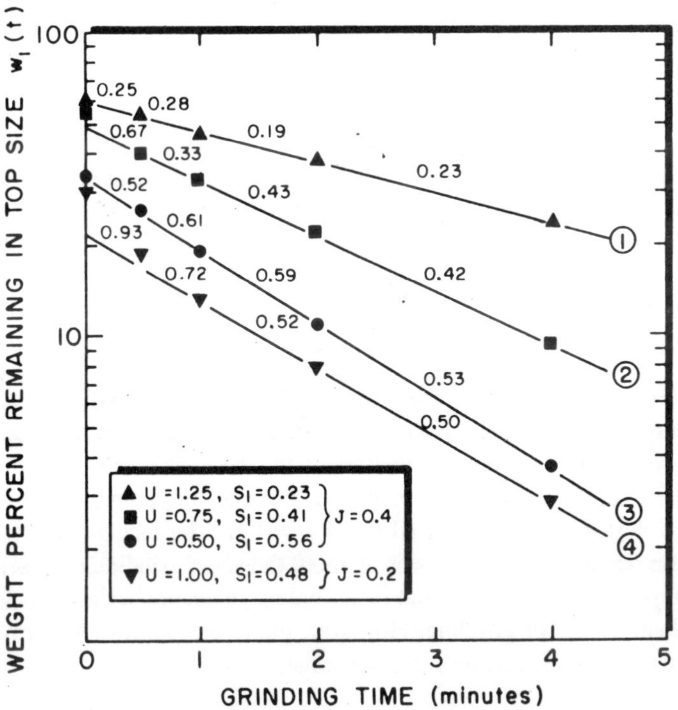

Fig. 1. Demonstration of first-order disappearance kinetics for 20 × 30 mesh quartz (after Ref. [20])

Further, it is to be noted that the choice of the "best fit" value is consistently closest to the rate value obtained for the last time-interval (i.e. 2-4 min period). As can be seen from the data presented in Table 6, for plot no. 2, 3 and 4, during the first two minutes more than 60% material breaks out of size-interval 1. Therefore, the *S* value obtained for the material left in size-interval 1 after two minutes of grinding (material which should belong mostly to the lower half size interval) can hardly be expected to provide a good approximation to the true *S* value for this size interval (i.e. the *S* value corresponding to a natural distribution of particle population over the size-interval).

Table 6: Representation of the data presented in Fig. 1 in the terms of the % reduction in M_1

Plot No.	% reduction in M_1				S_1^a	S_1^b	S_1^c
time (min):	0.5	1.0	2.0	4.0		(min $^{-1}$)	
1.	11.6	23.2	36.5	60.2	0.23	0.23	0.23
2.	28.6	39.5	60.7	83.1	0.50	0.43	0.41
3.	22.8	43.1	68.5	89.2	0.56	0.59	0.56
4.	37.2	56.1	71.9	90.5	0.93	0.72	0.48

a: refers to the first 40% reduction in M_1.

b: refers to the 40-60% reduction range.

c: refers to the value reported by Smaila [20].

Finally, it should be pointed out that the variation of grinding rate with time (or degree of size reduction) is not uniformly the same in all cases. In fact, besides the magnitude, the nature of variation also differs considerably from case to case. From the three sets of S values given in Table 6, it is expected that the correlations developed based on each set of data will be quite different from each other.

The data presented in Table 7 [20] provides a good illustration of some of the points discussed above. It can be seen that upto about 1 min grinding time, which corresponds to 60% reduction of the feed size, the grinding rate is practically independent of the ball load. Beyond this point, material breaks faster in the case of smaller ball load (J = 0.3). The fact that the first 60% reduction occurs in both the cases at about the same grinding rate is certainly more important than the fact that faster breakage occurs in the case of $J = 0.3$ in the later stages of grinding. However, the two S values given in Table 7 emphasize only the latter less important fact.

Table 7: A comparison of the disappearance kinetics for 20 × 30 mesh quartz ground under two different ball and powder loading conditions (after Ref. [20])

J	U		$M_1(t) \times 100$					S_1 [20] (min^{-1})
		t (min):	0.0	0.5	1.0	2.0	4.0	
0.3	0.75		55.5	41.0	33.3	17.5	6.6	0.53
0.4	0.75		55.3	39.5	33.4	21.7	9.3	0.41

Table 8 presents some disappearance kinetics data reported by Malghan [14]. It will be seen that most of the observations made above are valid for this data set also.

Table 8: Variation of grinding rate with degree of size reduction for 8/10 mesh dolomite ground under different operating conditions (after Ref. [14])

D (cm)	d (cm)	ϕ_c	J	U	S_1^a (min^{-1})	S_1^b (min^{-1})	S_1^c (min^{-1})	$S_1^{[14]}$ (min^{-1})
12.7	5.1	0.6	0.43	0.8	0.36	0.34	0.48	0.32
25.4	2.5	0.6	0.26	0.8	1.23	1.03	1.20	1.06
25.4	2.5	0.5	0.34	0.8	0.81	0.65	0.58	0.66
50.8	2.5	0.6	0.43	0.8	1.35	1.22	0.90	1.10

a: refers to the first 50% reduction in M_1.

b: refers to the 50-75% reduction range.

c: refers to the 75-90% reduction range.

LINEAR BREAKAGE KINETICS

It is widely believed that for given milling conditions, the S values are independent of the size distribution of the particulate material [1, 2, 6, 13]. Three main arguments have been given in support of this contention:

1. For a one-size-fraction starting feed, the S_1 value remains unchanged with time, though the size distribution of the particulate material changes considerably as grinding proceeds [1].
2. Radiotracing experiments show that the intermediate sizes also disappear at a constant rate in an environment in which the amounts of both smaller and larger sizes change with time [1].
3. Size distributions generated by batch grinding narrowly sized starting feeds can be accurately simulated using eqn. (1) with time-independent S and B values [1, 2, 6, 13, 16].

As shown earlier, the S estimates obtained from the disappearance kinetics data do not correspond to the natural size distribution within the size interval under consideration, and these estimates can be quite erroneous in some cases. Also, it can be seen from the published data [1, 2, 6, 8-11, 14, 20] that the size distributions generated by grinding narrowly sized feeds are not representative of the industrial milling conditions, and the size distribution range usually covered corresponds to a very small volume in the n-dimensional particle-size space. Systematic studies have been carried out at our laboratory to understand the dependence of the S values on the size distribution of the particle charge [28]. It has been found that, depending on the nature of the material being ground, the S values can vary considerably with the size-consist of the particle charge [28]. In the context of the main theme of this paper, two brief examples will suffice. Tables 9 and 10 give the relevant data for two materials : limestone and quartz. It may be mentioned that the feed charges were prepared by mixing

14/20 mesh material with –20 mesh material in different proportions. Experimental and computational procedures are described elsewhere [3, 27]. From the data presented in Table 10, it can be seen that the same 14/20 mesh size-fraction grinds in the natural size-distribution environment (corresponding to lower $M_{14/20}(0)$ values) at a considerably higher rate than that observed for it in the case of narrowly sized feed charge (corresponding to $M_{14/20}(0) = 0.89$). Similar results have been reported by Fuerstenau and Abouzeid [29].

Table 9: Size distribution of the –20 mesh feed charge

Material	$M_i \times 100$						
	Size interval (mesh):	20/28	28/35	35/48	48/65	65/100	–100
Limestone		30	22	9	9	9	21
Quartz		30	21	10	9	9	21

Table 10: Variation of grinding rate with size distribution for 14/20 mesh size-fraction

Material	$S_{14/20}$ (min^{-1})					
	$M_{14/20}$ (0):	0.89	0.47	0.32	0.17	0.09
Limestone		0.57	0.67	0.72	0.78	0.83
Quartz		0.47	0.51	0.55	0.60	0.60

DISCUSSION AND CONCLUSIONS

Precise determination of the rate parameter values is essential for the development of reliable correlations. This point is best illustrated by some examples given in Table 11 (see also Table 3). It will be seen that for most of the correlations S values must be estimated to an accuracy of at least ± 2%.

Table 11: Effect of error in the S values on the value of the constant 2.35 in Corr. 12, constant 1.2 in Corr. 6, exponent Γ in Corr. 7, and exponent θ in Corr. 9.

% error in S	Absolute error in the estimated value of the constant/exponent: Corr. 12*			Corr. 6			Corr. 7 or Corr. 9		
	(J_2-J_1):	0.1	0.2	(U_2-U_1):	0.5	1.0	$\frac{D_2}{D_1}$ or $\frac{d_2}{d_1}$:	1.5	2.0
2.5		0.50	0.25		0.10	0.05		0.12	0.06
5.0		1.00	0.50		0.20	0.10		0.25	0.12
10.0		2.00	1.00		0.40	0.20		0.49	0.25

* Corr. 12, an excellent approximation to Corr. 6, is:
$S \propto e^{-(1.2\,U + 2.35\,J)}$

The first-order disappearance kinetics hypothesis based rate parameter estimation scheme cannot be recommended for the purpose of correlation development because of three main reasons. First, it has been shown in this paper that the said hypothesis is not correct. Very significant variations can occur in grinding rate with time. In some cases, these variations are observed to be cyclic in nature, while in others they are systematic. Second, as it has been shown earlier, the S values estimated by different researchers based on the first-order disappearance hypothesis show a definite personal bias. Generally, the lowest S value observed during the batch grinding test is selected as the true value for which no rational justification exists. Third, when the size fraction of interest is ground as the top-size for a long time, the strength distribution and the size distribution of particles within this size-interval change quite significantly along with the overall size-distribution of the particulate charge. Obviously, the grinding rate value determined under these conditions has no relevance to the actual operating conditions. These important aspects of parameter estimation are discussed in detail elsewhere [27].

The data presented in Table 10 clearly shows that the S value obtained by batch grinding a nearly mono-size feed can be significantly lower than that obtained by grinding the same size-fraction in its more natural particle size-distribution environment (the one corresponding to the contents of the industrial mill). Further, it may be emphasized that the effect of some design and operating variables is significantly particle size-distribution dependent. For these reasons, the ball mill scale-up and design cannot be based on the S values determined by one-size-fraction method.

NOMENCLATURE

$B_{i,j}$	breakage distribution parameter
d	ball diameter
d°	a constant in Correlation 8
D	mill diameter
f_c	formal fractional filling of the mill volume by the particle charge based on particle-bed porosity of 0.4
J	formal fractional filling of the mill volume by the ball charge based on ball-bed porosity of 0.4
M_i	mass fraction in size interval i
P	net mill power
S_i	grinding rate parameter for size class i
t, t^*	grinding time
U	formal fractional filling of intersticies of ball-bed at rest
W	particle load

β an exponent in Correlation 8

Γ an exponent in Correlation 7

θ an exponent in Correlation 9

ϕ_c mill speed expressed as fraction of the critical speed

REFERENCES

1. Austin, L.G., Klimpel, R. and Luckie, P.T. : Process Engineering of Size Reduction: Ball Milling. New York : SME/AIME 1984.
2. Herbst, J.A., and Fuerstenau, D.W. : Int. J. Miner. Process. 7 (1980) 1.
3. Gupta, V.K., Zouit, H., and Hodouin, D. : Powder Technol. 42 (1985) 199.
4. Gupta, V.K. : Ph.D. Thesis. Indian Institute of Technology, Kanpur 1975.
5. Gupta, V.K., and Kapur, P.C. : Dechema Monographien 79, Nr. 1576-1588. Weinheim : Verlag Chemie 1076, p. 447.
6. Mika, T.S., Berlioz, L.M. and Fuerstenau, D.W. : Dechema Monographien 57, Nr. 993-1026. Weinheim : Verlag Chemie 1067, p. 205.
7. Gupta V.K., and Kapur P.C. : Powder Technol. 10 (1974) 217.
8. Le Houillier, R., and Marchand, J.C. : Powder Technol. 17 (1977) 101.
9. Austin, L.G., Shoji, K., and Luckie, P.T. : Powder Technol. 14 (1976) 71.
10 Shoji, K., Lohrasb, S., and Austin, L.G. : Powder Technol. 25 (1980) 109.
11. Shoji, K., Austin, L.G., Smaila, F., Brame, K., and Luckie, P.T. : Powder Technol. 31 (1982) 121.
12. Gupta, V.K. : The effects of ball and particle loading on grinding rate in dry ball milling (unpublished work).
13. Herbst, J.A., and Fuerstenau, D.W. : Trans. AIME 2523 (1972) 169.
14. Malghan, S.G. : D. Engg. Thesis. University of California, Berkeley 1976.
15. Austin, L.G. : Ind. Eng. Chem. Process. Des. Develop. 12 (1973) 121.
16. Malghan, S.G., and Fuerstenau, D.W. : Zerkleinern, Dechema Monographien 79, Nr. 1576-1588. Weinheim : Verlag Chemie 1976, p. 613.
17. Austin, L.G., Klimpel, R.R., Luckie, P.T., and Rogers, R.S.C. : In Mular, A.L., and Jergensen, G.V.(Eds.) : Design and Installation of Comminution Circuits. New York: SME/AIME 1982, p. 301.
18. Gupta, V.K., Hodouin, D., Remillard, M., and Vachon, R. : Ball mill scale-up in the context of pilot plant test work (unpublished work).
19. Gupta, V.K. : In Finch, J. (Ed.) : Proceedings of Symposium on Challenges in Mineral Sciences, 23rd Annual Conference of Metallurgists. Quebec : Canadian Institute of Mining and Metallurgy 1984, p. 1.
20. Smaila, F.M. : M.S. Thesis, The Pennsylvania State University 1982.
21. Gupta, V.K. : In Proceedings of the 17th Canadian Mineral Processors Conference Ottawa : Canadian Institute of Mining and Metallurgy 1985, p. 639.
22. Gupta. V.K. : The effect of mill speed on grinding rate in dry ball milling (unpublished work).
23. Herbst, J.A. and Fuerstenau, D.W. : Trans. AIME 254 (1973) 343.

24. Hiyame, H., Majima, M., and Fujinaka, Y., : CIM Bull. September (1974) 62.
25. Herbst, J.A., Siddique, M., Rajamani, K., and Sanchez, E. : Trans. AIME 272 (1982) 1945.
26. Gupta, V.K. : A reappraisal of various empirical correlations for the scale-up design of ball mills (unpublished work).
27. Gupta, V.K. : Determination of breakage rate and distribution parameters for scale-up design of ball mills (unpublished work).
28. Gupta, V.K. : The effect of particle size distribution on grinding rate in dry ball milling, 116th AIME Annual Technical Meeting. Denver : AIME 1987, Preprint No. 87-152.
29. Fuerstenau, D.W., and Abouzeid, A.Z.M. : Int. J. Miner. Process. 31 (1991) 151.

Feldspar Flotation: Theory and Practice

K. HANUMANTHA RAO and K.S.E. FORSSBERG
Division of Mineral Processing,
Luleå University of Technology, S-97187 LULEÅ, Sweden.

ABSTRACT

Over the years a considerable amount of study has been devoted to the flotation principles governing the separation of feldspars from quartz. Several reviews have been published on the flotation of silicate minerals as such and to the authors' knowledge, there is not single comprehensive review dealing with the subject of feldspar flotation. The present paper is therefore, an attempt to provide an indepth analysis of the literature on the flotation studies of the isomorphous group of alkali feldspars and plagioclases.

In general, feldspars are separated in fluoride ion medium using cationic collectors in acidic circuit. There is no general agreement on how the fluoride ions activate feldspar surfaces for amine cation adsorption and not quartz surfaces. A critical analysis on the role of fluoride, based on chemical thermodynamic equilibria, has been presented. The use of HF is no longer acceptable due to the associated environmental and health problems, and quite a few new reagents schemes have been reported. The available evidences show that the success of these new reagent schemes is very limited. A new reagent scheme which is environmentally safe and cost effective for the separation of feldspars from quartz is yet to be resolved.

INTRODUCTION

Feldspars are found under a large variety of geological conditions, and are more widespread than any other mineral group. They occur in granites and pegmatites, together with quartz and to some extent mica minerals. Feldspars are composed of same elements differing only in the relative amounts of potassium, sodium and calcium phases. They exist mostly as isomorphous mixtures of $NaAlSi_3O_8$, $KAlSi_3O_8$ and $CaAl_2Si_2O_8$ molecules. Feldspars are used as raw materials in glass making, ceramics and fillers. The separation of feldspars into individual phases is of great importance as a source of scarce potash feldspar to the industry. However, it is very difficult to separate feldspars into pure K, Na and Ca phases by flotation due to their identical crystalline structures and close similarities in their physical and chemical properties. Hardly any flotation studies have been reported in the literature involving the separation of individual phases. For practical purposes, the separation of feldspars consisting of one of the phases as a predominant component is acceptable. The ratio of

$K_2O : Na_2O$ in feldspars is often used as a raw material specification by the industry. For example, a ratio of 2 : 1 to 3 : 1 is required for ceramic products whereas for high-voltage insulators, abrasives and electrodes production, the ratio must be higher [1].

The separation of feldspars from quartz has been the subject of numerous studies since 1930. In these mineral systems, the traditional methods often fail because of overlapping densities and magnetic susceptibilities. The flotation process so far proves to be the most suitable method for the separation of feldspar from quartz [2]. In a conventional process, feldspar is separated from quartz using a cationic amine collector and hydrofluoric acid as an activator for feldspar. The use of HF suffers from the growing awareness of environmental and health problems, in addition to its inherent high cost and corrosive action on flotation cells. Several new reagent schemes not involving the use of hydrofluoric acid have been reported in the literature.

Over the years a considerable amount of study has been devoted to the flotation principles governing the separation of feldspar from quartz. The studies dealing with the conventional process and the role of HF in the separation, have been discussed briefly in the past [3, 4]. In the present review, a major emphasis is placed on the new reagent schemes reported in the literature.

FELDSPAR MINERALS

Feldspars are by far the most abundant minerals in the earth's crust, constituting around 65% by volume. Feldspar is a general term applied to a group of minerals which are aluminium silicates, combined with potassium, sodium or calcium or a mixture of these elements. These are described as potassium feldspar (orthoclase, $KAlSi_3O_8$), sodium feldspar (albite, $NaAlSi_3O_8$) and calcium feldspar (anorthite, $CaAl_2Si_2O_8$). Any feldspar containing more than 5% of the third component may be called ternary. Anorthoclase is albite containing up to one-third K-feldspar (and some Ca-feldspar). The nomenclature of homogeneous feldspars in the ternary system forming two binary series, Na-K feldspars (alkali feldspars) and Ca-Na feldspars (plagioclases) is shown in Fig. 1 [5].

The structure of feldspars is based on a frame work of SiO_2, where each oxygen is ideally shared between two silicon atoms. When all four oxygen atoms are tetrahedrally bounded to one silicon atom, they are also shared with other silicon atoms. Hence, every oxygen atom in the tetrahedra is bound to two silicon atoms by one valency bond. Thus, the electrical neutrality is achieved throughout the resulting three-dimensional network structure of quartz mineral.

Feldspar mineral structures are composed of corner-sharing AlO_4 and SiO_4 tetrahedra linked in an infinite three-dimensional array. When Al^{3+} ions replace

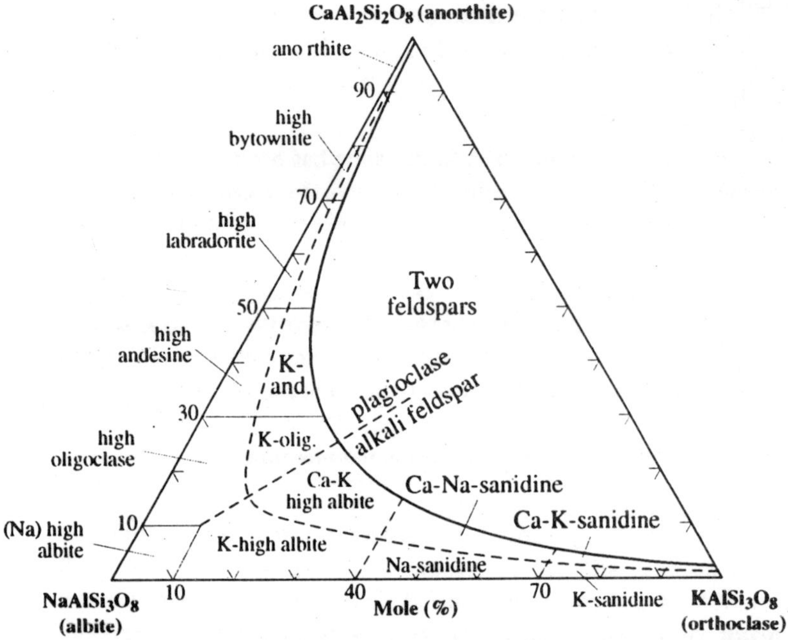

Fig. 1. Nomenclature of homogeneous feldspars in a high structural state of the ternary system of albite, orthoclase and anorthite (after Ref. [5])

Si^{4+} ions in the oxygen tetrahedron, cations must be included in the interstices within the framework for charge equilibrium. The replacement of approximately 25% of the silicon by aluminium and the introduction of alkali metal ions to balance the charge gives rise to feldspar minerals. A schematic diagram of the structure of feldspars is shown in Fig. 2 [1]. Aluminium and silicon ions are situated in the tetrahedron centers while oxygen ions being at the apex. The Na, K and Ca cations occupy the cavities of the tetrahedron framework to compensate the anion charges. The general formula AT_4O_8 designates the stoichiometry, where T represent Al, Si and A is a divalent (Ca or Ba) for $Al_2Si_2O_8$ alkaline-earth feldspars, or a monovalent (Na or K) for the $AlSi_3O_8$ alkali feldspar series of solid solutions and mixed crystals. A complete range of compositions is observed in the plagioclase feldspar series, $Na_yCa_{1-y}Al_{2-y}Si_{2+y}O_8$ ($0 < y < 1$) [6].

When feldspars are ground, the surfaces consisting of silicon and aluminium atoms exposed are unsaturated due to anionic tetrahedron breakage. In the presence of water, the surface atoms are saturated with hydroxyl groups. The dissociation of surface hydroxyl leads to a negatively charged surface of AlO^- and SiO^-. Similarly, a negatively charged surface of SiO^- occurs in the case of quartz. Thus, feldspars containing AlO^- and SiO^-, and that of quartz containing SiO^- differ only in cation composition. The difference in the surface cations of feldspar and quartz could possibly be utilized for selective adsorption of appropriate collector molecules and thus effecting selective flotation separation. However, the dissolution of feldspars leads to either K/Na/Ca-deficient or aluminium deficient surface layer. Hence, a knowledge of the dissolution/sorption chemistry of surface atoms is required to gain a better understanding of the physico-chemical aspects of feldspar flotation.

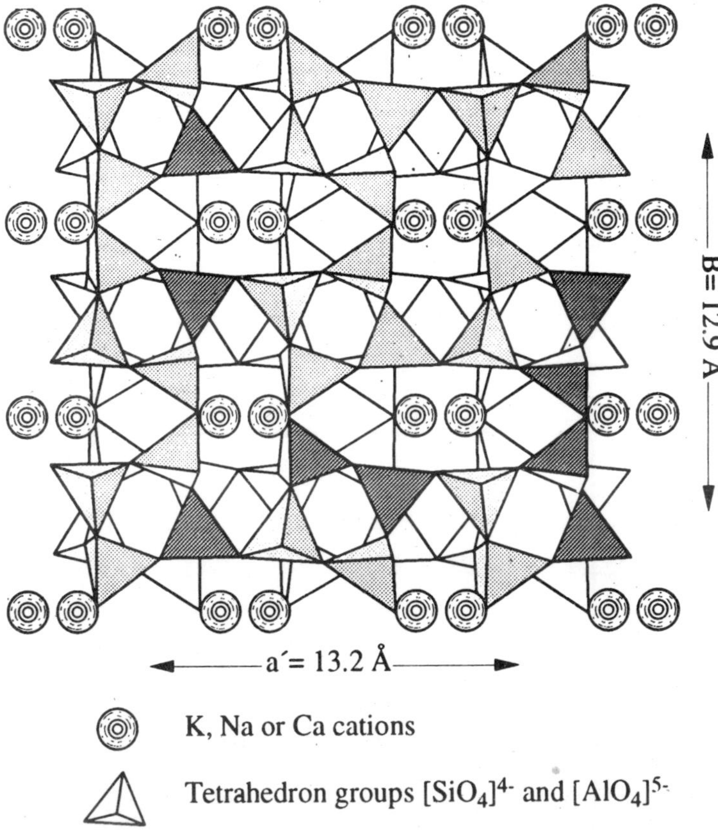

Fig. 2. Schematic diagram of the structure of feldspars (after Ref. [1]).

DISSOLUTION STUDIES

Feldspar weathering rates have been extensively studied in connection with environmental and geochemical issues and the works have been reviewed by Velbel [7] and, Schott and Petit [8]. In general, the kinetics of feldspar dissolution is found to comprise o.ᶠ two stages. In the early stages, high dissolution rates with a non-stoichiometric release of alkalis and alkaline earth metal ions relative to silica and aluminium are observed. At this stage, a thin layer of hydrolyzed aluminosilicate is thought to be formed on the surface due to hydrogen ion exchange for alkalis and alkaline earth metal-ions. As such, the composition of surface layer is pH dependent. For example, a K-feldspar contains a 6 : 1 ratio of SiO_2 and Al_2O_3, and this ratio is maintained between pH 6-10, whereas at pH 3 the ratio is 12 : 1 and 8 : 1 at pH 11 [9]. This means that Al_2O_3 dissolves more readily on the acid side below pH 4 and on the alkaline side above pH 10. The altered residual surface layer is about two unit cells thick (1.5-2.5 nm) to a few tens of nanometer thick. From the loss of potassium in the outer portion of a K-feldspar which is treated with HF/H_2SO_4, the thickness of the leached layer is calculated to be about 6.0 nm [10]. Sodium losses from a leached albite is found to be negligible at distances greater than 30-40 nm from the mineral surface [11]. Thus, a hydrogen-exchange cation depleted surface layer exists and it is at most a few tens of nanometer thick. After the initial stage of dissolution, further dissolution is found to be stoichiometric, with resultant release of all constituents in solution. Here, the rate of dissolution of the outer surface of the residual layer is similar to the rate at which alkalis are replaced by hydrogen at the interface between the fresh feldspar, and the residual layer. A schematic representation of an albite grain in mildly acidic solutions is shown in Fig. 3 [12].

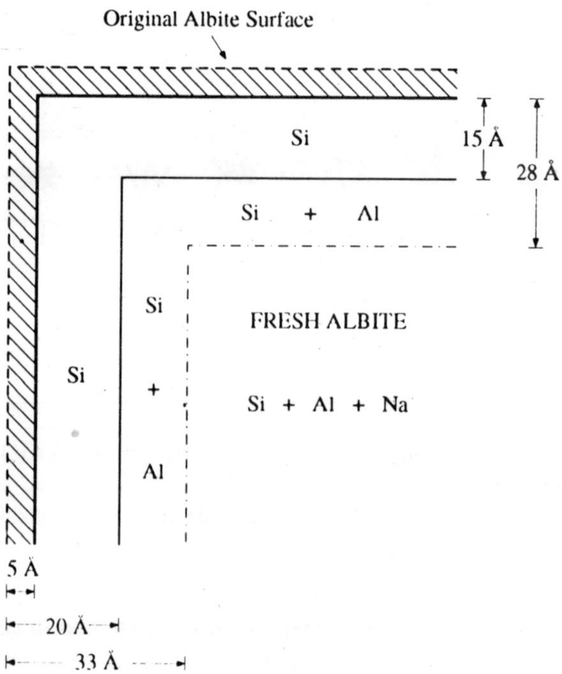

Fig. 3. Schematic representation of an albite grain after 20 days of reaction in mildly acidic aqueous solutions (after Ref. [12])

Thus, the dissolution reactions of feldspar are incongruent. When feldspars are exposed to aqueous solutions, there is an initial rapid exchange of alkali ions for H^+ ions followed by a slow dissolution in which the elements are released to the solution, or to the formation of a secondary solid phase. This process may be represented by [13] :

$$\text{Cation Al silicate (s)} + H_2O = \text{Cation} + OH^- + H_4SiO_4 + \text{Al silicates}$$

Only a few authors have carried out the dissolution of feldspar minerals with a view to evaluate its effect on the flotation of such minerals. Joy *et al.* [14] treated albite with perchloric acid and only aluminium was found in the washed liquors. At low pH values, aluminium ions are preferentially leached out and this process is aided by the presence of fluoride. The dissolved Al ions at pH 3 correspond to all of the surface-adjacent aluminium atoms, which made these authors to believe that the surface aluminium sites are not responsible for the collector adsorption mechanism. With microcline, both potassium and silica are detected in solution. The dissolution of potassium is suppressed in the presence of potassium chloride [15], in accordance with the formation of hydrolyzed surface layer mentioned earlier.

The dissolution of microcline in the presence and absence of fluoride as reported by Warren and Kitchener [16], and Read and Manser [17] is shown in Fig. 4. Although, the equilibration period in both of these studies was 24 hours, the difference in the solids content and the particle size used might affect the concentrations of dissolved species. In water and at low solids content, the concentrations of Si and Al are increased as the pH value decreases to 5 and this effect is extended to pH 3 in the case of higher solids concentration. At higher solids content, the concentrations of dissolved Si and Al are found to be the same as when the pH is decreased to 3, whereas, a preferential dissolution of Al can be seen at low solid concentrations. Even if the dissolved ions concentrations are the same, preferential dissolution of aluminium can be noted from the results since the aluminium/silica ratio in the microcline is 0.35 whereas, it varies from 1 to 2 in the solution. In the presence of fluoride, the concentrations of Si and Al are increased considerably. Below pH 3, there is hardly any effect on the concentrations of either Si or Al. The increased concentrations at pH 3 are attributed to the presence of molecular HF which attacks the surface and degrades the lattice. As the increase in the concentration of undissociated HF is marginal below pH 3, no further effect on the surface is observed. The dissolution of microcline/quartz mixture with increasing fluoride concentrations at pH 3 as reported by Read and Manser [17] is shown in Fig. 5. Both Si and Al concentrations increase with increasing amount of fluoride. Above 10 mM fluoride, the Al/Si ratio in solution is found to be the same as that of the microcline itself, and thus made the authors to suggest that higher levels of fluoride create fresh feldspar surfaces.

ELECTROKINETIC STUDIES

In the process of grinding feldspar or quartz, the Si/O and Al/O bonds are fractured due to their framework structure. The compensating surface (Na^+, K^+ and Ca^{2+}) cations are readily soluble in aqueous medium leaving behind a negative lattice. Hence, the PZC of feldspar and quartz is low and occurs in the

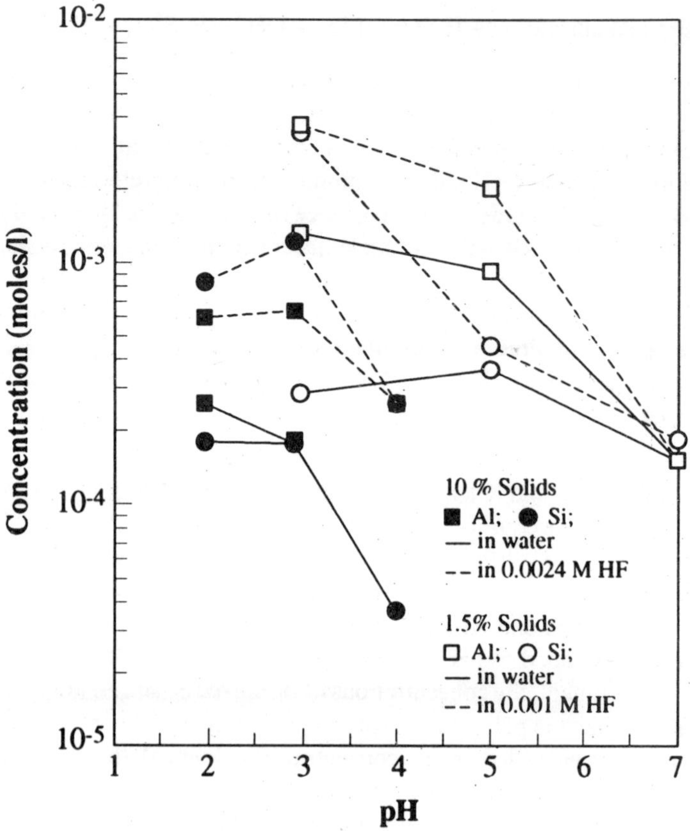

Fig. 4. Dissolution of microcline in the presence and absence of fluoride as a function of pH after 24 hours equilibration period (after Refs. [16, 17])

region of pH 2. Several authors have reported the PZC's of feldspar and quartz minerals. Typical values are shown in Table 1, as summarized by Fuerstenau and Fuerstenau [4].

Table 1: Point of zero charge of feldspar and quartz minerals (After Ref. [4])

Mineral	Chemical formula	PZC
Microcline	$K(AlSi_3O_8)$	pH 1.9, 1.7, 1.8
Orthoclase	$K(AlSi_3O_8)$	pH 1.7, 1.4
Sanidine	$K(AlSi_3O_8)$	pH 1.7
Albite	$Na(AlSi_3O_8)$	pH 1.9, 2.3, 1.9
Anorthite	$Ca(Al_2Si_2O_8)$	pH 2.4, 2.0, 3.6
Oligoclase	Na,Ca Feldspar	pH 1.5
Anorthoclase	$(K,Na)[AlSi_3O_8]$	pH 1.6
Quartz	SiO_2	pH 2.2, 1.8, 1.4, 2.3

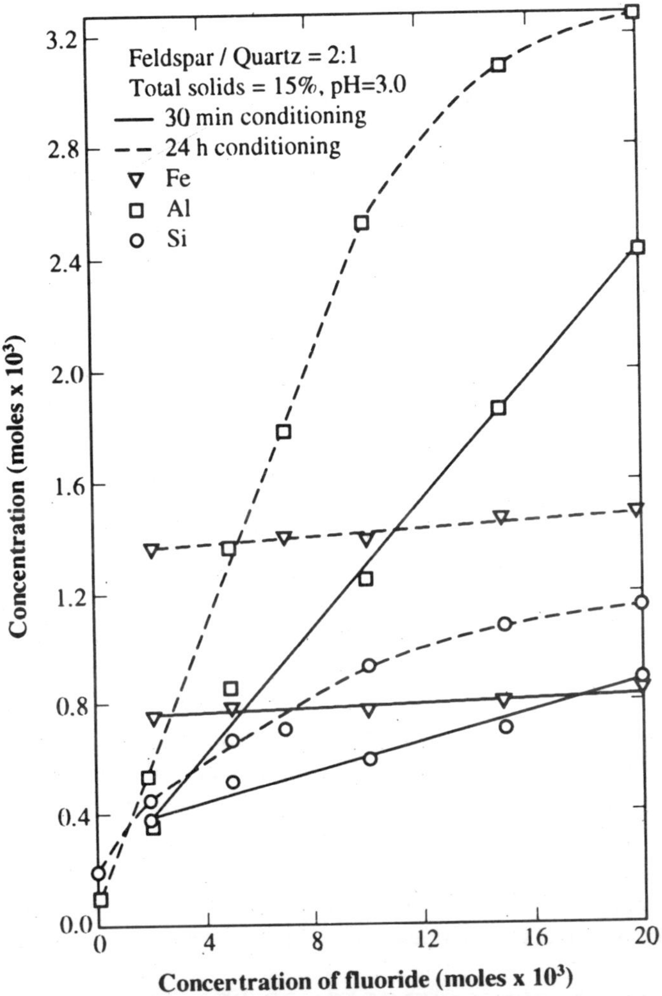

Fig. 5. Dissolution of microcline/quartz mixture as a function of fluoride concentration at pH 3.0 after 30 min and 24 hours equilibration periods (after Ref. [17])

Ney [18], presented the zeta potential versus pH curves for various feldspar minerals using perchloric acid as a pH modifier. The PZC's of many minerals from various locations were found to be around pH 2, as shown in Table 1. However, microcline, orthoclase and sanidine showed negative potentials in the entire pH region, down to pH 1.

The zeta potentials of quartz and microcline under simulated flotation conditions were reported by Warren and Kitchener [16]. At pH 3, the zeta potential of quartz is not changed when the pH was adjusted with either HCl or HF. The magnitude of negative potential is however reduced, when the potential was measured in feldspar liquors (1 mM HF), which is attributed to the specific adsorption of Al^{3+} extracted from feldspar at pH 3. In the case of microcline, the presence of fluoride increased the magnitude of negative potential. Thus, fluoride is stated to cause quartz depression and activation of feldspar in cationic amine collectors systems by decreasing and increasing the negative potentials, respectively. From the adsorption of fluoride ions on feldspar and the magnitude of negative potentials, maximum activation of feldspar is claimed to occur in dilute equilibrium concentrations of fluoride and at pH 5.

Similar electrokinetic behaviour of increasing and decreasing magnitude of negative zeta potentials of feldspar and quartz respectively, in the presence of fluoride (0.47 mM NaF) was also observed by Read and Manser [17]. These authors investigated the effect of fluoride through streaming potential measurements between the pH region of 4 and 10. When the minerals were washed with HF, the negative potentials are increased in both the cases, but with the greater effect being displayed by microcline (– 60 mV). The large increase in the magnitude of the negative potential with HF washed feldspar is thought to be due to the adsorption of alumino or silico-fluoride complexes formed during the HF leaching which were not removed by subsequent washing operation. Microcline further exhibited a very strong negative potential at pH 2.5 (40 mM HF), where the leached solution showed a near stoichiometric release of both Al and Si ions. This result made the authors to suggest that one of the principal action of fluoride is to remove amorphous silica from the surface and to create a fresh feldspar surface. The variation in the zeta potential of acid-treated and untreated feldspar was found to be small by Shimoiizaka *et al.* [19]. However, the amount of dodecyl amine adsorbed on HF treated feldspar is greater than that found on the untreated feldspar or on quartz. Thus, whether negative fluoride complexes are formed or not, the concentrations of HF used during the treatment could affect the magnitude of zeta potentials.

STUDIES IN THE PRESENCE OF FLUORIDE IONS

Flotation Studies

Flotation separation of feldspars from quartz was first accomplished in 1936 [20]. However, the selective activation effect of fluoride ions on feldspars in strongly acidic conditions was first reported by O'Merra *et al.* [21] while using lauryl amine collector in the presence of hydrofluoric acid and fluorine-bearing salts. Subsequently, Dean and Ambrose [22] showed that the addition of HCl or HF, reduced the flotation of quartz to zero at pH 4. They also found that these two acids affect the flotation of feldspars differently. Feldspar recovery was a mere 10% in the presence of HCl, but 97% in the presence of HF with dodecyl amine as a collector. Since the time of these publications, the use of HF in feldspar-quartz separation with cationic collectors has proved to be the most suitable reagent scheme and as a result, it became a routine method in plant practice.

Several authors have studied the flotation behaviour of feldspar and quartz in the presence of fluoride using single minerals and their mixtures. Studies by Dean and Ambrose [22] reveal that the concentration of fluoride is critical depending on the type of feldspar mineral being treated. Buckenham and Rogers [23] observed that the additions of NaF from 0.5 to 5 mM had no significant effect on the flotation behaviour. Hallimond tests showed that quartz was depressed below pH 4 in the presence of fluoride (5 mM), whereas the region of high recoveries of feldspar was extended. Similar tests by Suliin and Smith [24] with concentrated HCl and HF pretreated microcline show that HF pretreatment activated while HCl pretreatment depressed flotation. The effect of HF concentrations on different size fractions of feldspar-quartz mixtures was investigated by Tröndle *et al.* [25]. A greater selective action of HF over HCl is observed at pH values < 3.5. A minimum in the feldspar yield with increasing HF concentrations is noticed at pH 3.5 with all size fractions of orthoclase, but the minimum tends to be displaced with increasing fineness, towards higher HF concentrations or lower pH values. No such minimum values have been observed however with microcline but a sharp increase in enrichment occurs up to 10 mM of HF at a pH value of approximately 2.5. The minimum in the flotation behaviour is persistent with almost all types of feldspars studied by Uhlig [26], shown in Fig. 6. This figure shows principally the same characteristic relationship between various feldspar recoveries and concentrations of HF. All the feldspars are activated between pH 2.7 and 1.8, which corresponds

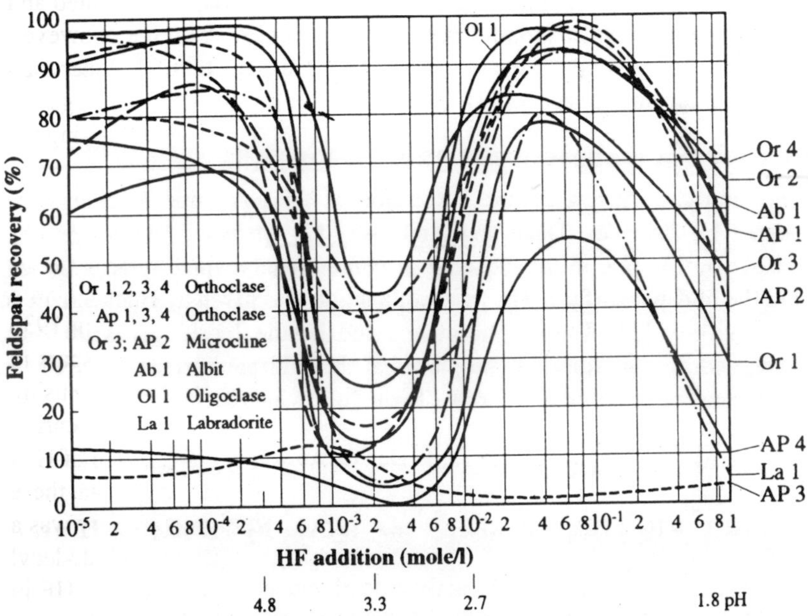

Fig. 6. Dependence of feldspar recovery on HF addition and pH, respectively, for various pegmatitic ores (after Ref. [26]).

to a very high level of HF addition. As quartz no longer floats below pH 3, the results suggest that the separation of feldspar can be made as long as the fluoride concentration is sufficiently high. The studies on the effect of particle size on the requirement of HF indicated that a dosage of 1 to 10 kg/t of HF was necessary for the size fraction 0.1 - 0.315 mm. The separation of very fine size fractions (5-20 μm) is also possible, but at a very high acid consumption.

With natural feldspar ores, O'Meara *et al.* [21] in their very first publication showed that the addition of HF up to 1.5 kg/t of the ore increased both grade and recovery of feldspar. On the other hand, the recovery decreases with increasing HCl addition. For example, the recovery of feldspar decreased from about 95 to 50% with HCl additions from 0.025 to 0.2 kg/t. Pegmatite ores with mica inclusions were studied by several workers who developed flotation flowsheets for the individual concentrated products of mica, feldspar and quartz [27-32]. One such flotation flowsheet for the continuous processing of mica, heavy minerals, feldspar and quartz is shown in Fig. 7 [29]. In general, mica and iron minerals are removed, prior to the separation of feldspar from quartz. Initially, mica is floated in acidic circuit with sulfuric acid and using tallow amine acetate together with either fuel oil or kerosine as a collector. This is followed by an anionic flotation also in acidic circuit where iron minerals and residual mica are removed with petroleum sulfonate. The separation of feldspar from quartz is carried out by activating feldspar with hydrofluoric acid (at a dosage of 1-2 kg/t) and using tallow amine cationic collector. Laboratory flotation tests on coarse pegmatite show an optimum dosage of 2.5 kg /t HF for the separation of potash-soda-feldspar from quartz [33]. A similar optimization study on a different pegmatite ore showed a HF dosage of 1.5 kg/t for a good feldspar grade and recovery [17].

Mechanism of Feldspar Activation by Fluoride Ions

Surprisingly, very few authors have studied the activation mechanisms of feldspar by fluorides. Dean and Ambrose [22] attributed the activation of feldspar by fluoride to the cleaning of contaminants which retards flotation from the feldspar surface. The first explanation was given by Buckenham and Rogers in 1954 [23]. These authors suggested that the fluoride gives rise to sites of charged aluminofluoride complexes at the feldspar surface on which the cationic collector adsorbs according to the following reactions:

$$(\equiv \mathrm{Al} - \mathrm{OH})_{surf} + 2\,\mathrm{HF\,(aq)} \Leftrightarrow (\equiv \mathrm{Al} - \mathrm{F_2^-})\,\mathrm{H^+}_{surf} + \mathrm{H_2O}$$

$$(\equiv \mathrm{Al} - \mathrm{F_2^-})\,\mathrm{H^+}_{surf} + \mathrm{RNH_3^+} \Leftrightarrow (\equiv \mathrm{Al} - \mathrm{F_2^-})\,\mathrm{RNH_3^+} + \mathrm{H^+\,(aq)}$$

The reason they gave for the activation of feldspar but not quartz is that the fluoride complexes of the type $M_3(AlF_6)$ are almost insoluble in water whereas the silicon tetrafluoride (formed by the action of HF on silica) hydrolyses and forms very stable fluorosilicic acid in solution so that there is no possiblity for

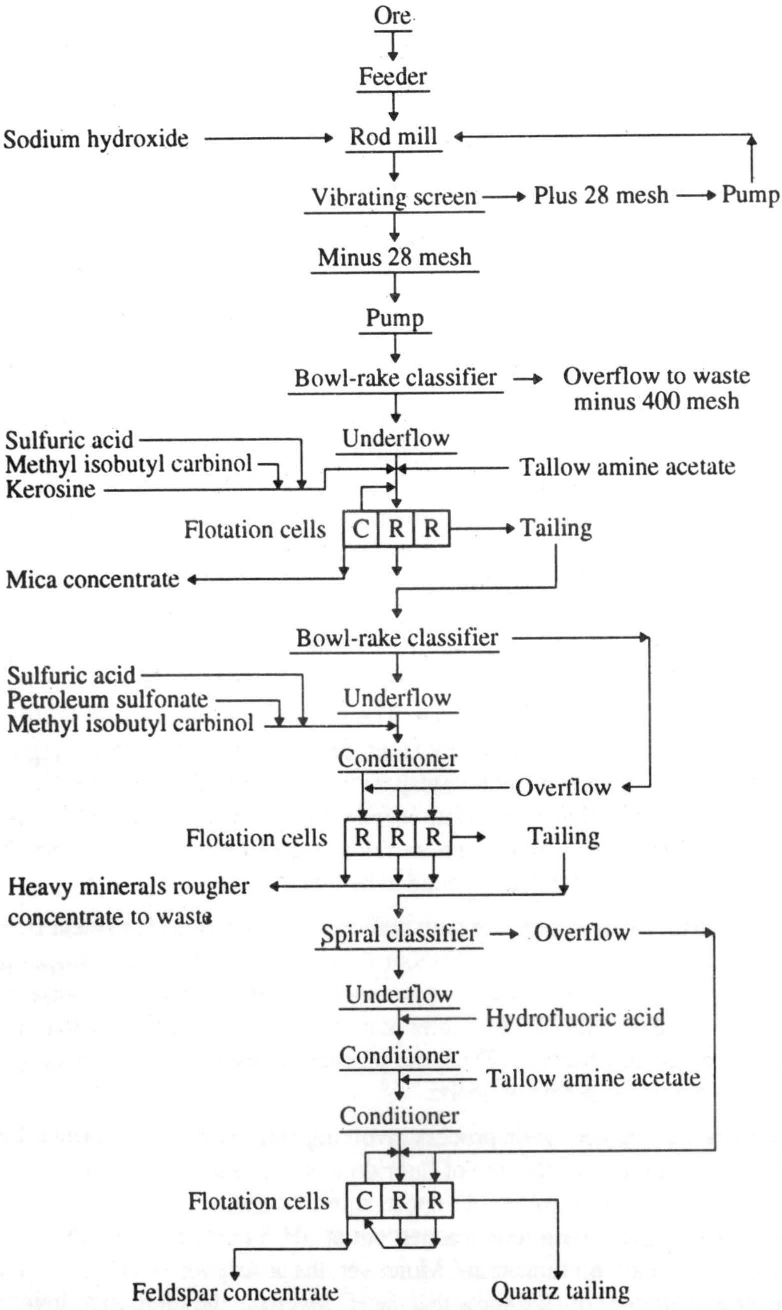

Fig. 7. Flotation flowsheet for the continuous processing of mica, heavy minerals, feldspar and quartz (after Ref. [29])

the formation of an insoluble fluoride on the surface of quartz. They also noted that the pH increases when fluoride is added to a suspension of feldspar which is consistent with the scheme given above. The adsorption of HF_2^- complexes on aluminium sites at pH < 3 is also proposed by Schubert and Abido [34], and Abido [35].

According to Smith [36], HF initially attacks the surface silicic acid thereby forming hexafluo-silicate ions. These SiF_6^{2-} ions adsorb onto aluminium sites creating negative sites favouring the adsorption of amine cations as follows:

$$\equiv Al - OH + SiF_6^{2-} \Leftrightarrow \equiv Al - SiF_6^- + OH^-$$

$$\equiv Al - SiF_6^- + RNH_3^+ \Leftrightarrow \equiv Al - SiF_6 NH_3 R$$

The role of SiF_6^{2-} ions in the activation of feldspar is further substantiated by contact angle, infrared and Hallimond tube flotation studies [3, 24, 37]. From the infrared spectra recorded after treating the microcline in HF (1N ≈ pH 1.8) containing dodecylamine, they showed the presence of potassium, sodium and amine fluosilicates on or near the surface of microcline. In addition, the flotation studies showed an activation of microcline with the compound $(RNH_3)_2SiF_6$. Shimoiizaka *et al.* [19] also observed sodium or potassium fluosilicate on the surface of the feldspar from infrared studies. These authors assumed a reaction between the amine and the surface sodium/potassium fluosilicate to give a hydrophobic character to the mineral. The formation of K_2SiF_6 on feldspar surface in HF solution is reported by Sclar [38] in connection with geological field tests. It is well known that K-feldspar must be exposed either to HF fumes or be immersed in HF in order to stain it yellow with sodium cobaltinitrite, and that only the HF reaction product on the feldspar surface is stained but, not the feldspar. The reaction product is found to be hieratite (cubic K_2SiF_6).

The activation of microcline with NaF in acidic solutions is evident from the results shown in Fig. 8 [3]. Without fluoride ions, the contact angles of quartz and feldspar are the same in the presence of dodecyl amine. The presence of fluoride increased the contact angles of microcline below pH 5 but no effect on quartz could be observed. Thus, the process is described as activation of feldspar and not depression of quartz.

Contrary to the activation process involving surface aluminium sites, Joy *et al.* [14] postulated that the role of fluoride was one of depressing quartz and that the aluminium sites were not important in feldspar activation. This study shows that the aluminium ions leached out at pH 3 correspond to all of the surface-adjacent aluminium atoms. Moreover, the adsorption of H^+ ions in the presence and absence of NaF show that the H^+ loss from the solution following fluoride adsorption is in excess of the fluoride loss by a factor of 6.9 at pH 4.5, falling as the pH is increased. This strong pH dependence led the authors to suggest that hydrogen fluoride is the active species since the proportion of fluoride ions present as HF exhibits similar behaviour with increase in pH. The depression of quartz is thought to be caused by the exchange of surface silicon hydroxyl group for fluoride ions thereby reducing the number of surface negative sites. The fluoride loss from the solution is found to be so small as to preclude the adsorption of SiF_6^{2-} ions. From the analysis of the concentration of various fluoride species at which maximum adsorption of fluoride occurs on microcline, Warren and Kitchener [16] observed that the concentration of hexafluosilicate ions above pH 4 are of the order of only 1 to 10 μM. These authors considered the adsorption of fluoride ions by aluminium sites on feldspar surface.

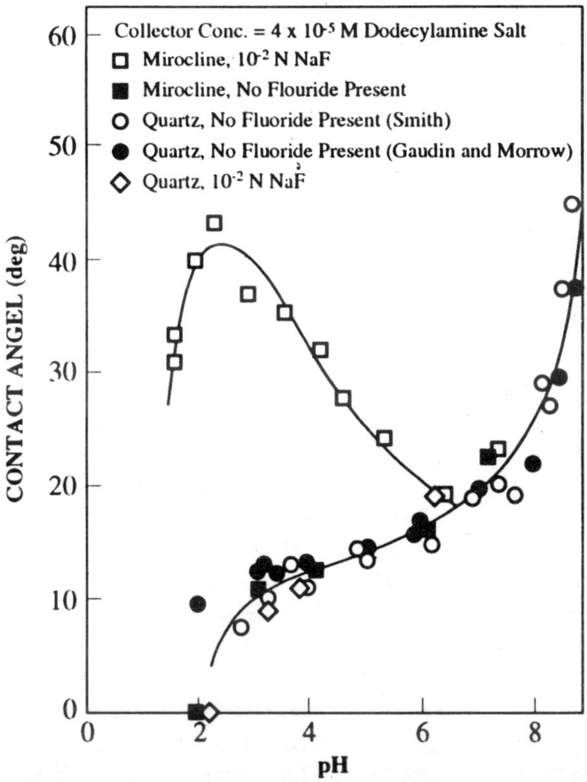

Fig. 8. Wettability of microcline and quartz in aqueous dodecyl amine solutions as a function of pH in the presence and absence of sodium fluoride (after Ref. [3]).

At the best flotation conditions of pH < 3, the HF attacks feldspar and more so quartz, thereby producing significant amounts of silicon in solution. Hence, the concentrations of HF used and its effect on the minerals to be leached out of the lattice ions depend on the distribution of various amounts of silicon and aluminium fluoride complexes in solution. It should be pointed out that the authors who disagree about the role of SiF_6^{2-} ions invariably used a very low amount of HF, whereas Smith and co-workers used 10 mM of HF (Fig. 8) and this amount corresponds to the dosage used in either laboratory batch flotation tests or in plant practice, for the separation of feldspar from natural ores. The effect of HF in actual flotation of feldspar from a pegmatite ore was analysed by Read and Manser [17], who also provided a complete chemical analysis of the pulp liquid at the time of flotation. The equilibrium composition of the solid and liquid species in solution were calculated using HALTAFALL computer program. This program is only capable of handling not more than six components, sixteen complexes and six solid phases. We have re-examined the equilibria while using a more powerful software SOLGASWATER taking into account all possible solution and solid species shown in Table 2.

Table 2: The components and species used in the theoretical calculations

Species No.		Log K_f	Stoichiometry matrix (components)				
			H^+	F^-	Fe^{3+}	Al^{3+}	$Si(OH)_4$
Solution							
1	H^+	0.000	1.00	0.00	0.00	0.00	0.00
2	F^-	0.000	0.00	1.00	0.00	0.00	0.00
3	HF	3.170	1.00	1.00	1.00	0.00	0.00
4	HF_2^-	3.670	1.00	2.00	0.00	0.00	0.00
5	Fe^{3+}	0.000	0.00	0.00	1.00	0.00	0.00
6	$Fe(OH)^{2+}$	−2.190	−1.00	0.00	1.00	0.00	0.00
7	$Fe(OH)_2^+$	−5.700	−2.00	0.00	1.00	0.00	0.00
8	$Fe(OH)_4^-$	−21.600	−4.00	0.00	1.00	0.00	0.00
9	$Fe_2(OH)_2^{4+}$	−2.900	−2.00	0.00	2.00	0.00	0.00
10	$Fe_3(OH)_4^{5+}$	−6.300	−4.00	0.00	3.00	0.00	0.00
11	FeF^{2+}	5.180	0.00	1.00	1.00	0.00	0.00
12	FeF_2^+	9.130	0.00	2.00	1.00	0.00	0.00
13	FeF_3	11.900	0.00	3.00	1.00	0.00	0.00
14	Al^{3+}	0.000	0.00	0.00	0.00	1.00	0.00
15	$Al(OH)^{2+}$	−4.990	−1.00	0.00	0.00	1.00	0.00
16	$Al(OH)_2^+$	−9.300	−2.00	0.00	0.00	1.00	0.00
17	$Al(OH)_3$ (aq)	−15.000	−3.00	0.00	0.00	1.00	0.00
18	$Al(OH)_4^-$	−23.000	−4.00	0.00	0.00	1.00	0.00
19	$Al_2(OH)_2^{4+}$	−7.700	−2.00	0.00	0.00	2.00	0.00
20	$Al_3(OH)_4^{5+}$	−13.900	−4.00	0.00	0.00	3.00	0.00
21	AlF^{2+}	7.000	0.00	1.00	0.00	1.00	0.00
22	AlF_2^+	12.600	0.00	2.00	0.00	1.00	0.00
23	AlF_3	16.700	0.00	3.00	0.00	1.00	0.00
24	AlF_4^-	19.100	0.00	4.00	0.00	1.00	0.00
25	AlF_5^{2-}	19.400	0.00	5.00	0.00	1.00	0.00
26	AlF_6^{3-}	19.800	0.00	6.00	0.00	1.00	0.00
27	$Si(OH)_4$	0.000	0.00	0.00	0.00	0.00	1.00
28	$SiO(OH)_3^-$	−9.860	−1.00	0.00	0.00	0.00	1.00
29	$SiO_2(OH)_2^{2-}$	−23.270	−2.00	0.00	0.00	0.00	1.00
30	$Si_2O_3(OH)_4^{2-}$	−18.810	−2.00	0.00	0.00	0.00	2.00

(contd.)

Table 2: (contd.)

31	$Si_2O_2(OH)_5^-$	−8.010	−1.00	0.00	0.00	0.00	2.00
32	$Si_2O(OH)_6$	1.200	0.00	0.00	0.00	0.00	2.00
33	$Si_3O_6(OH)_3^{3-}$	−27.990	−3.00	0.00	0.00	0.00	3.00
34	$Si_3O_5(OH)_5^{3-}$	−26.960	−3.00	0.00	0.00	0.00	3.00
35	$Si_4O_7(OH)_5^{3-}$	−24.980	−3.00	0.00	0.00	0.00	4.00
36	Si_4^{4-}	−35.420	−4.00	0.00	0.00	0.00	4.00
37	SiF_4	20.660	4.00	4.00	0.00	0.00	1.00
38	H_2SiF_6	31.398	6.00	6.00	0.00	0.00	1.00
39	SiF_6^{2-}	26.847	4.00	6.00	0.00	0.00	1.00
Solids							
40	$Fe(OH)_3$	−3.200	−3.00	0.00	1.00	0.00	0.00
41	$Al(OH)_3$	−8.500	−3.00	0.00	0.00	1.00	0.00
42	SiO_2	2.740	0.00	0.00	0.00	0.00	1.00

Figures 9 and 10 show the distribution of various alumino and silicon fluoride complexes with respect to fluoride concentration at pH 3. The total aluminium and silicon concentrations used are 1 and 5 mM, which are of the same order as shown in Fig. 5. It can be seen that aluminium ions and positive aluminium fluoro-complexes predominate in the solution below 1 mM of fluoride and above which, AlF_3, F^-, HF_2^- and AlF_4^- are the predominant

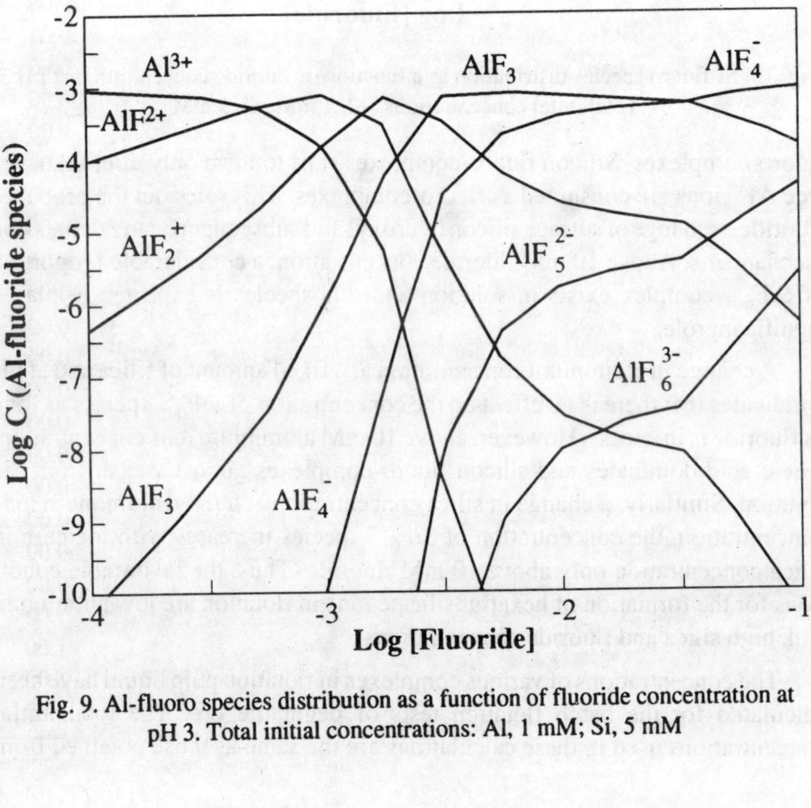

Fig. 9. Al-fluoro species distribution as a function of fluoride concentration at pH 3. Total initial concentrations: Al, 1 mM; Si, 5 mM

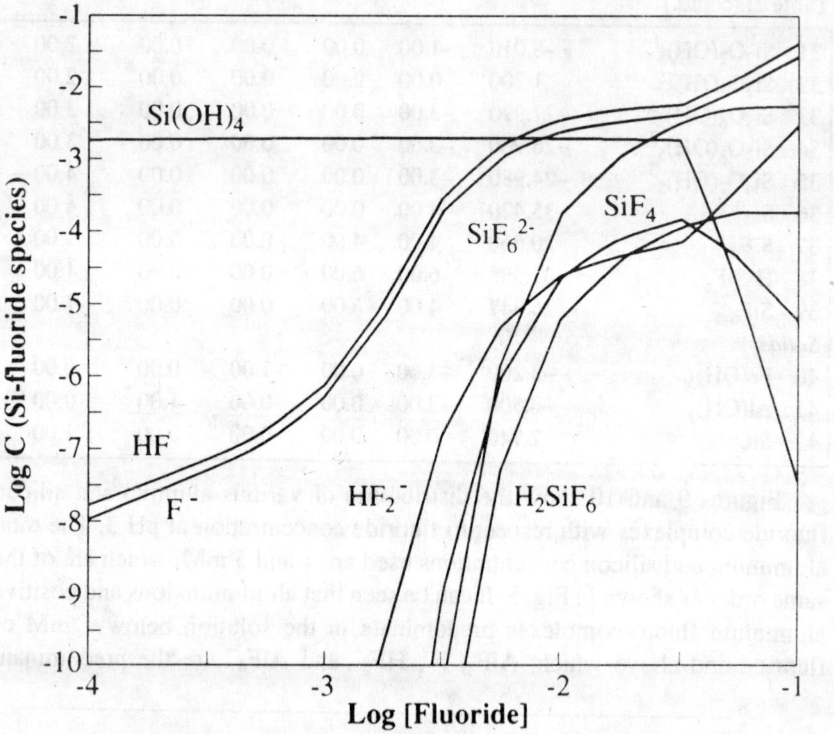

Fig. 10. Si-fluoro species distribution as a function of fluoride concentration at pH 3. Total inital concentrations: Al, 1 mM; Si, 5 mM

fluoro-complexes. Silicon fluoro-complexes start to form only after all of the free Al^{3+} ions are consumed as fluoro-complexes. This rules out the proposed fluoride exchange of surface silicon hydroxyl and subsequent quartz depression mechanisms. Above 10 mM fluoride concentration, a considerable proportion of SiF_6^{2-} complex exists in solution and this species is expected to play a significant role.

A change in aluminium concentration at a fixed amount of silica and at pH 3 indicates that there is no effect on the concentration of SiF_6^{2-} species as long as fluoride is in excess. However, above 10 mM aluminium ions concentration, silicic acid dominates and silicon fluoro-complexes almost vanish from the solution. Similarly, a change in silica concentration at a fixed aluminium ions concentration, the concentration of SiF_6^{2-} species increases with increase in silica concentration only above 10 mM fluoride. Thus, the favourable conditions for the formation of hexafluosilicate ions in flotation are low aluminium and, high silica and fluoride concentrations.

The concentrations of various complexes in flotation pulp liquid have been calculated for the batch flotation tests of pegmatite ore. The total initial concentrations used in these calculations are the same as those obtained from the analyses results of the aqueous pulp phase at the time of collector addition [17]. The results are presented in Table 3.

Table 3: Analyses of aqueous phase of pulp liquid at the time of collector addition of batch flotation tests of pegmatite ore at pH 2.5 (after Ref. [17])

HF	Analysis of pulp liquid, mM				Flotation result of feldspar	
kg/t	F	Al	Si	Fe	Grade, %	Recovery, %
0.0	0.00	0.37	0.57	0.57	No float	
0.5	12.10	0.30	0.93	0.57	89	37
1.5	41.00	0.26	12.10	0.38	90	87
2.5	59.00	0.48	15.00	0.38	92	85
5.0	137.00	0.44	25.00	0.54	94	88

Concentrations of calculated fluoride solution species at pH 2.5 with total initial concentrations as per the analysis shown above:

HF	Calculated F-solution species, mM			Flotation result of feldspar	
kg/t	F^-+HF	HF_2^-	SiF_6^{2-}	Grade, %	Recovery, %
0.0	0.00	0.00	0.00	No float	
0.5	7.61	0.03	0.23	89	37
1.5	30.92	0.44	0.88	90	87
2.5	45.28	0.94	1.10	92	85
5.0	107.99	5.35	1.85	94	88

Although, the dissolved ions could originate from other minerals apart from feldspar, some inference can be made from the results shown in Table 3. The activation of feldspar is seen at 1.5 kg/t HF concentration. At this level, both SiF_6^{2-} and HF_2^- species are stable and their concentrations are nearly 1 and 0.5 mM, respectively. The respective concentrations of AlF_3 and AlF_4^- are around 0.11 and 0.15 mM, while the concentrations of the species AlF_5^{2-} and AlF_6^{3-} are negligible at 1.5 kg/t HF addition. The surface aluminium can be in the form of fluoro-complex, but it is not certain that the surface will be negatively charged. With SiF_6^{2-} as the predominant species at the best flotation conditions, the responsible species for the activation of feldspars could be the same while HF_2^- species might aid the activation process at best. The criticism that the surface of feldspar will be devoid of aluminium atoms has to be considered while proposing the adsorption of any species on aluminium sites. At high dosages of HF and practical flotation conditions (1.5 kg/t $\approx 1-2 \times 10^{-2}$ M), the dissolution could degrade the whole surface lattice and result in a near stoichiometric ratio of Al/Si as depicted in Table 3. Thus, the feldspar surface could contain all the surface elements under practical flotation conditions [17].

The dosage of HF needed to float feldspar should be considerably greater than that required to complex either Fe^{3+} or Al^{3+} ions in solution. The effect of ferric ion concentration at a fixed aluminium (1 mM), silica (5 mM) and fluoride (10 mM) is shown in Fig. 11. Above 1 mM ferric ion, the concentrations of SiF_6^{2-} and HF_2^- species decrease and positive iron-fluoro complexes predominate in the system. This explains why the iron-rich minerals are separated as much as possible by desliming, magnetic separation or using a special flotation stage, before the separation of feldspar from quartz by flotation is attempted.

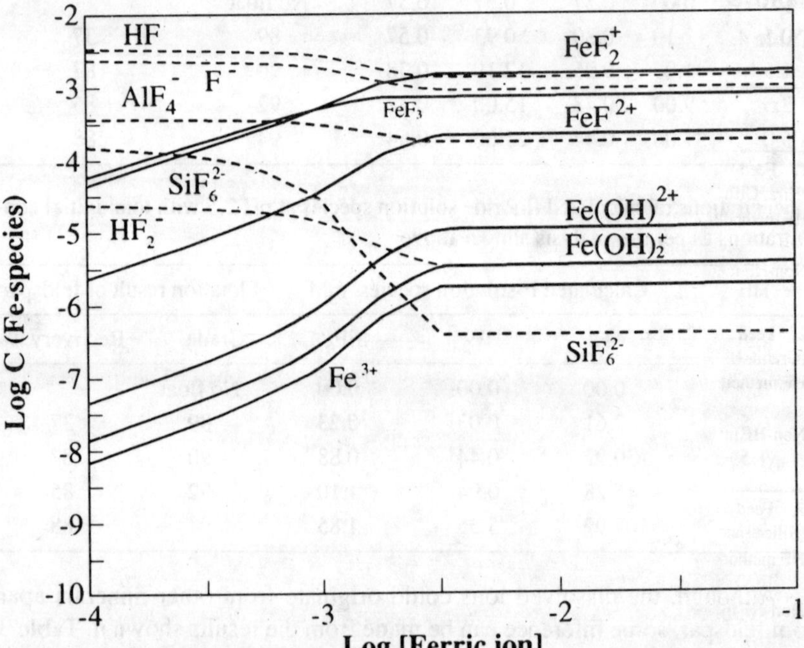

Fig. 11. Fluoro species distribution as a function of ferric ion concentration at pH 3. Total initial concentrations: Al, 1 mM; Si, 5 mM; F, 10 mM

STUDIES IN THE ABSENCE OF FLUORIDE IONS

Flotation Studies

Katayanagi [39] patented a flotation process in 1974, which involves using a mixture of cationic and anionic collectors in acidic circuit at about pH 2 for the separation of feldspar from quartz in the absence of fluoride ions. Alkyltrimethylene diamine acetate together with sodium petroleum sulfonate are used as collectors and the pH is controlled by using either hydrochloric or sulfuric acid. This process is reported to be applied in industrial practice [40]. Malghan [41] has reported a process which does not use HF medium, but is based on a similar reagent scheme as that of Katayanagi (a combined cationic and anionic collector, Duomeen TDO, N-tallow 1, 3-propylene diamine dioleate with carboxylate anionic group instead of sulfonate). Both of these reagent schemes have been tested with a number of mineralogically different ores and compared with the conventional HF method. The results as presented in Table 4 [40, 42] show that the method of using mixed cationic and anionic collectors is as effective as the HF reagent system in obtaining high grade feldspar concentrates. However, studies indicate that the floatability of feldspar and the size of the particle are interdependent and when the feed contains plus 600 μm, the HF system performed better than the non-HF reagent system [42]. This limitation is partly overcome by split conditioning of the feed whereby improved selectivity can be obtained with a more efficient usage of collector.

Table 4: Flotation results on natural samples in fluoride medium (HF method) and non-fluoride medium (non HF method) (after Refs. [40, 42])

		Assay, %					Distribution
		Al_2O_3	K_2O	Na_2O	CaO	Fe_2O_3	(%)
1. Feed: (Kira, Gifu, Japan)		7.88	4.15	0.20		0.27	
HF method:	Feldspar. conc.	18.02	12.60	1.08		0.28	30
	Quartz conc.	0.80	(0.10)			0.04	70
Non-HF method:	Feldspar. conc.	18.00	12.61	1.10		0.27	30.5
	Quartz conc.	0.90	(0.10)			0.04	69.5
2. Feed: (Granit, Hiroshima, Japan)		13.62	4.00	3.77		0.91	
HF method:	Feldspar. conc.	22.00	6.70	5.65		0.10	67
	Quartz conc.	0.09	(0.10)			0.02	33
Non-HF method:	Feldspar. conc.	18.50	6.70	5.33		0.09	68
	Quartz conc.	0.11	(0.10)			0.02	32
3. Feed: (Silica sand, Shimane, Japan)		6.00	4.18	0.88		0.13	
HF method:	Feldspar. conc.	17.17	11.00	1.51		0.48	12
	Quartz conc.	1.62	(0.10)			0.08	88
Non-HF method:	Feldspar. conc.	17.62	11.00	1.42		0.30	12
	Quartz conc.	1.40	(0.10)			0.07	88
4. Feed: (Kings Mountain, N. Carolina)		5.97	3.57	0.56	0.05		
HF method:	Feldspar. conc.	18.50	11.00	2.30	0.14		27.3
	Quartz conc.	0.63	0.09	0.03	0.02		72.7
Non-HF method:	Feldspar. conc.	18.70	12.20	1.80	0.14		28.6
	Quartz conc.	0.88	0.11	0.06	0.02		71.4
5. Feed: (Lillesand, Norway)		13.30	4.50	4.01	0.95		
HF method:	Feldspar. conc.	18.80	6.40	5.90	1.10		68.8
	Quartz conc.	0.82	0.11	0.10	0.04		31.2
Non-HF method:	Feldspar. conc.	18.70	6.40	5.70	1.35		69.9
	Quartz conc.	0.78	0.10	0.09	0.03		30.1
6. Feed: (Feldspar Corp., Alabama)		12.72	4.63	4.00	0.34		
HF method:	Feldspar. conc.	19.90	7.20	6.10	0.53		64.4
	Quartz conc.	0.39	0.08	0.06	0.03		35.6
Non-HF method:	Feldspar. conc.	19.80	7.30	6.30	0.53		62.6
	Quartz conc.	0.82	0.22	0.22	0.04		37.4

While the feldspar recovery and concentrate grades are relatively unaffected below pH 2, a loss in selectivity is observed above pH 2. An optimum pH region of 1.6 to 1.9 has been reported which means that a very high acidic pulp conditioning (pH < 2) has to be maintained to achieve selectivity in non-HF reagent system.

A comparative study by Bolin [43] using a coarse pegmatite from Limbergsbo, Forshammar, Sweden, and a milling waste from a sulphide ore concentrator at Åmmeberg, Sweden, showed that the HF reagent system gave higher grades and better yields than the combined collector system of Duomeen TDO. The best results obtained are shown in Fig. 12. Although, an improvement in the results with Duomeen TDO is reported at a lower pH value of 1.9, any substantial improvement is stated to be small in the absence of fluoride ions. This study indicates that the method of combined collectors is not effective for all types of materials. The significant difference in the materials studied by Bolin is the presence of iron oxide impurities, which affected the results during the first stage mica flotation. Fe-ions present during the follow up feldspar flotation may cause a reduction in the negative charge on the minerals. In other

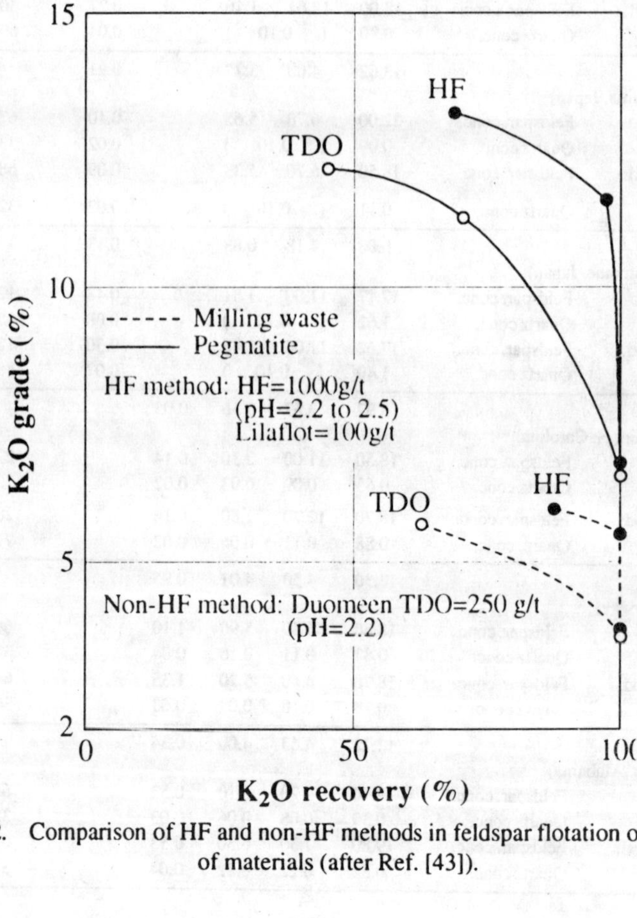

Fig. 12. Comparison of HF and non-HF methods in feldspar flotation of two types of materials (after Ref. [43]).

words, it could be only a charge difference between feldspar and quartz that led to a feldspar flotation in combined collector system. Iron-bearing raw marine sand consisting of quartz (88%) and feldspar (10%) was treated by Mathieu and Sirois [44] using the same combination of anionic-cationic promotors. The iron minerals were first floated with either petroleum sulfonate at pH 2.1 or with a commercial iron mineral collector (CX-62) at pH 6.8. After the first stage flotation, feldspar flotation was carried out at pH 2.5 with tallow amine acetate or propane diamine dioleate. Alternatively, feldspar-iron flotation was carried out in a single stage at pH 2.4 by using a combination of petroleum sulfonate and propane diamine dioleate. In either case, the quartz and feldspar concentrates were subsequently upgraded to various market specifications by acid leaching and/or magnetic separation.

Tang *et al.* [45, 46] reported the flotation of feldspar minerals from natural silica sand while using the same reagent scheme as that of Katayanagi, but surprisingly, at natural pH values. These authors stated that a prior preparation of the material, i.e., grinding the material in basic pH and desliming, made it possible to float feldspar minerals selectively at a natural pH of 7.8. The feldspars were selectively floated at a collector ratio of sulfonate/diamine greater than 6 but all previous studies were made using a ratio not exceeding 3. At higher ratios, the anionic collector is claimed to depress quartz but not feldspar due to its Al-positive charge centers, although both minerals are negatively charged at natural pH values. It is however hard to believe the suggested desorption of diamine from quartz surfaces which is attributed to the high pulp densities employed during the flotation as described by the authors. Nevertheless, a thorough study of the characteristics of mixed reagents in solution, and the interaction between reagents and with minerals at different ratios needs immediate attention, since better flotation results are reported with even several natural ores.

Recently, several authors have reported the selectivity in feldspar-quartz flotation system using new reagent schemes without fluoride ions. However, most of these studies have been limited to either a single mineral flotation or artificial mixtures. Jakobs and Dobias [47] used a polycondensation product of dicyandiamide-formaldehyde (Melflock) together with a alkyl aryl sulfonate or a carboxylic acid. The polyelectrolyte is stated to activate feldspar around pH 2 and subsequent flotation with an anionic collector. Even if, Melflock preferentially adsorbs on feldspar surfaces due to protonized amine groups in acidic pH, the advantage of using this product over alkyl diamine is not clear, since Melflock does not impart any hydrophobicity due to the lack of C-C bonds. The floatability in this system is thus, governed by subsequent adsorption of an anionic collector. The optimum ratio of cationic Melflock and anionic sulfonate collector concentrations is noted where a maximum amount of turbidity appeared in a system with increasing concentration of anionic sulfonate at a constant Melflock concentration.

Liu *et al.* [48] reported a double-collection flotation process, in which the first step involves conditioning of the pulp in the presence of anionic collector (sodium oleate or hydroxamic-acid sodium salt) for one minute, followed by two minutes conditioning with sodium hexametaphosphate (HMP). In the second step, cationic dodecyl amine-hydrochloride (DAC) is used to float feldspar selectively after three minutes conditioning. Initially, oleate ions are stated to adsorb on positively charged areas of quartz and feldspar and the follow up conditioning with HMP results in the exchange of all the oleate ions on quartz surface but, partially on feldspar surface. The oleate adsorption layer on feldspar enhances subsequent flotation with DAC collector. Single mineral flotation results did indicate the selectivity between feldspar and quartz when the pulp pH is controlled below 5 or 6. Such a separation has been demonstrated on artificially mixed samples, though the process is yet to be proved on natural ores.

El Salmawy *et al.* [49] floated quartz from feldspar while employing a non-ionic surfactant, cetyl ether polyoxyethylene (20 oxyethylene units, trade name Brij 58). It is reported that alkali feldspar cannot be floated at all, irrespective of the pH or collector concentration. The flotation of quartz through hydrogen bonding between the silanol groups of quartz and the ether groups of polyoxyethylene moiety was described. To prove their contention, the feldspar was subjected to leaching to remove the alkali cations from the surface, and as a result, such a feldspar was made to respond to a nonionic flotation. Though, the Hallimond tube recovery data presented by the authors may suggest that quartz be floated from feldspar at a low pH using nonionic surfactants, at least theoretically, the depletion of the alkali cations on the surface of feldspar at low pH values is to be considered. The success of this method would then depend on the dissolution kinetics of feldspar in relation to the time needed to bind nonionic surfactants with the silanol groups of quartz. In other words, the effect of conditioning period in flotation needs to be thoroughly investigated.

Mechanism of Feldspar Collection in Mixed Collector Systems

The adsorption mechanism of mixed cationic and anionic collectors in feldspar flotation has not been clearly defined. However, some mechanistic explanations have been suggested based on the studies made by Buckenham and Schulman [50] on sulfide and quartz minerals. In their work on mixed cationic and anionic collectors, using xanthates as the anionic collector and alkyl trimethyl ammonium bromide (TAB) or iodides as the cationic collector, they suggested the attachment of mixed xanthate/TAB complex with xanthate ion over the positive sites of sulfide mineral surface. Similarly, the adsorption of TAB ion over the negative sites on quartz followed by association and interaction between TAB and xanthate ion is suggested to explain the collecting properties of xanthate/TAB mixtures on quartz. The reagents which gave a combined dissociating complex produced enhanced flotation than the individual reagents. They stated that the mechanism (Fig. 13) which requires the concept of dissociating molecular complexes is supported by their preliminary work on fatty acid/TAB, fatty acid/amine, and the nondissociating long chain sulfate/TAB systems.

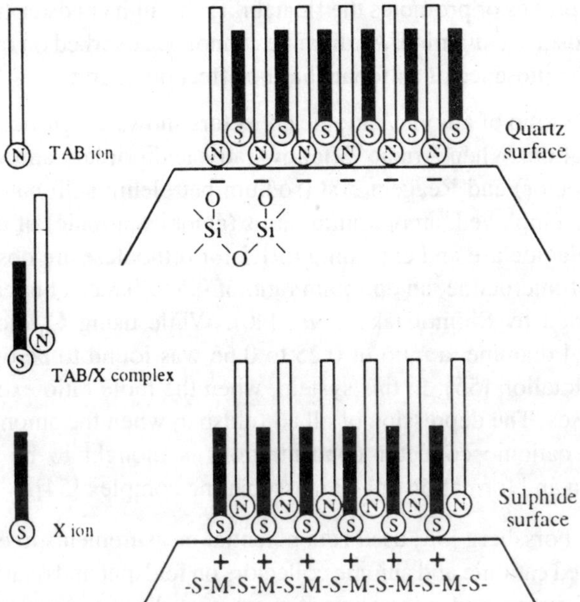

Fig. 13. Schematic representation of attachment of anionic / cationic collectors on negative and positively charged mineral surfaces (after Ref. [50])

For cationic-anionic combinations, there is a high probability of complex formation and precipitation through charge neutralisation at comparable ratios [51]. Monolayer and surface tension work by Buckenham and Schulman [50] showed that there is a strong association between xanthate and TAB ions. Intermolecular and ionomolecular complexes possess high hydrophobic capabilities and the role of complex formation is claimed to increase flotation recovery due to the significantly higher surface acivities of these complexes [52]. The high surface activities of bifunctional complexes compared to those of the species involved in the complex formation have been deduced by several authors from surface tension reduction efficiency [53].

Shimoiizaka *et al.* [40] made an attempt to describe the role of a mixed collector system in feldspar-quartz flotation through adsorption, zeta potential and flotation studies. In acidic pH 3 or less, diamine is not adsorbed on quartz, irrespective of the presence of an anionic sulfonate. On the other hand, the amount of diamine adsorbed by feldspar increases in the presence of sulfonate. Although, the adsorption of sulfonate in the presence of diamine is not measured, coadsorption of anionic sulfonate on partially covered feldspar surfaces with diamine is suggested. The coadsorption is thought to result in the formation of amine salt and this assumption is supported by the correlation of the increase in flotation recovery to that of the precipitation region of the mixed diamine and sulfonate collectors. In essence, the adsorbed diamine alone cannot produce sufficient hydrophobicity to float feldspar but, the simultaneous use of sulfonate improves or promotes the floatability through coadsorption with the previously adsorbed diamine. As diamine cannot be adsorbed on quartz at low pH values, the presence of sulfonate has no effect on quartz.

The mole ratio of anionic/cationic collectors shows a significant effect on feldspar flotation. When Armac T (tallow-fatty acid primary amine acetate as cationic collector) and Reagent 308 (sodium petroleum sulfonate as anionic collector) are employed, an optimum ratios (anionic/cationic) of 0.5 to 1.0 in the case of plagioclase and extending to 1.5 for orthoclase are observed [54]. In the case of microcline, an optimum ratio of 0.5 to 0.8 can be read from the results reported by Shimoiizaka *et al.* [40]. While using C_n-dioxyethylene sulphate/alkyl diamine, a ratio of 0.25 to 0.66 was found to be optimum for microcline flotation [55]. In this system, when the mole ratio exceeds unity, flotation ceases. The depression of alkali feldspar, when the anionic collector exceeds the cationic collector concentration, is thought to be due to the neutralization and formation of a nondissociating complex [54].

Rao and Forssberg [55] used zeta potential measurements to evaluate the effect of mixed cationic and anionic collectors on feldspar and quartz. The zeta potentials of quartz and microcline at a constant diamine concentration but increasing anionic concentration at a specified pH and at different pH values, are shown in Figs. 14 and 15, respectively. Figure 14 shows the zeta potentials at pH 2.5 where both minerals are negatively charged. The initial high positive potentials of quartz and microcline are due to the presence of 5×10^{-5}M diamine and its adsorption on the minerals. With increasing anionic sulfate concentrations, the zeta potentials are seen to decrease either due to coadsorption or complexation of anionic sulfate with diamine at the surface. Further increase in the concentration led to a sudden charge reversal as observed at 1 : 1 mole ratio and above, while the potentials varied slightly between –30 to – 40 mV. As these negative potentials are nearly the same as those of the minerals in water at pH 2.5, it is difficult to establish whether these potentials are due to the excess anions adsorbed on their surfaces or that the surfaces are devoid of any collector species. It is possible that when the concentration of the anionic collector exceeds that of the cationic collector, the amine cation cannot be adsorbed on the surface due to neutralizing effect and precipitation in the solution. An increase in flotation recovery is observed below the charge reversal collector concentration while above it, flotation ceases. This shows that the particles are hydrophilic above the charge reversal. If the negative potentials are assumed to be caused by the excess anions adsorbed, then they should be in reverse orientation in this concentration region. The zeta potentials as presented in Fig. 14, show that the mechanism of mixed collectors on feldspar and quartz is the same at pH 2.5 where both minerals exhibit negative potentials.

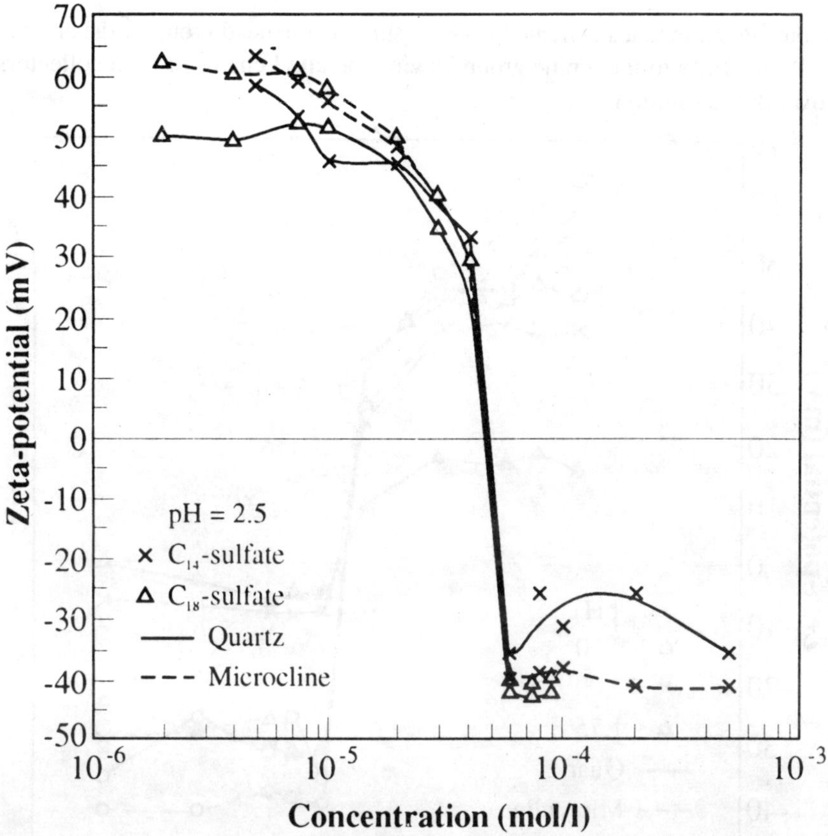

Fig. 14. Effect of anionic collector concentration on the zeta-potentials of microcline and quartz in the presence of 5×10^{-5} M diamine at pH 2.5 (after Ref. [55])

The reduction in the positive potential observed on quartz in the presence of diamine when the pH is decreased from 2.5 to 1.75 is attributable to the corresponding decrease in diamine adsorption (Fig. 15). Since the PZC of quartz is around pH 2, the diamine can no longer be adsorbed below this pH value. But, microcline still exhibits negative potential at pH 1.75 and hence the high positive potentials due to adsorbed diamine. Thus, it is the inherent charge differences of quartz and feldspar below pH 2 which is responsible for the selective flotation of feldspar in a mixed collector system. Initially, cationic diamine are adsorbed on the negatively charged feldspar at low pH values due to electrostatic attraction but not on quartz having almost no charge. The partial hydrophobicity thus obtained by feldspar is increased in the presence of anionic collector either through coadsorption or by complexation with the previously adsorbed diamine. The adsorption mechanism is similar to the representation shown in Fig. 13. This representation possibly suits better the alkyl diamine used in most of the studies, i.e., alkyltrimethylene diamine. As the second protonated amine also situated near the surface, the head group of the anionic collector fixes to the amine group leaving the alkyl chains of both collectors towards the solution.

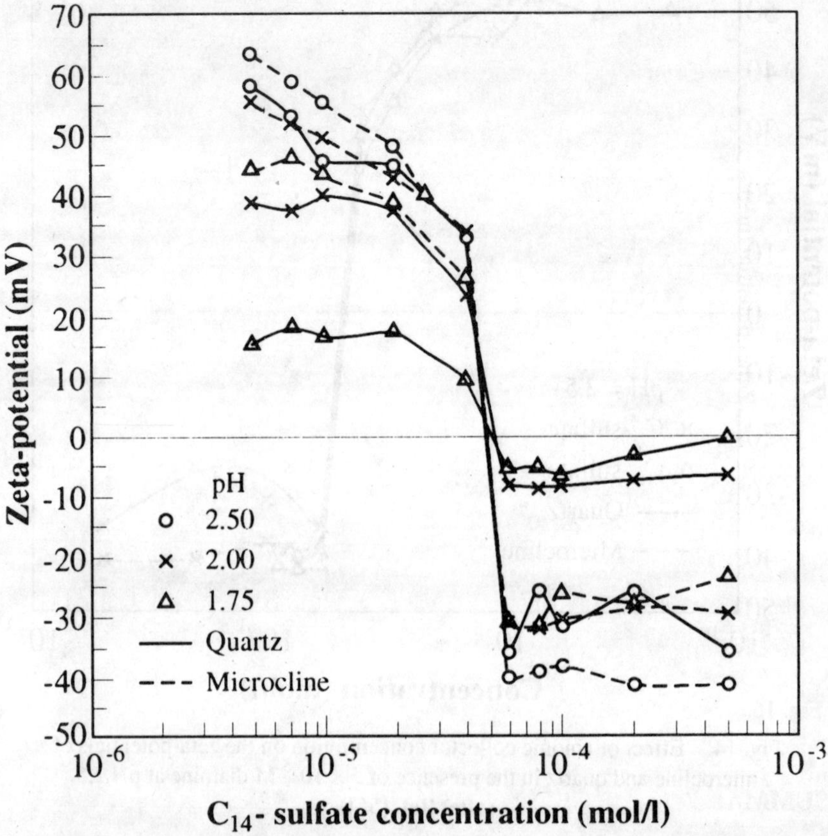

Fig. 15. Effect of anionic collector concentration on the zeta-potentials of microcline and quartz in the presence of 5×10^{-5} M diamine at various pH values (after Ref. [55])

The studies of Bolin [43] also indicate that the success of a mixed collector system depends mainly on the charge difference of feldspar and quartz in flotation pulps. The flotation results as presented in Fig. 12 show that the HF method performed better than the combined collector system (non-HF method). The zeta potential measurements of feldspar and quartz in deionized water and in water from the same flotation experiments are shown in Fig. 16. The best flotation results were obtained where there is a large difference in the zeta potentials of quartz and feldspar. In the presence of fluoride ions, the difference is enlarged as more negative charges are placed on the feldspar surfaces for the amine adsorption. With Duomeen TDO collector and in the absence of fluoride ions, there is no big charge difference even at very low pH values (due to the presence of Fe-ions in the pulp liquid) and as a result poor flotation results are obtained.

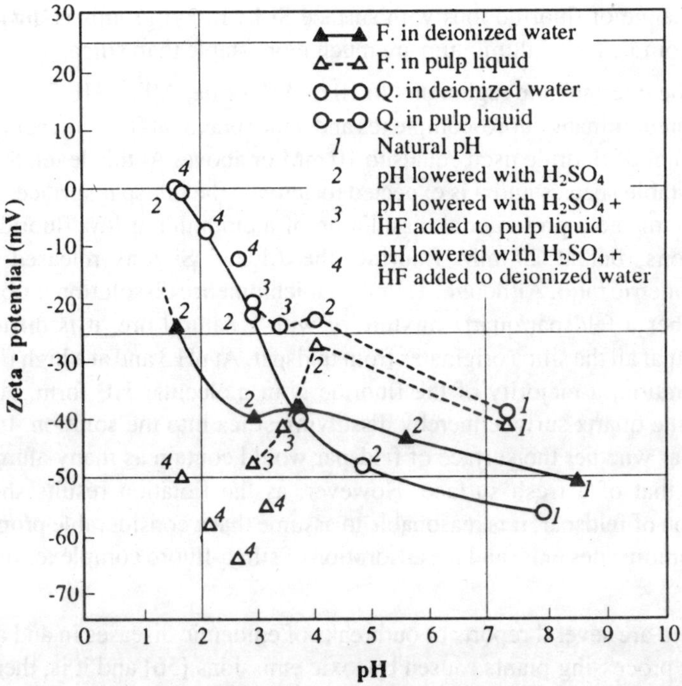

Fig. 16. Zeta-potentials of agate-milled feldspar and quartz in deionized water and pulp liquid (after Ref. [43])

SUMMARY

Despite the success of the HF-method in feldspar-quartz flotation system, only a few authors have attempted to explain the role of fluoride ions in the activation mechanism of feldspar at low pH values. The mechanisms proposed are conflicting and the studies can be classified mainly into two groups, one at low concentrations of fluoride and the other at high concentrations. The authors who used low concentrations proposed the adsorption of either fluoride ions or hydrogen fluoride complex on aluminium sites, whereas the group who employed high concentrations similar to the practical conditions, explained the role of hexafluosilicate ions in feldspar activation. Figure 6 clearly shows two distinct regions of feldspar flotation, one occurring at low concentrations and the other at very high concentrations. At low concentrations (< 1 mM), AlF_2^+, AlF^{2+} and AlF_3 are the predominant fluoro-complexes. It is expected that the surface aluminium would complex with fluoride ions, but it is not certain that such complexation can impart a negative charge. As the zeta-potentials show, at increased negative charge even at low fluoride additions, fluoride ions could exchange the surface Al-hydroxyl groups. It should be mentioned here that the low concentrations used correspond to a pH region of 4 to 5 at which, free fluoride ions exist in the solution while HF_2^- species are almost non-existent. An exchange of fluoride ions with surface Si-hydroxyl groups is difficult as fluoro-complexes of aluminium are much more stable than silica.

In the intermediate region of 1 to 10 mM fluoride, AlF_4^-, HF_2^- and SiF_6^{2-} are the predominant fluoro-complexes and under practical flotation conditions, the amount of fluoride used equals to 10 mM or above. At this level, SiF_6^{2-} is a more stable species and it is expected to activate the feldspar surface. All the studies indicate a preferential dissolution of aluminium at low fluoride concentrations, but at 20 mM fluoride, the Al and Si ions released are in stoichiometric ratio. Although, a similar stoichiometric dissolution is observed with either a feldspar/quartz mixture or with a natural ore, it is difficult to assume that all the silica originates from feldspar. At pH 3 and at a high fluoride concentration, a majority of the fluoride is in molecular HF form. HF also attacks the quartz surface thereby dissolving silica into the solution. It is not very clear whether the surface of feldspar would contain as many aluminium sites as that of a fresh surface. However, as the flotation results show an activation of feldspar, it is reasonable to assume that a considerable proportion of aluminium sites exist and the adsorption of silico-fluoro complexes on these sites.

There are several reports of outbreaks of epidemic diseases in and around feldspar processing plants caused by toxic emissions [56] and it is, therefore, imperative to search for new reagent schemes. Although, the method of using mixed cationic and anionic collectors for feldspar flotation without HF at a very low pH value appears to be promising, studies have so far showed only a limited success. This method rests principally on the inherent charge difference of feldspar and quartz around pH 2. At this pH, quartz possess near zero charge while feldspar is slightly negative so that diamine adsorbs on feldspar but not on quartz. The partial hydrophobicity so obtained by feldspar could thus be increased with anionic collector either through co-adsorption or complexation with a previously adsorbed diamine. However, it is very difficult to maintain these charge differences due to the existence of various metal cations in flotation pulp liquid. Moreover, the PZC's of several feldspar minerals are nearly the same as that of quartz. Hence, an effective reagent scheme for feldspar-quartz flotation system is not yet established.

The dissolution of feldspars are stoichiometric between pH 6 and 10 with respect to Si and Al. At this region, feldspars differ from quartz in surface composition. There is an extensive literature on the use of chelating agents and n-octyl imino-bis-methylene phosphonic acid, where no flotation of silica is observed, but garnet and tourmaline show strong floatability below pH 7 and 6 respectively [57]. It could be worthwhile to try these types of reagents. The adsorption of metal cations could be different on feldspar and quartz. It is possible that the metal cations act as either selective depressants in cationic flotation or selective activities in aniomic flotation. With a metal ion depression in cationic flotation, encouraging results were obtained in the separation of two feldspar minerals [1].

ACKNOWLEDGEMENT

The financial support from the Swedish Mineral Processing Research Foundation (MinFo) is gratefully acknowledged.

REFERENCES

1. Revnivtzev, V.I., Putrin, A.M., and Archangelskaya, I.N. : Proce VIII Int. Miner. Process. Congr., Leningrad, 1968, Vol. D-7, pp 1-8.
2. Fuerstenau, D.W., and Raghavan, S. : Proc. XII Int. Miner. Process. Congr., Sal Paulo, Brazil, 1977. Sao Paulo : Nacional Publicacoes e publicidade S/C Ltda., 1980, Vol. II, pp 368-415.
3. Smith, R.W., and Akhtar, S. : In Fuerstenau, M.C. (Ed.) : Flotation. Baltimore : AMIE 1976, Vol. 1, pp 87-116.
4. Fuerstenau, D.W., Fuerstenau, M.C. : In King, R.P. (Ed.) : Principles of Flotation. Monograph Series No. 3, Johannesburg : SAIMM, 1982, pp 109-158.
5. Smith, J.V. : Feldspar Minerals-Crystal structure and physical properties. Berlin : Springer-Verlag 1974, p 447.
6. Ribbe, P.H. : In Brown, W.L. (Ed.) : Feldspars and Feldspathoids. Dordrecht : D. Reidel Publishing Company 1984, pp 1-54.
7. Velbel, M.A. : In Davis, J.A., Hayes, K.F. (Eds.) : Geochemical Processes at Mineral Surfaces. ACS Symp. Series 323, 1986, pp 615-634.
8. Schott, J., and Petit, J-C. : In Stumm, W. (Ed.) : Aquatic Surface Chemistry. New York : John Wiley & Sons 1987, pp 293-315.
9. Correns, C.W. : Clay Minerals Bulletin 4 (1961). 249. Cited in : Lidström, L. : Acta Poly. Scandinavica, Chemistry Metallurgy Series No. 66, 1967, 112 pp.
10. Perry, D.L., Tsao, L., and Gaugler, K.A. : Geochim. Cosmochim. Acta. 47 (1983) 1289-91. Cited in reference 7.
11. Della Mea, G., Dran, J-C., Petit, J-C., Bezzon, G., and Rossi-Alvarez, C. : Paper presented at 6th Ion Beam Analysis Meeting, Tempe, Arizona, 1983. Cited in reference 7.
12. Chou, L., and Wollast, R. : Am. J. Science 285 (1985) 963-993.
13. Stumm, W., and Furrer, G. : In Stumm, W. (Ed.) : Aquatic Surface Chemistry. New York : John Wiley & Sons 1987, pp 197-219.
14. Joy, A.S., Manser, R.M., Lloyd, K., and Watson, D. : Trans. IMM 75 (1966) C81-C86.
15. Bustamante, H., Sparroe, G.J., and Warren, L.J. : Chemistry in Australia 48 (10) (1981) 375-379.
16. Warren, L.J., and Kitchener, J.A. : Trans. IMM 81 (1972) C137-C147.
17. Read, A.D., and Manser, R.M. : The action of fluorides as a modifying agent in silicate flotation. Mineral Processing Information, Note No. 8, Stevenage : Warren Spring Laboratory 1975, 42 pp.

18. Ney, P. : Zeta-Potentiale und Flotierbarkeit von Mineralen. Wien : Springer-Verlag 1973, pp 149-189.
19. Shimoiizaka, J., Nagai, R., Taira, Y., and Nakatsuka, K. : J. Min. Metall. Inst. Japan 80 (918) (1964) 1054-1058.
20. Ladoo, R.G., and Meyers, W.M. : Nonmetallic Minerals. New York : McGraw-Hill Book Co. Inc. 1951, p 216.
21. O Meara, R.G., Norman, J.E., and Hammond, W.E. : Bull. Am. Ceram. Soc. 18 (8) (1939) 286-292.
22. Dean, R.S., and Ambrose, P.M. : Development and use of certain flotation reagents. US Bureau of Mines 1944, Bulletin No. 449.
23. Buckenham, M.H., and Rogers, J. : Trans. Inst. Min. Metall. 64 (1954) 11-30.
24. Suliin, D.B., and Smith, R.W. : Trans. IMM 75 (1966) C333-C336.
25. Tröndle, H.M., Clement, M., and Brehler, B. : Interceram (3) (1970) 185-192.
26. Uhlig, U. : Proc. XVth Int. Miner. Process. Congr. France : Societe de L' industrie Minerale 1985, Vol. 2, pp. 265-277.
27. Lutjen, G.P. : Engg. Min. J. 154 (5) (1953) 92-95.
28. Baarson, R.E., Ray, C.L., and Treweek, H.B. : In Fuerstenau, D.W. (Ed.) : Froth Flotation. New York : Trans. AIME 1962, pp 427-453.
29. Eddy, W.H., Browning, J.S., and Hardemon, J.E. : Selective flotation of minerals from North Carolina mica tailings. US Bureau of Mines 1969, Report Investigations No. 7319, 10 pp.
30. Weiss, R. : Interceram 19 (2) (1970) 112-114.
31. Batchelor, R.W. : Trans. J. Brit. Ceram. Soc. 72 (1) (1973) 7-10.
32. Redeker, I.H. : Erzmetall 30 (12) (1977). 566-572.
33. Rao, D.V.R., and Hohmann, E.H. : Chemical Age of India 20 (3) (1969) 195-198.
34. Schubert, H., and Abido, A.M. : Bergakademie 19 (10) (1967) 601-605.
35. Abido, A.M. : Technical Journal, Nat. Metall Lab. India 11 (2) (1969) 25-29.
36. Smith, R.W. : Trans. SME-AIME 232 (1965) 160-168.
37. Smith, R.W., and Smolik, T.J. : Trans. SME-AIME 232 (1965) 196-204.
38. Sclar, C.B. : American Mineralogist 57 (1972) 287-291.
39. Katayanagi, T. : Flotation separation of feldspar. US Pat. No. 3,844,939, 1974.
40. Shimoiizaka, J., Nakatsuka, K., and Katayanagi, T. : In Weiss, A. (Ed.) : World Mining and Metals Technology. Baltimore : Port City Press 1976, pp 423-438.
41. Malghan, S.G. : Selective flotation of feldspar in non-fluoride medium. SME-AIME Fall Meeting 1976, Denver, CO, Preprint No. 76-H-331.
42. Malghan, S.G. : Mining Engineering (1981) 1616-1623.
43. Bolin, N.J. : Erzmetall 36 (9) (1983) 429-432.
44. Mathieu, G.I., and Sirois, L.L. : in Jones, M.J., and Oblatt, R. (Eds.) : Reagents in the Minerals Industry. London. : Inst. Min. Metall. 1984, pp 57-67.
45. Tang, J., Mao, J., and Sun, B. : In Forssberg, E. (Ed.) : Proc. XVI Int. Miner. Process. Congr. Amsterdam : Elsevier 1988, Vol. 10B, pp 1529-1536.
46. Tang, J., Sun, B., Cheng, Z., Chen, X., Zhao, H., and Zhang, S. : Proc. XVIII Int. Miner. Process. Congr. Parkville : Australian Inst. Min. Metall. 1993, Publication Series No. 3/93, Vol. 4, pp 851-856.

47. Jakobs, U., and Dobias, B. : New aspects in the flotation separation of feldspar and quartz. XVII Int. Miner. Process. Congr., Dresden 1991, Poster session.
48. Liu, Y., Gong, H., Qui, J., and Zhang, K. : Proc. XVIII Int. Miner. Process. Congr. Parkville : Australian Inst. Min. Metall. 1993, Publication Series No. 3/93, Vol. 4, pp 857-862.
49. El Salmawy. M.S., Nakahiro, Y., and Wakamatsu, T. : Proc. XVII Int. Miner. Process. Congr. Parkville : Australian Inst. Min. Metall. 1993, Publication Series No. 3/93, Vol. 4, 845-849.
50. Buckenham, M.H., and Schulman, J.H. : Trans. SME-AIME 226 (1963) 1-6.
51. Elworthy, P.H., Florence, A.T., and Macfarlane, G.B. : Solubilisation by Surface Active Agents and Its Application in Chemistry and Biological Sciences. Suffolk: Chapman and Håll Limited 1968.
52. Somasundaran, P. : Int. J. Miner. Process. 3 (1) (1976) 35-41.
53. Rosen, M.J. : In Scamehorn, J.F. (Ed.) : Phenomena in Mixed Surfactant Systems. Washington : ACS Symposium Series 311, American Chemical Society 1986, pp; 144-162.
54. McEwen, R., Hansen, G.W., and Lee, G.F. : Trans. SME-AIME 260 (1976) 97-100.
55. Rao, K.H., and Forssberg, K.S.E. : Proc. XVIII Int. Miner. Process. Congr. Parkville : Australian Inst. Min. Metall. 1993, Publication Series No 3/93, Vol. 4, pp 837-844.
56. Lercher, P., and Kofler, W. : In Seemayer, N.H., and Hadnagy, W. (Eds.) : Environmental Hygiene. Berlin : Springer-Verlag 1988, pp 175-178.
57. Collins, D.N., Wright, R., and Watson, D. : In Jones, M.J., and Oblatt, R. (Eds.): Reagents in the Minerals Industry. London : Inst. Min. Metal. 1984, pp 1-12.

Flotation of Complex Sulfide Ores

PRADIP, K. K. DAS and R. SINGH
Tata Research Development and Design Centre,
1 Mangaldas Road, Pune 411001, India

ABSTRACT

The flotation of complex sulfide ores containing primarily copper-lead-zinc ore minerals continues to be a challenging area of current research interest. The complexity of processing such ores is mainly due to: the fine dissemination and intimate association of ore minerals in the matrix; the presence of clay and other hydrophobic gangue minerals (talc, graphite etc.); the presence of excessive amounts of pyrite - pyrrhotite minerals; the occurrence of oxide and oxidized ore minerals and due to the presence of minor minerals of major economic value such as those containing precious metals, cadmium, mercury, arsenic, bismuth, antimony, molybdenum, rhenium, nickel, cobalt etc. Some of the significant recent advances in the field are reviewed in this paper under five broad categories, namely, characterization tools, reagent chemistry, flotation machines, flotation circuits design and pulp chemistry & grinding media wear. Selected case studies based on the authors' own work on the processing of Indian complex sulfide ores are also presented in order to illustrate a few possible scientific approaches to deal with the problems through well designed systematic investigations.

INTRODUCTION

The flotation of complex sulfide ores containing copper-lead-zinc minerals remains one of the most important challenges in base metals metallurgy. The complexity of processing such ores stems from one or more of the following :

- Fine dissemination of ore minerals in the matrix.
- Complex mineral association, that is, a number of sulfide minerals and sulfo-salts in close association with each other such that the liberation of individual minerals and hence selective separation in order to produce clean concentrates becomes difficult.
- Presence of clay and other gangue mineral slimes (e.g. sericite) adversely affecting the selectivity of separation.
- Occurrence of naturally flotable (hydrophobic) gangue minerals such as talc, pyrophyllite, graphite and mica interfering in achieving the desired selectivity.
- Presence of excessive amounts of pyrite and pyrrhotite causing problems in separation due to difficulty in the depression of iron sulfides and due to

the enhanced propensity of the ore minerals to oxidize (galvanic interactions).

- Occurrence of oxide and oxidized minerals along with sulfide minerals leading to problems of losses in recovery during flotation with conventional reagents.
- Presence of minor minerals of economic importance containing precious metals, cadmium, mercury, arsenic, antimony, bismuth, molybdenum, rhenium etc.; recovery of these minerals is important but often difficult and involves tradeoff between conflicting requirements. Cyanide for example improves selectivity but leads to lower recovery of precious metals.
- Ore type variability in mineralogical composition, head grade, degree of oxidation and dissemination affecting its metallurgical response; presence of a variety of valuable minerals having different flotation behaviour as well as different liberation sizes requires a complex flowsheet in order to achieve optimum performance.

It is well recognized that any alteration of mineral samples either due to the geological origin of the ore deposit and/or a consequence of environmental factors during mining, storage and transport causes problems during the flotation of sulfide ores. The process flowsheets and the reagent mix employed for the beneficiation of such complex ores therefore tend to become more complicated. It may involve several stages of cleaning, regrinding and recirculation steps. The techno-economic feasibility of utilizing such complex ore deposits thus depends on a proper design of comminution-flotation circuits. We present in this paper some of the possible problem areas and approaches to overcome them. A few case studies based on our work on Indian base metals ore deposits are discussed in greater detail to illustrate the salient features of complex ore processing.

RECENT ADVANCES IN PROCESSING OF COMPLEX SULFIDE ORES

Significant advances have been made in the industrial flotation practice so as to be able to process complex sulfide ore deposits. These are briefly reviewed in this section under the following broad categories: (i) better techniques to characterize ores, concentrates and rejects, (ii) design, development and application of improved reagent chemistry tailor-made for specific separation problem, (iii) improved design of flotation machines, (iv) computer aided design of optimum flotation circuit configurations, and (v) better understanding of pulp chemistry and its effect on flotation. Several recent monographs and proceedings volumes on the processing of complex sulfide ores are available and the readers are referred to those for details [1-14, 85, 86].

Characterization of Flotation Feeds, Concentrates and Rejects

Conventional assessment of the metallurgical performance of a flotation circuit is based on the assumption that each mineral is perfectly liberated. On the other hand, the separation efficiency of a circuit treating complex ores depends on the behaviour of the middling particles (composites), particularly with respect to their split in various separation steps (rougher, cleaner, recleaner etc.). The advent of recent semi-automatic equipment like QEM-SEM have made it possible to provide quantitative measures of liberation in a given sample [15]. It is thus possible to establish as to what extent the poor liberation limits the plant performance. The results of Frew and Davey's [16] work on Elura ore for example, illustrate the usefulness of such measurements. Liberation measurements with respect to each component under study (e.g. galena, sphalerite and pyrite) on four to seven size fractions each of the various stream samples collected in the plant could be conveniently made with the help of QEM-SEM. The combined effect of size as well as liberation could thus be examined not only for the whole circuit but for each unit operation in the plant that is, roughing, scavenging, cleaning and regrinding separately [17]. It was established, for example, that the effect of regrinding was to increase the recovery of sphalerite by reducing the amounts of sphalerite in less liberated sizes rather than by increasing the liberation of sphalerite within a given size fraction [16].

Reagent Chemistry

Considerable progress has been made in our understanding of flotation reagent chemistry for treating complex sulfide ores [10, 12, 13, 14, 18-23, 87]. Current research activity in this area is directed towards finding alternate reagent schemes which (i) are less dependent on pH control for optimum effectiveness, (ii) offer enhanced recoveries of small amounts of precious metals and other minor but valuable constituents such as Au, Ag, Ni, Co, Mo, Rh, Pt etc. (iii) yield more selective flotation separation over pyrite, pyrrhotite, talc, graphite etc. - the components of considerable nuisance value present in complex sulfide ores, (iv) improve recovery of relatively coarse and partially liberated particles - these can be subsequently reground and refloated, if necessary but such an approach can lead to enormous savings in crushing/grinding costs, (v) enhance recovery of oxidized minerals during the beneficiation of mixed sulfide-oxide ores, (vi) perform more effective differential flotation among sulfide minerals (Cu/Mo, Cu/Zn, Pb/Zn, Cu/Pb, Cu/Ni etc.) so as to achieve the desired quality of concentrates obtained by processing complex ores, (vii) are effective in reducing entrainment of fine gangue during flotation - new frother chemistry, for example, so as to control the froth structure and froth drainage, (viii) can substitute the environmentally hazardous but commonly used inorganic modifiers such as cyanides, dichromates, ferro/ferricyanides, sulfides, hydrosulfides, and hypochlorites.

A few examples of relatively new collectors for treating complex sulfide ores are listed in Table 1. Some of these reagents are already being used in commercial operations whereas others have been tested extensively in laboratory and pilot plant scale. In China, for example, conventional xanthate collectors are reported to be increasingly replaced with dibutyl dithiophosphate,

Table 1: Some examples of relatively new flotation collectors for sulfide ores

Reagent	Structure	Applications	Ref.
Sodium Diisobutyl-dithiophosphinate (Aerophine 3418A)	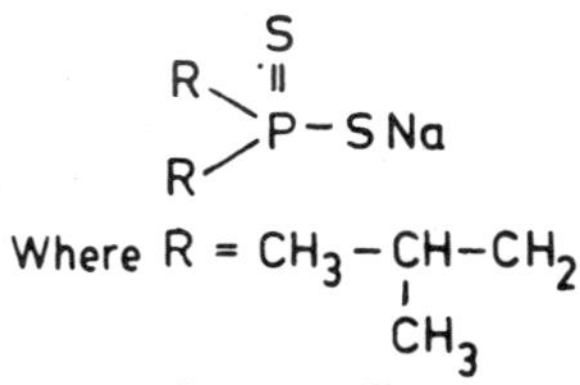	Improved selectivity against Zn and iron sulfides, increased recovery of precious metals, rapid flotation kinetics	24
Alkoxy Carbonyl Alkyl Thiono-carbamates	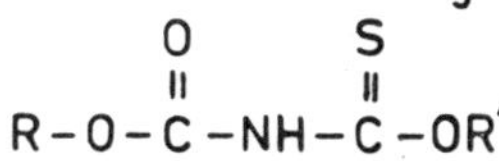	Excellent for copper rich minerals like bornite, covellite & chalcocite	25
Alkoxy Carbonyl - Alkyl Thioureas	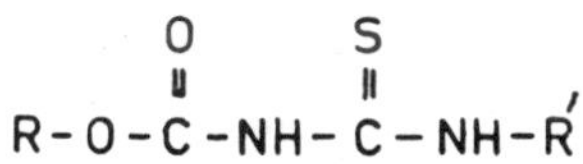	Excellent for chalcopyrite flotation	25
Dialkyl or Diaryl Monothio-phosphates	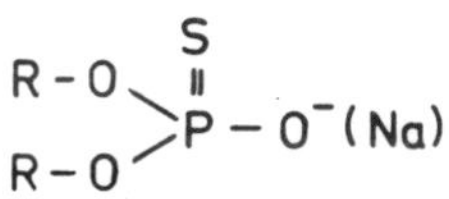	Effective Acid Circuit Collectors	25
Dialkyl Monothio-phosphinates	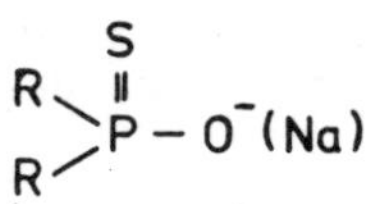	Effective in neutral and alkaline circuits	25
Alkyl Dithiophosphates	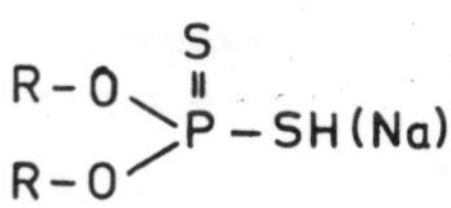	Effective at pH > 9	25
Dicresyl Monothiophosphate	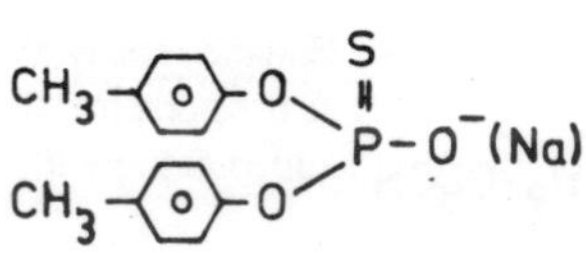	Excellent for selective gold flotation	26
1, hydroxy-ethyl-2-heptadecenyl glyoxalidine (AMINE 220)	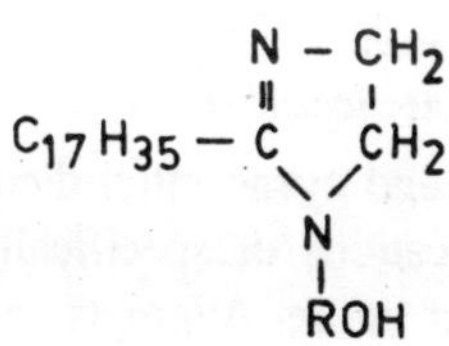	Better than conventional sulfide collectors	27

(Contd)

Table 1 (*contd.*)

Reagent	Structure	Applications	Ref.
Alkyl Derivatives of Mercaptobenzo-thiazole Aminothiophenol Mercaptobenzo-oxazole	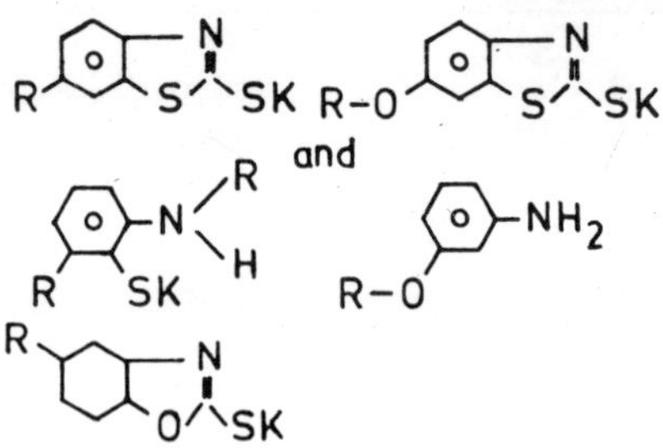	Excellent for the beneficiation of oxidized lead and zinc ores	28, 29
Alkyl Hydroxamates	R – C – O H – N – OH	Excellent for mixed sulfide-oxide ores (particularly copper)	30, 31, 32
F Series	$R_1-X-(CH_2)_n-N\langle^{R_2}_{R_3}$ e.g. $C_6H_{13}S(CH_2)_2NH$	Excellent for chalcocite, sphalerite & precious metals	33
S Series	R_1-S-R_2 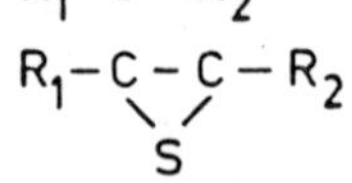 e.g. $C_8H_{17}SC_2H_5$	Excellent for chalcopyrite, molybdenite, pentlandite	33
Alkyl Derivatives of Salicylaldoxime (SALO)	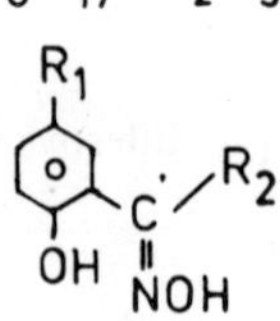	Excellent for oxidized copper, lead, zinc minerals	34
Modified (Dithio)* carboxylic acids	$R_nX-C(=S)-S-CH_2-COOH$	Enhance recovery of copper & precious metals when added along with xanthate	35

* e.g. x=O O-alkyl xanthato acetic acid (n=1)
x=N (N,N dialkyldithiocarbamate) acetic acid (n=2)

benzene ammonium dithiophosphate and cyano ethyl diethyl dithio carbamate [36]. The need and hence the applications of specifically designed flotation collectors is likely to grow in the near future. Attempts are also being made to develop the science of reagents design based on molecular recognition mechanisms so that one can engineer the molecules for specific applications [14, 21].

Several attempts made to find substitutes for environmentally hazardous depressants for the beneficiation of sulfide ores have been successful. For example, natural polysaccharides and their derivatives are found to be quite promising [37]. Starch, carboxymethyl cellulose (CMC), dextrin, guargum, for example, are known to be effective gangue depressants. Laskowski and coworkers [38] have recently demonstrated their utility in differential flotation amongst sulfide minerals. The adsorption of dextrin was found to be maximum at pH 10-11 on galena and at pH 4-9 for chalcopyrite, the pH values coinciding with the isoelectric points (iep) of the corresponding hydroxides that is lead, copper and iron respectively. Based on this observation, it was possible to selectively float chalcopyrite from galena at pH 11-11.5 using xanthate and chalcocite from heazlewoodite (nickel sulfide) in the range of pH 11.5-12, both with xanthate as well as with diphenyl guanidine collectors [39, 40]. Dextrin addition was found beneficial in enhancing selectivity both during rougher as well as cleaner stages.

Sodium sulfite is another reagent which is found to enhance selectivity in the sequential flotation scheme (first copper, then lead, followed by zinc) for the beneficiation of complex sulfide ores. Houot and Duhamet [41] recently reported the role of metal ions (Ca^{2+}, Ba^{2+}) or corresponding metal hydroxides in the depression of galena using sulfite. The proposed depression mechanism involves two steps: the adsorption of metal ions followed by reaction with sulfite ions to form insoluble metal sulfites on the surface which prevents collector adsorption.

Carboxymethyl cellulose (CMC) is a common depressant for gangue minerals like talc and pyrophyllite. The mechanism of enhancing selectivity in case of a nickel transition ore from Western Australia containing pentlandite, pyrrhotite, violarite and pyrite was found to be slime dispersion rather than depression [42]. The main cause of high magnesite content of the concentrate during flotation was found to be the slime coating of fine magnesia containing minerals. An interesting observation reported by the investigators was that the flotation of this ore in a saline pulp (10% NaCl) resulted in a better flotation response. An inexpensive reagent like sodium chloride could thus substitute for CMC.

Conventional cyanide-dichromate reagent scheme for beneficiating Cu-Pb-Zn ores suffers from a number of drawbacks - the environment hazards posed by cyanide, prohibitive amounts of secondary sulfide minerals causing problems in selectivity with cyanide, loss of precious metals due to cyanide complexing and high alkalinity of cyanide flotation with its viscous froth causing problems in separation [36]. In China, several alternate reagent schemes have been employed commercially [36] for example, bulk flotation with Na_2CO_3-$ZnSO_4$ - sodium thiosulphate depressant system followed by copper-lead separation with dichromate. Copper-lead separation can be carried out with dichromate-sulfite or hydrogen peroxide-ferrosulfate-sulfite or 2-carboxylic cellulose sulfite or a simple sulfite system with an increased temperature of flotation (60-70°C). Enhanced temperature of flotation can also be used for achieving Pb-Zn separation. An interesting study reported by Reddy *et al.* [43] indicates that preactivation of cyanide depressed sphalerite with formaldehyde and nickel sulfate solutions helps reduce the consumption of copper sulfate. The use of formaldehyde was found to improve the grade and recovery of not only sphalerite from the ore sample obtained from Rajpura-Dariba, India but also increase the silver content in the zinc concentrate.

Conventional industrial practice of treating mixed sulfide-oxide ore deposits involves sulfidizing the oxidized minerals with sodium sulfide or hydrosulfide before flotation. There have been considerable research efforts devoted to finding alternate sulfidizing agents because sulfide ions, if not regulated, also depress sulfide flotation. Oxidized copper minerals of commercial importance are malachite and chrysocolla. A substantial improvement in flotation was reported recently for malachite sulfidized with sodium tetrasulfide instead of conventional hydrosulfide ions [44]. The detrimental effect of sulfidization residue is due to the presence of copper species as well as thiosulfate ions which consume a part of collector thereby decreasing recovery. Tetrasulfide solution was observed to be a much better sulfidizing agent because, (a) the sulfidized coating is more uniform and adherent and, (b) the amounts of thiosulfate and solution copper species generated are smaller [45].

Calcium sulfide, produced by reduction roasting of gypsum with coal is being used in China as a sulfidizing agent for oxidized copper ores [46]. Zhang and Poling [47] have shown that prior or simultaneous addition of ammonium sulfate can not only enhance the activating effect of hydrosulfide ions on the xanthate flotation of malachite but it can also eliminate the detrimental effect of excess residual hydrosulfide. Similarly, ethylene diamine phosphate (EDP) is also found to be an excellent sulfidization-promoter for oxidized copper minerals. In China, with an addition of 82-156 g/t of EDP, the recoveries of oxide copper could be increased by 6-11% with a reduced consumption of butyl xanthate by 43% and sodium sulfide by 34% [46].

Several recent research publications on reagent testing have clearly demonstrated the need to look at both the reagent kinetics as well as the reagent selectivity while conducting laboratory scale testing, particularly with respect to replacing one reagent with another. The measurement of flotation rate constant, k, is crucial in the evaluation of reagent performance. The scale up factors for k going from laboratory to pilot plant and plant scale are unfortunately not unique [18, 48, 49]. There is, thus, an urgent need to understand the scale up problems so that meaningful predictions regarding plant performance can be made based on bench scale laboratory data.

Flotation Machines

Over the last three decades, in response to the realization that conventional mechanical cells are not suited for the recovery of relatively coarse and fine particles, several new designs of flotation machines have emerged on the scene.

Some of the machines which have been commercialized in the processing of complex ores are briefly discussed in this section.

Column flotation has now become widely accepted in the mineral processing industry in general and processing of complex sulfide ores in particular. The column has been found particularly attractive for applications involving multiple cleaning stages since it can upgrade the feed in a single stage compared to several stages of conventional flotation cells. The air sparger unit is the most crucial aspect of column design. Several designs of air sparger units are now available such as Cominco air sparging system [50], US Bureau of Mines (USBM) design [52], Virginia Polytechnic Institute Microbubble flotation column [53] and Deister Flotaire Bubble generator system [51]. Several variations on the design of column flotation cell have also been attempted. For example Wemco-Leeds column having a number of horizontal barriers arranged vertically one above the other thus separating the pulp into a series of vertical compartments is claimed to yield better grades than conventional column [50]. Michigan Tech University (MTU) packed column flotation technology [54], Jameson Cell [50], AFT Sprayflot Cell [55], air sparged hydrocyclone [56] and EKOF's pneumatic flotation cell technology [50] are also being commercially tried.

Recently Outokumpu has introduced the new HG machine which is claimed to provide the metallurgical performance comparable to column cells. It is similar to conventional Outokumpu cells except that the rising froth is forced to pass through a tapered channel where froth is water sprayed with the help of special nozzles. Outokumpu's Skim-Air flash flotation machine for the flotation of relatively coarse particles are installed in a number of sulfide ore processing plants all around the world. The feed to these machines is essentially the cyclone underflow from the comminution circuit such that coarse, liberated particles of flotable sulfide minerals are selectively captured before being reground, thus resulting in savings in energy, reduction in overall slime losses, increasing the capacity of the flotation circuit and improvements in the ease of handling of the final concentrate (since it is coarser) with respect to dewatering/drying. Introduction of flash flotation has also resulted in better recovery of precious metals such as silver and gold during flotation of complex sulfide ores [57, 58].

Flotation Circuit Design

Commercial flotation circuits are currently designed and evaluated on the basis of laboratory scale semi-batch flotation tests, locked cycle tests and pilot plant trials. An important input in the design of such circuits is a set of more or less empirical/heuristic rules that have evolved over the years for improvement in circuit performance based on experience and intuition [59]. It is now realized however, that these rules are only indicative of the general direction for improvement and, in general, those cannot really optimize the circuit configuration for some given constraints. For a synthesis of an optimum circuit, it is important to make a detailed consideration of several factors such as ore feed and its liberation characteristics, reagent chemistry and flotation rate constants, process variables (e.g. pulp density, aeration rate, froth volume, residence time), design parameters like the number and size of cells per bank and interconnections between them and the objective or profit function which may range from simple aim of maximizing recovery to highly complex and non-linear function of recovery and grade. For a multi-component, multi-stage flotation circuit as required for the beneficiation of a complex sulfide ore, involving a number of recycle streams, only a comprehensive strategy based on an integral approach to circuit optimization is meaningful. A number of optimization algorithms have now been developed which along with the efficient and fast computing power available today, can be utilized to arrive at the best circuit configuration. The utility of such an integrated approach to circuit optimization has been demonstrated for a number of large-scale complex multi-stage industrial flotation circuits. Kapur *et al.* [60] has shown that utilizing the raw plant data, the given circuit in the plant can be simulated to a remarkably precise degree through such techniques - the flotation can be mathematically described in terms of a well defined spectrum of flotable species having certain rate constants. This approach thus opens up a whole new set of possibilities of optimizing flotation circuit configuration in a scientifically rigorous fashion [61].

Another interesting development in the area of flotation circuit design is the concept of multi-feed flotation circuits [36]. It turns out, for example, that if the rougher flotation concentrate is mixed with the incoming new feed and floated again in a bank (second stage), better flotation performance is observed. Such multi-feed flotation circuits where froth from the first stage of rougher flotation is partially or completely combined with the feed from second part of flotation have been in commercial operation in China for processing a number of low grade, finely disseminated complex ore deposits including sulfides [36]. The main advantages of such circuits are the reduction of reagent consumption by 20-30%, better metallurgical performance due to a thicker froth in the second stage resulting in better drainage of unwanted gangue and the autogenous carrier effect of relatively coarse particles of the first froth on the recovery of fine and ultrafine particles in the second stage leading to enhanced recovery of fines [36]. Such innovations in circuit design through modelling and optimization must be explored in order to exploit the difficult-to-process complex sulfide ore deposits.

Grinding Media Wear, Galvanic Interactions and Its Effect on Flotation of Complex Sulfide Ores

It is well documented that galvanic interactions between two or more sulfide minerals as well as those between a sulfide mineral and the iron grinding media in aqueous medium play an important role in determining the flotation behaviour of a complex sulfide ore [62-74]. With their rest potentials being much higher than that of iron, sulfide minerals act as a cathode while the iron balls act as anode in flotation pulps. Hence with xanthate collector this cathodic potential of sulfides will lead to reduced floatability [64]. Influence of galvanic contact between various grinding media and pyrrhotite on flotation was studied by Pozzo and Iwasaki [65, 66]. Rest potential measurements of pyrrhotite (PO), austenic stainless steel (ASS), martensitic stainless steel (MSS), mild steel (MS), zinc and magnesium showed that it decreases in the order

$$PO > ASS > MSS > MS > Zn > Mg$$

Thus, pyrrhotite will act as cathode and the grinding media as anodes on contact with each other. The oxidation products formed on the surface of pyrrhotite and responsible for reduced floatability was analysed to be iron sulfate. In general, grinding in steel mills leads to suppression of floatability of sulfide minerals.

Similarly galvanic interaction between two sulfide minerals leads to the preferential dissolution of the active mineral (anode) in the couple and the cathodic reduction of oxygen on the surface of the noble minerals [67]. The electrochemical reactions could be represented as follows:

$MS \rightarrow M^{2+} + S + 2\,e$ Oxidation of active sulfide mineral

$1/2\,O_2 + H_2O + 2\,e \rightarrow 2\,OH^-$ Cathodic reduction of oxygen on noble sulfide

The change in the flotation response of a sulfide mineral could be due to the following possible reasons [63]:

- Surface Passivation (Tarnishing): due to oxygen reduction on the cathodic surface leading to the formation of the constituent metal ion oxyhydroxide species on the surface.
- Migration of Metal Ions: generated due to the dissolution of anodic sulfide mineral which could react with hydroxyl ions generated at the cathode surface leading to coating of metal hydroxides and its attendant influence on the flotation behaviour:
- Activation Effect: due to the dissolved metal ions (as a result of galvanic interactions) activating an otherwise nonfloatable mineral. For example, activation of sphalerite surface when contacted with chalcopyrite and/or galena.
- Area Effect: due to the differences in the relative surface area of the two electrodes in a galvanic couple. A lesser amount of active mineral in the mixture (smaller anode in contact with relatively larger cathode) is known to favour increased anodic oxidation.

Thus, it is always the floatability of the noble sulfide mineral in the couple which is adversely affected due to galvanic interactions; the flotation of active mineral (anode) is not significantly impaired.

The presence of oxygen enhances galvanic interactions and hence grinding media wear during grinding of complex sulfide ores with attendant deleterious effects. Cheng and Iwasaki [68], for example, reported that the floatabilities of chalcopyrite and pyrrhotite during flotation of their synthetic mixtures were not affected significantly at neutral to acidic pH when the grinding was carried out in nitrogen atmosphere, whereas the recovery of pyrrhotite was considerably higher if the mixture was ground in an oxygen atmosphere.

Similarly Natarajan [69] found through a detailed study on the grinding of Indian complex sulfide ores from Malanjkhand Copper Deposit in Madhya Pradesh and Rajpura-Dariba Lead-Zinc-Copper Deposit, Rajasthan that maintaining a nitrogen atmosphere during wet grinding of such ores leads to significant reduction in the wear rate. The combined effect of oxygen as well as the grinding media wear on flotation of a copper ore (chalcopyrite) is illustrated in Table 2.

Table 2: Flotation recovery (denver laboratory flotation) of a copper ore after wet grinding (hyper steel balls as grinding media) (after Ref. [67])

Nature of Conditioning	Percent Recovery			
	Dry Grinding	Wet Grinding		
		N_2	Closed Mill	O_2
Frother alone	50.8	16.1	9.5	4.5
Frother + Collector	80.5	34.1	18.9	13.1
10^{-5} M Collector*	–	38.0	22.9	18.3
10^{-4} M Collector*	–	53.4	45.1	35.6

* Added during grinding; xanthate collector

As expected, the wet grinding particularly in presence of oxygen atmosphere leads to flotation depression. Addition of xanthate collector during grinding seems to help mitigate this adverse effect of galvanic interactions, even though the mechanisms of this interesting effect is not clearly understood [67].

Galvanic interactions, in general, thus not only enhance grinding media wear but also affect the flotation response adversely. The magnitude of this influence, of course, depends on the kind of sulfide minerals present in the ore and the type of grinding media used during grinding. During the beneficiation of complex multi-metal sulfide ores, particularly those containing relatively large quantities of pyrite-pyrrhotite minerals, the role of galvanic interactions and hence its adverse effect on flotation behaviour needs to be established so that proper remedial strategies could be worked out.

SPECIFIC CASE STUDIES

A few specific case studies based on our work on Indian ores where some of the above-mentioned problems associated with the processing of complex sulfide ores were successfully overcome after detailed investigations, are presented in this section.

COMPLEX MINERALOGY (MIXED SULFIDE-OXIDE ORES)

A detailed petromineralogical study on the ore is the first step in the test work on complex ores. Quantitative information about the association as well as liberation size of each mineral constituent provides crucial guidelines to the mineral engineers regarding the fineness of grind, the quality of concentrates and the potential problems anticipated during flotation.

For example, a critical review of plant results and experience at Malanjkhand Copper Deposit (India), suggested that it is essential to establish the role of mineralogy, particularly the presence of secondary sulfides and oxidized minerals on the flotation plant performance. Five ore samples collected from the mines in consultation with plant engineers and geologists were first studied mineralogically and then subjected to extensive bench scale flotation testing for this purpose. Extremely important clues regarding unsatisfactory plant performance were thus established.

The results of mineralogical, liberation and bench scale flotation studies on these five samples are summarized in Fig. 1. Depending on the location in the mine, one can thus have an ore feed quite different from the chalcopyritic ore for which the plant is designed. The feed richer in secondary sulfide minerals like chalcocite and covellite (SS & Tr. CHP) can still be processed by a judicious control of xanthate (SIPX) reagent. The feed richer in oxidized minerals like malachite and cuprite however requires sulphidization or an auxiliary collector. The sulphidization test work indicated that the control of pulp Eh is essential though difficult in practice to obtain satisfactory results [32]. Alternatively, we found copper chelating hydroxamate reagent to be quite effective. As illustrated in Fig. 1, replacing 50% of xanthate with hydroxamate enhanced the copper recovery considerably for both the high oxide (HO) samples. It is interesting to note that an increase in xanthate dosage alone will not work due to kinetic factors. While the equilibrium recovery R_∞ for HO(I) could be increased to levels comparable to sulfide minerals, the kinetics of flotation was observed to be much slower at higher dosages, the kinetics of sulfide minerals was in fact reduced considerably. This is the reason why higher xanthate dosages during plant trials led to an apparent depression of sulfide recovery. Xanthate-hydroxamate (SIPX-KHXM) combination, on the other hand, was found to be ideal. Subsequent pilot plant trials conducted in collaboration with plant engineers corroborated our findings (See Table 3).

Table 3: A comparison of sodium isopropyl xanthate (SIPX) with xanthate-hydroxamate (SIPX-KHXM) combination on malanjkhand copper ore (after Ref. [75])

Feed Grade 1% Cu
Pine oil dosage: 25 g/t
Number of sets of pilot plant campaigns: 17

	SIPX Alone	SIPX-KHXM Combination
Collector Dosage, (g/t)	48.6 ± 7	48.6 ± 7 SIPX 24.2 ± 3.2 KHXM
Total Copper Recovery, (%)	77.2 ± 11.2	85 ± 6.3
Concentrate Grade		
% Cu	31.2 ± 4	31 ± 5.9
% Insolubles	13.2 ± 8	16 ± 7.9

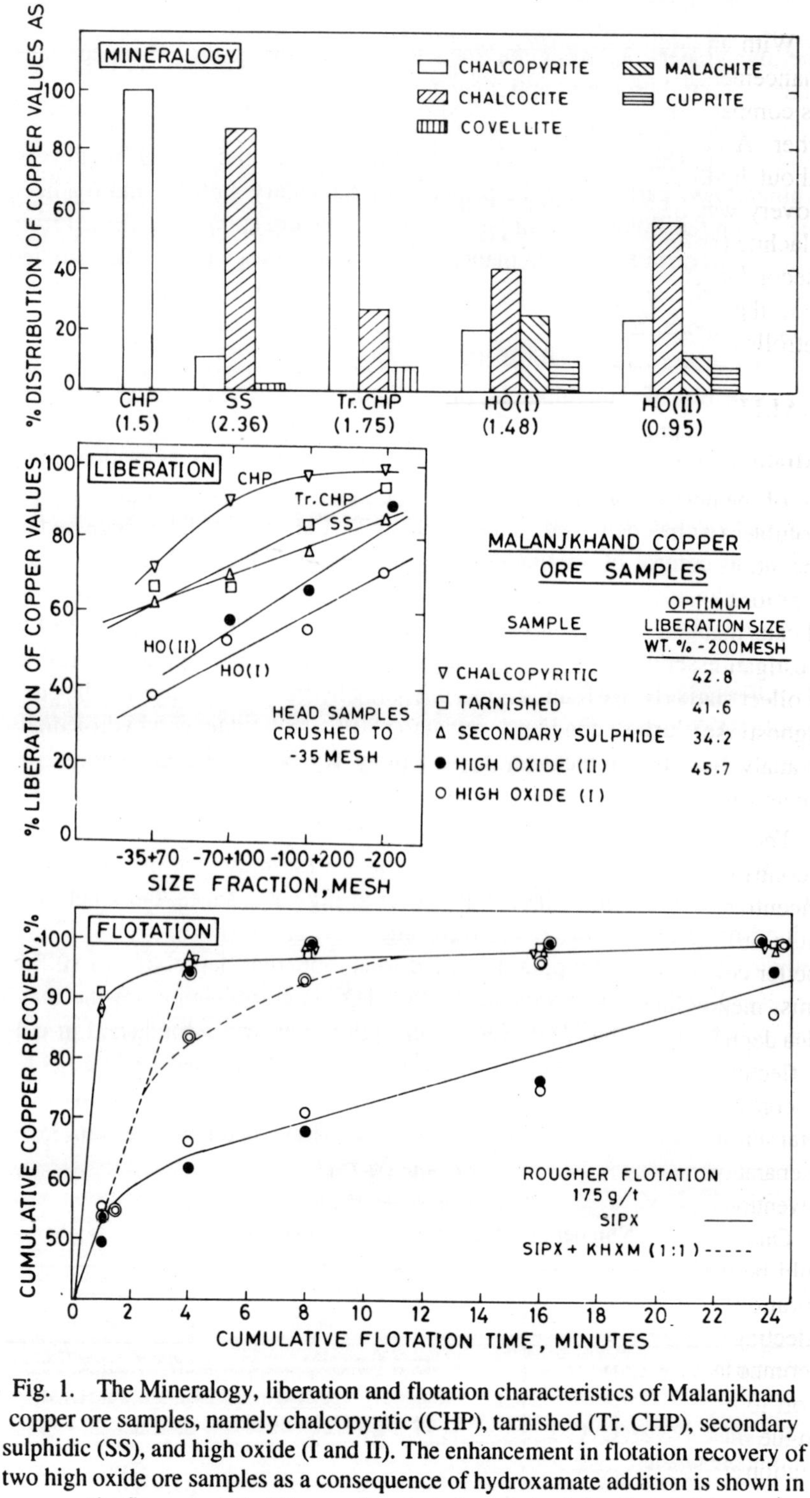

Fig. 1. The Mineralogy, liberation and flotation characteristics of Malanjkhand copper ore samples, namely chalcopyritic (CHP), tarnished (Tr. CHP), secondary sulphidic (SS), and high oxide (I and II). The enhancement in flotation recovery of two high oxide ore samples as a consequence of hydroxamate addition is shown in the figure in dotted lines and encircled symbols. (after Ref. [32])

With an additional dosage of 24 g/t of hydroxamate on an average, an enhancement in total copper recovery of about 8% could be achieved. The grade was comparable except that with hydroxamate, insolubles content was a little higher. A careful mineralogical examination of the concentrates with and without hydroxamate collector confirmed that the enhancement in copper recovery was due to the additional recovery of oxide copper in the form of malachite (not recoverable with xanthate alone in plant) and higher recovery of secondary sulfide minerals [32]. The relatively higher insolubles content was due to the fine silicate gangue being entrained during flotation, which can be controlled by other means as discussed in the following section.

QUALITY OF FINAL CONCENTRATES

Reduction of Silica

One of the common concerns in base metal ores flotation is the reduction of insolubles (% silica) in final concentrates. Unusually high silica content in the concentrates causes problems in subsequent smelter and hydrometallurgical operations. Our extensive investigations on a variety of Indian base metal ores indicate that the solution to this problem lies in scientific and systematic investigations to first establish its cause. Predesigned plant campaigns in order to collect sizewise data followed by mass balancing and analysis with a plant diagnostic software (for example, PREDICT, developed by us) and mineralogical analysis of concentrate products, can be used to pinpoint the source of silicate gangue in the concentrate.

The silicate gangue (sometimes graphite/mica gangue) for example, could be contributed by well-liberated coarse particles reporting in the final cleaner concentrate. The reason could be the lack of an appropriate reagent combination that is sufficiently selective against gangue. The remedy could lie in selecting a better collector and/or suitable gangue depressants such as modified guar-gums, metal-silicate hydrosols and carboxymethyl cellulose derivatives for silica depression. It is important to recognize that these gangue depressants will be effective only at relatively coarse sizes. Fines entrained along with water cannot be depressed effectively with this method. As illustrated in Fig. 2, the addition of metal silicate hydrosols led to general depression of flotation recovery. Selectivity of separation, in fact, was lower in presence of the depressant, as indicated by the conventional recovery-grade plot shown in the figure.

One of the obvious reasons for higher insolubles in the final concentrate could be the presence of composite particles in the feed (due to inadequate liberation). Composites may be collected in the concentrate due to strong collecting action of the added reagents. In such cases, therefore, the suppression of composite particles will lead to a corresponding drop in recovery. One may resort to a regrinding step so as to enhance liberation and hence rejection of gangue in subsequent flotation. The remedy is to optimize the comminution-flotation circuit, if it is established that composite particles are the cause of poor

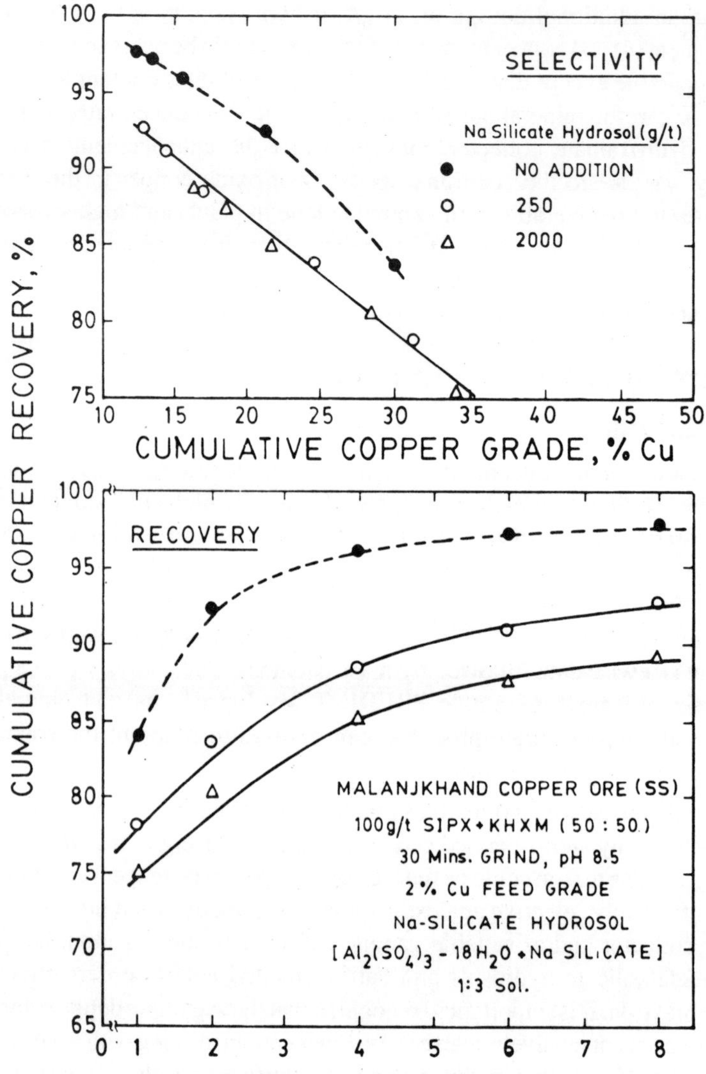

Fig. 2. Effect of a gangue depressant, sodium silicate-metal ion hydrosol on the flotation of a secondary sulphide ore sample (SS) from Malanjkhand, India using SIPX-KHXM reagent combination as a collector. (See Fig. 1 for details of mineralogy)

quality concentrates. Several plants all around the world including Rampura-Agucha of Hindustan Zinc Ltd. (HZL) have a regrinding step, that is regrinding the middlings (scavenger concentrate and the cleaner tails) and recirculate those in the flotation circuit.

Most often, however, the gangue particles report in the froth in the form of extremely fine particles. Entrainment of fine gangue particles could occur due to the kind of froth structure existing in the cells. By a judicious control of frother dosage as well as the type of frother, one can improve the froth drainage and hence better rejection of unwanted gangue. The effect of frother dosage on the flotation performance is clearly illustrated in Fig. 3. Both the recovery as well as the selectivity of separation are found to be quite sensitive to the frother dosage and hence a need to control it in the plant.

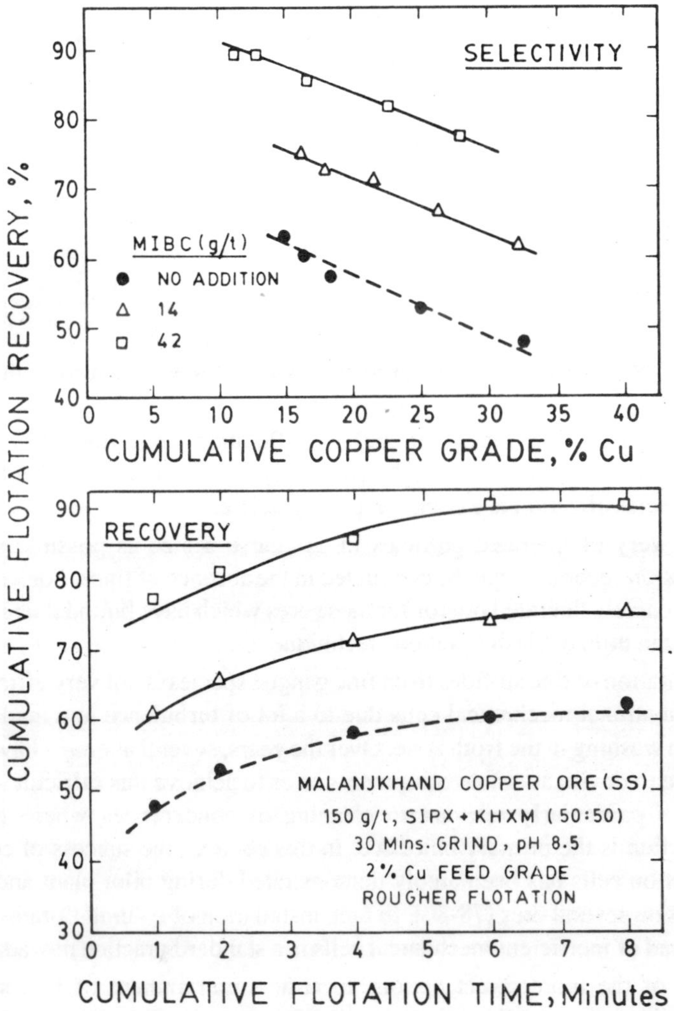

Fig. 3. Effect of frother dosage on the flotation performance of a secondary sulphide ore sample (SS) from Malanjkhand, India with SIPX-KHXM reagent combination. (See Fig 1 for detailed mineralogy of the sample)

A direct correlation between the entrainment of ultrafine quartz particles with the recovery of water in the concentrate during sulfide flotation is reported by a number of researchers in the field [76, 77]. In order to overcome the problem of entrainment, one could choose from one or more of the following options (Fig. 4):

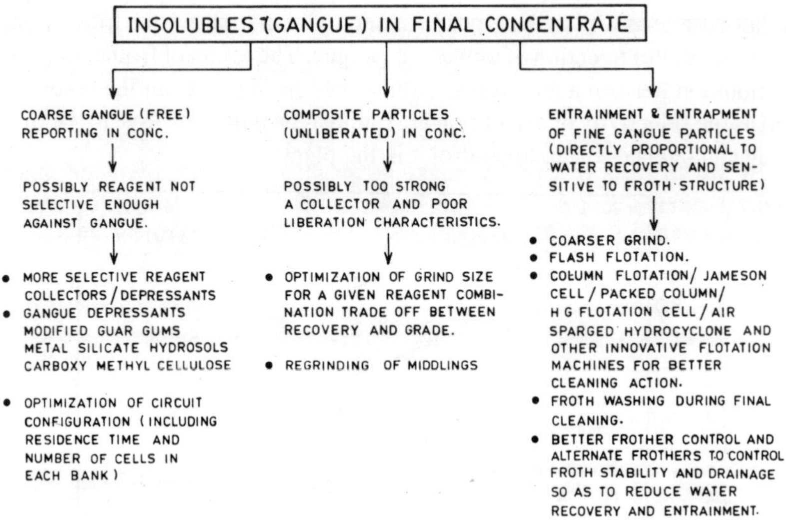

Fig. 4. The possible causes and remedies for higher insolutble content in the final concentrates

- Reduction of fine slimes (gangue) in the flotation feed (which may or may not be possible always) through better design of comminution circuits. Two stage classification is a common plant practice.
- Recovery of liberated particles in as coarse a size as possible so that flotation separation can be conducted in the absence of fines. For instance, for coarsely liberated ores or for those ores which have bimodal liberation, one can utilize "flash flotation" technique.
- Separation of fine sulfides from fine gangue species is not very efficient in conventional mechanical cells due to a lot of turbulence and inadequate froth washing in the froth zone. Over the years, several attempts have been made to devise alternate equipments so as to achieve this difficult separation - particularly relevant in cleaning of concentrates where gangue rejection is the primary objective. In this context, the success of column flotation cells has been amply demonstrated during pilot plant and plant trials on several ores [78-80]. In fact, installation of column flotation cells instead of inefficient mechanical cells is a standard practice nowadays.
- One of the more direct options to reduce entrainment of fine silicate particles is to install water spray nozzles over the flotation cells so that whole of froth layer can be continuously washed. In certain installations, this technique has led to spectacular success, since the addition of extra water helps in the drainage of fine hydrophilic gangue particles [81].
- Even though it has long been known that the choice of frother is also an important parameter determining the success of froth flotation, not much work was available in literature on the design and the efficiency of frothers. It is only recently that the importance of frothers in determining the froth

structure of the concentrate and its consequences in film drainage (facilitating rejection of fine hydrophilic gangue) is being recognized. It has been, for example, demonstrated that by selecting an appropriate frother, one can improve the concentrate grade significantly [82].

In particular, for Malanjkhand copper ore beneficiation plant, our investigations indicated that in case of primary chalcopyritic (sulfide) ore both the ultrafines (– 400 mesh) as well as the relatively coarser sizes due to composite particles contribute to the total insolubles content of the final concentrate. Interestingly, for a complex ore containing secondary sulfides and oxidized minerals, the insoluble content is primarily due to ultrafines (Fig. 5). Appropriate remedies, as discussed above, were thus suggested to the plant engineers [75].

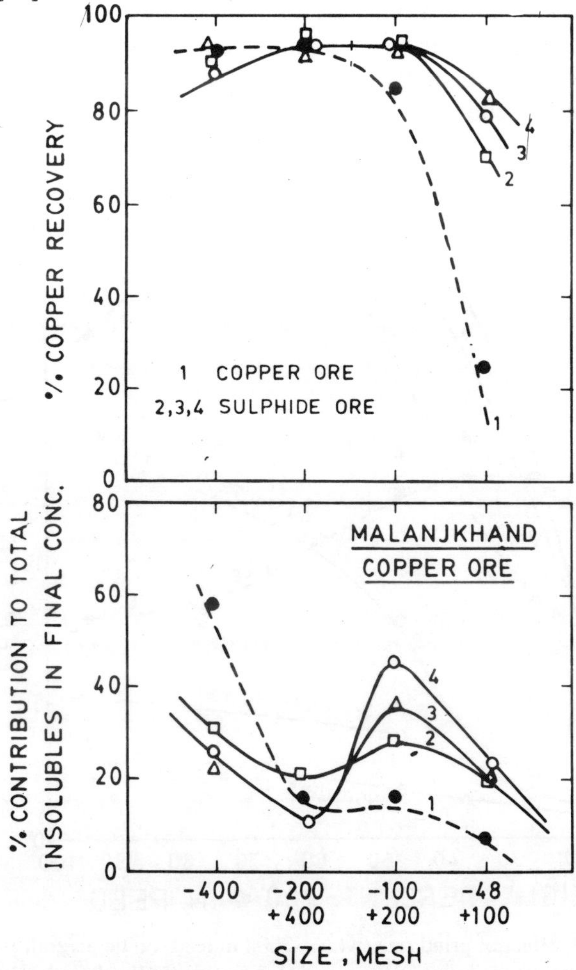

Fig. 5. Results of 'PREDICT' (A process diagnostic software) analysis on plant data collected during campaigns carried out a Malanjkhand: size-wise recovery as well as percent contribution of each size fraction to the total insolubles reporting in the final concentrate) (after Ref [61])

Reduction of Graphite Content

Rajpura-Dariba lead-zinc-copper ore (1.45% Pb, 6% Zn, 8.7% Fe, 1.6% graphitic carbon and 0.14% Cu) being processed by Hindustan Zinc Ltd. occurs mainly in calc-silicate (CS) rock and graphite-mica schist (GMS) zones. The beneficiation of this ore is complicated due to the high pyrite content, finely disseminated and amorphous nature of graphite and fine grained interlocking of galena and sphalerite. One of our important findings was the need to optimize the fineness of grind in order to reduce the graphite content of the lead concentrate. Our findings were subsequently confirmed during plant trials.

Laboratory flotation tests were carried out on a representative ore sample from the mines at different grind sizes following a standard procedure [83]. The recovery, kinetics and lead grade as well as the distribution and grade of misplaced zinc and graphite were monitored during the tests. The results are summarized in Fig. 6. R_∞ and K shown in the figure were calculated by fitting a first order rate equation

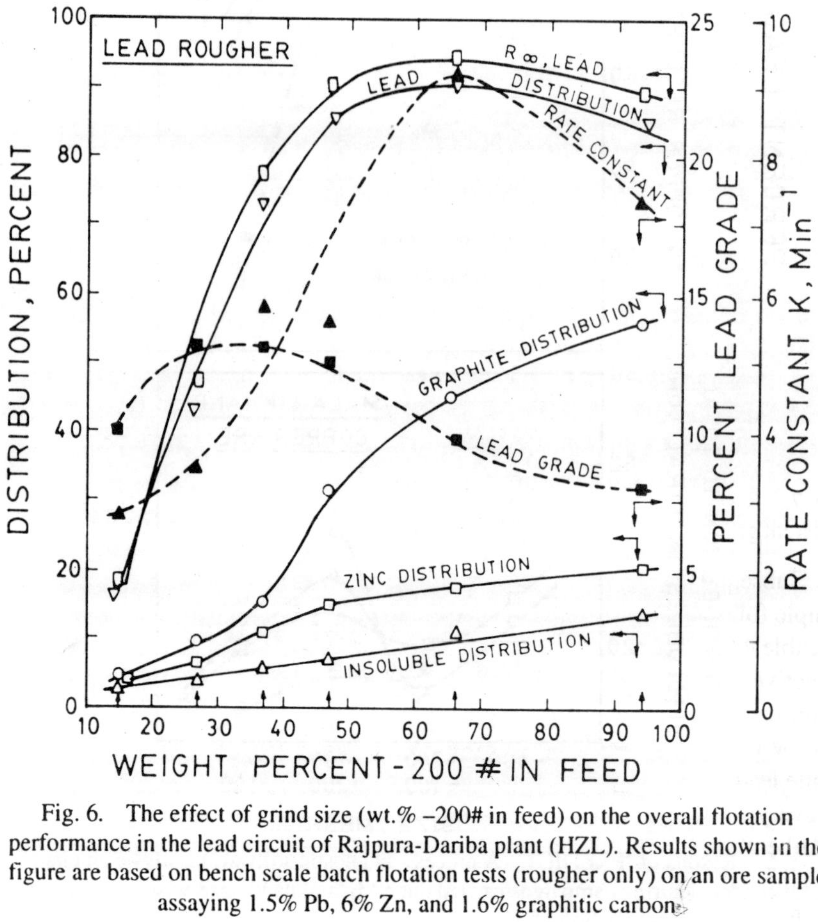

Fig. 6. The effect of grind size (wt.% –200# in feed) on the overall flotation performance in the lead circuit of Rajpura-Dariba plant (HZL). Results shown in the figure are based on bench scale batch flotation tests (rougher only) on an ore sample assaying 1.5% Pb, 6% Zn, and 1.6% graphitic carbon.

$$R = R_{\infty}[1 - (1/Kt)\{1 - \exp(-Kt)\}]$$

to the recovery vs time data obtained through batch flotation tests. The results indicate that the fineness of grind should be controlled, preferably below 66% minus 200 mesh to obtain best performance particularly with respect to graphite distribution in the lead concentrate. The recovery vs grade data plotted in Fig. 7 suggests the same.

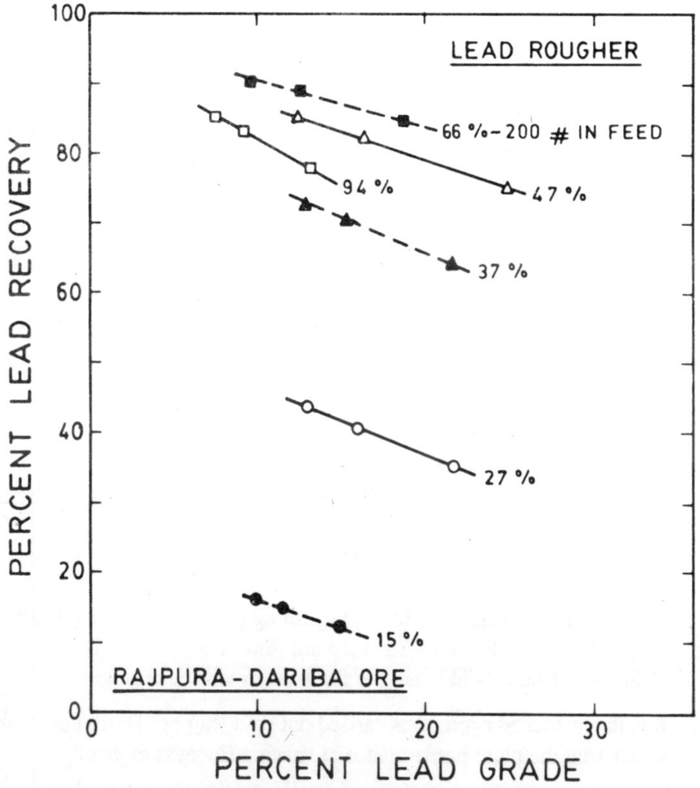

Fig. 7. Recovery vs grade plot for lead flotation at cumulative 1, 2 and 3 minutes at different grind size of feed. Feed assayed 1.5% Pb, 6% Zn and 1.6% graphitic carbon

Subsequently locked cycle tests were also performed on the same ore sample following the plant flowsheet as shown in Fig. 8. The results presented in Table 4, illustrate the effect of grind size on the flotation performance. The objective was to achieve at least 50% grade for both lead and zinc concentrates. Additionally, the graphite in lead concentrate should be less than 2%. It is clear that the recovery of lead and zinc values requires different fineness of grind. While lead recovery and grade is achieved at 58% - 66% minus 200 mesh, zinc recovery and grade is best at 78% minus 200 mesh. The detailed mineralogical and liberation studies carried out on the sample indicated that the liberation size for zinc values was certainly finer than lead and hence consistent with the observed flotation behaviour [83].

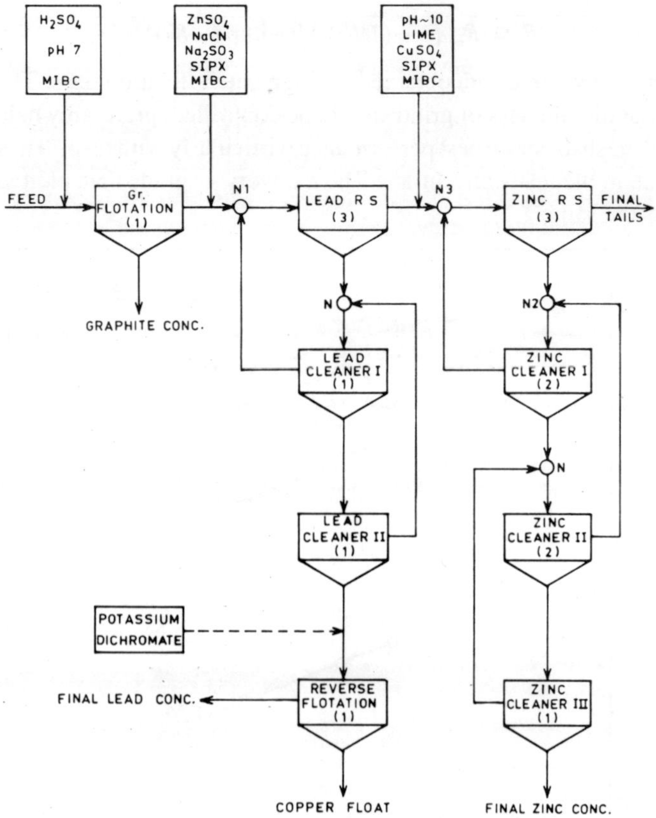

Fig. 8. Flow sheet for locked cycle tests (3 cleaning stages both for lead and zinc circuits) for simulating the Rajpura-Dariba plant flow sheet (HZL). The numbers shown in brackets indicate the flotation time at each stage

Table 4: Results of locked cycle tests carried out on a Rajpura-Dariba ore sample simulating the plant performance at varying fineness of grind

Feed assaying 1.45% Pb, 5.97% Zn, 8.8% Fe, 1.58% Gr.C, 59% Insol.

Grind Size	Procedure Details*	Lead circuit				Zinc circuit			
		Rec %	Grade, %			Rec%	Grade, %		
		Pb	Pb	Gr.C	Zn	Zn	Zn	Insol.	Pb
58	B	61.7	35.3	1.7	8	83.4	23.6	7.3	0.26
	C	68	49.6	1.5	9.3	78.8	39.2	2.3	0.23
66	C_3	66.5	47.8	1.9	8.9	74.5	47.4	2.0	0.22
	C_4	68.5	58.1	1.7	6.6	82.7	46.1	2.2	0.31
78	B	66.5	38.1	2.4	8.5	83.9	56.1	3.7	0.33

* B : 4 kg feed floated in 6.8 litres cell in each cycle

C : 2 kg feed in 6.8 litres cell floated twice in each cycle. Cleaner tails divided in two equal parts and reciculated along with each 2 kg new feed. C_3 represented three cleanings in lead circuit whereas C_4 experiments involved one additional cleaning.

The results therefore indicated the need to carry out grinding in two stages that is, primary coarse grind for lead recovery, followed by a finer grind for the recovery of zinc values. The next best option was to maintain a grind size around 66% minus 200 mesh, a compromise between the two. Our test work also indicated that finer grind leading to excessive slimes, were detrimental to both recovery as well as the quality of final concentrates. The plant trials conducted later on these lines confirmed our findings.

It was also interesting to find that graphite in Rajpura-Dariba ore should be floated out ahead of lead circuit (See Fig. 8). A proper control of frother dosage was observed to be necessary for achieving best results. Detailed studies indicated that graphite present in the Rajpura-Dariba ore floats well. All attempts, therefore, to prevent its flotation using common depressants such as Nigrosine were not successful. In addition to preflotation of graphite, the final lead concentrate when subjected to reverse flotation, (that is graphite flotation in presence of dichromate as galena depressant), the graphite content in the final concentrate could be reduced further. The desired quality of the final lead concentrate particularly with respect to the graphite content could thus be achieved by a judicious choice of grind size, preflotation of graphite ahead of lead flotation followed by reverse flotation of graphite during the final cleaning stage.

OPTIMIZATION OF PROCESS VARIABLES : STATISTICAL DESIGN OF EXPERIMENTS

The processing of complex ores commonly involves differential flotation of copper, lead and zinc minerals using a combination of reagents, namely collectors, frothers, depressants, activators and modifiers. In addition, process variables like pH, air flow rate, agitation intensity, conditioning time also influence the flotation performance. The optimization of such variables for a given objective function (a combination of recovery, grade and flotation kinetics) thus necessarily requires an efficient methodology. We have successfully employed the statistical techniques for this purpose, as illustrated by the results from one such study presented in the following paragraphs.

The differential flotation of a lead-zinc ore (1.5% Pb, 10% Zn) from Rampura-Agucha mines owned by Hindustan Zinc Ltd. requires addition of potassium ethyl xanthate (KEX) collector, zinc sulphate, sodium cyanide, sodium sulphite, Nigrosine (and/or Aero633) as depressants, copper sulphate as activator and sodium carbonate as pH modifier. Our investigations indicated that it was more prudent to grind the ore in two stages with flotation in between the two stages for maximum recovery. The experiments were thus conducted as per the standard procedure illustrated in Fig. 9. The grinding was carried out in a laboratory rod mill and the flotation experiments were conducted in a standard batch flotation cell. In order to monitor kinetics of flotation, the concentrate was collected at the end of 1, 2 and 4 minutes at each stage. The

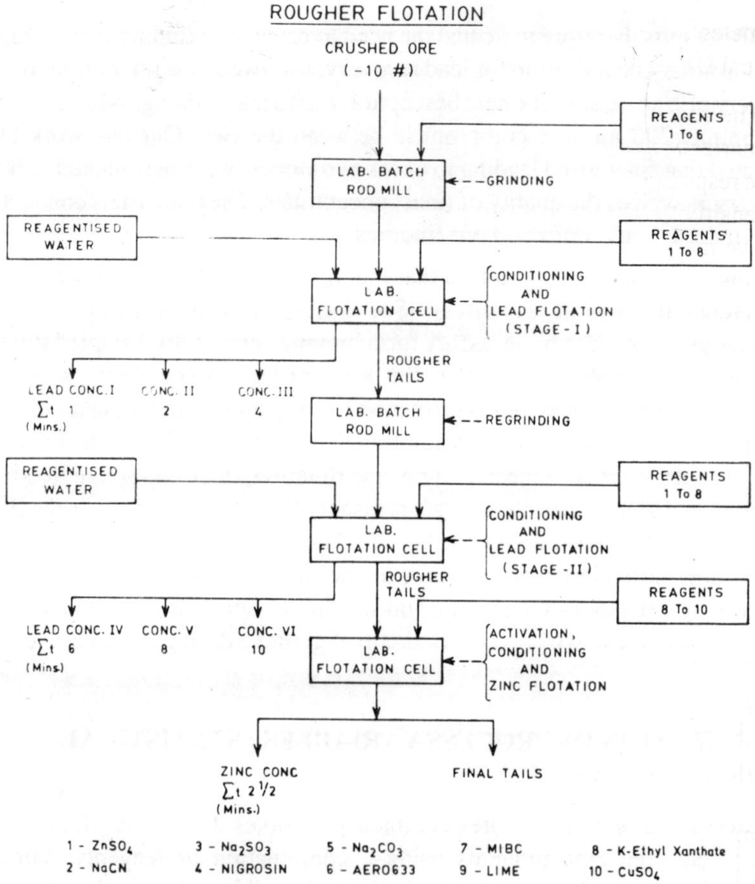

Fig. 9. Schematic flow sheet for conducting the statistically designed bench scale batch flotation (rougher) experiments on Rampura-Agucha ore assaying 1.5% Pb and 10% Zn

response parameters monitored during each experiment included grade and recovery of lead, distribution and grade of misplaced zinc as well as graphite and the flotation rate constant for lead. All the reagent dosages, the regrind time, pH and air flow rate, in all a total of 12 variables at three different levels each (the range selected on the basis of preliminary experimentation and plant data) were thus chosen. The experiments were conducted as per the L27 orthogonal array design [84]. The results were analysed using standard analysis of variance (ANOVA) methods, first for each response separately and then for a combined response parameter called desirability index. The desirability index (D.I.) for a given response parameter was computed as:

$$\text{D.I.} = (X_i - X_{i,m})/(X_i^* - X_{i,m}); \quad X_{i,m} < X_i < X_i^*$$

where X_i, X_i^*, $X_{i,m}$ are defined as the measured value, maximum and minimum possible value respectively of a response parameter X. For two or more response parameters, one can compute the overall D.I. as the geometric mean of individual DIs.

The results for lead recovery, grade and the rate constant in Stage I are summarized in Fig. 10 for illustration. The effect of each individual parameter on the response behaviour can be seen clearly. Furthermore, ANOVA indicated the sensitivity of each of these parameters. The objective of the study was to achieve maximum recovery of lead with minimum misplacement of zinc and graphite in the lead concentrate. The main factors contributing to each of these responses could thus be established (Table 5). It turned out that pH was the most important parameter affecting all the responses significantly. Other important parameters and their sensitivity is also seen in the table. The kinetics of

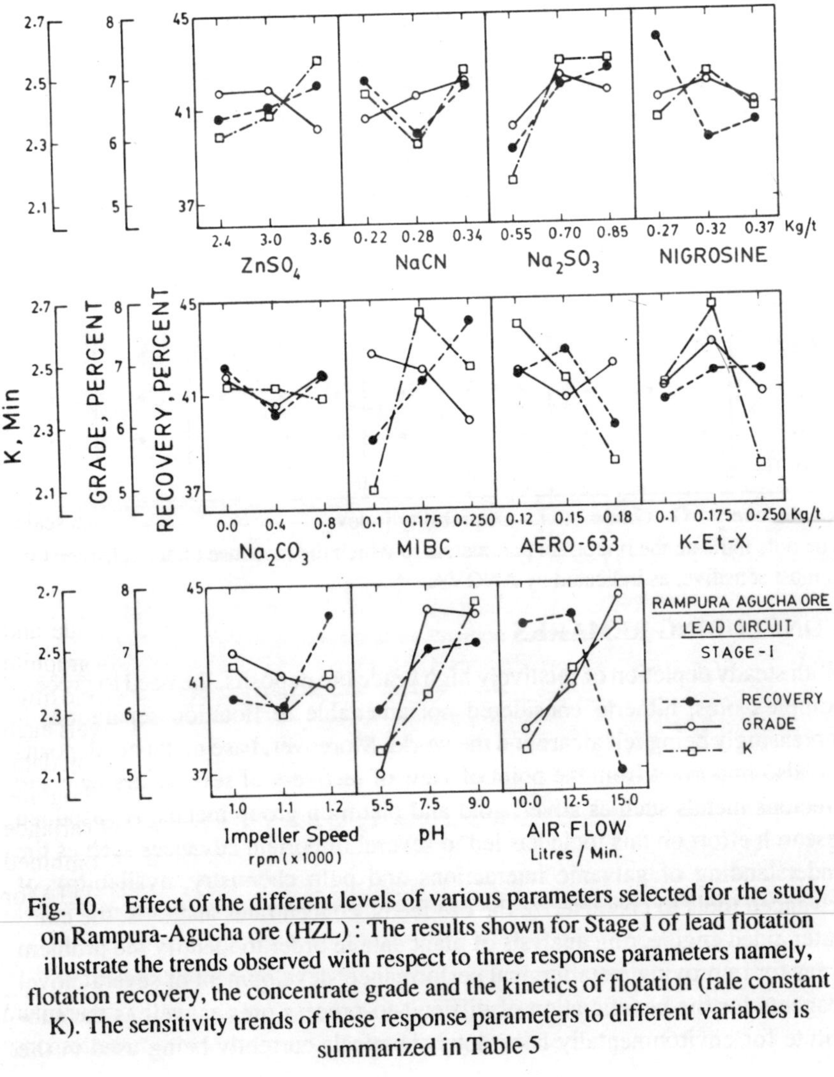

Fig. 10. Effect of the different levels of various parameters selected for the study on Rampura-Agucha ore (HZL) : The results shown for Stage I of lead flotation illustrate the trends observed with respect to three response parameters namely, flotation recovery, the concentrate grade and the kinetics of flotation (rale constant K). The sensitivity trends of these response parameters to different variables is summarized in Table 5

flotation, for example, is sensitive to the collector (KEX) dosage whereas $ZnSO_4$ influences zinc recovery. The variables not shown in the table were found to be unimportant in the range selected for this analysis. The recommendations based on this systematic test work were found useful to the plant engineers. Statistical design of experiments is thus an important tool to optimize such multi-component, multi-variable, multi-response problems particularly to identify the most sensitive parameters so that those could be controlled in the plant to optimize as well as to maintain a stable plant performance.

Table 5: List of most important factors with respect to individual response as well as combined response as defined by desirability index (D.I.) - Stage I flotation (as indicated by ANOVA of the results of statistically designed experiments)

Variable	Lead			Zinc			Graphite			Rate Const.	Combined
	R	G	DI_{Pb}	R	G	DI_{Zn}	R	G	DI_G	K_{Pb}	DI
KEX										•	•
Aero 633										•	
$ZnSO_4$				•		•					
Na_2SO_3			•							•	
MIBC	•									•	
pH		•	•	•	•	•	•	•	•	•	•
Air-flow rate	•	•	•	•		•				•	•

R - Recovery G - Grade D.I. - Desirability Index
The dots indicate the important parameters to which the response under consideration is most sensitive, as indicated by ANOVA.

CONCLUDING REMARKS

With steady depletion of relatively high grade ore deposits, the need to process complex ores, hitherto considered not amenable to flotation separation, is increasingly being felt all around the world. Moreover, base metal ore deposits are also important from the point of view of recovery of several strategic and precious metals such as silver, gold and platinum group metals. A sustained research effort on this topic has led to several important advances such as the understanding of galvanic interactions and pulp chemistry, availability of advanced tools to characterize the ore feeds, concentrates and tails, the computer aided engineering analysis of plant data in order to identify the problem areas for improving metallurgical performance, development of several novel reagents for the beneficiation of difficult-to-process ores as well as the substitute for environmentally hazardous chemicals currently being used in the plants, and the innovations in the design of flotation machines to extend the size range of efficient flotation separation. Research efforts in the coming years will be directed towards finding ways and means to improve metallurgical performance, to enhance energy efficiency, to mitigate environmental concerns through waste recycling and to recover as many byproducts as possible in order to improve the overall economic viability of the complex ore processing plants.

ACKNOWLEDGEMENTS

The authors express their sincere gratitude to Prof. E. C. Subbarao for his constant interest, encouragement, help and support for this work. We are grateful to the staff, engineers and management of Hindustan Copper Ltd. (HCL) and Hindustan Zinc Ltd. (HZL) for providing sustained cooperation and help during the course of several investigations mentioned in the paper. Financial support received over the years for the continuing research programme at our laboratories on the processing of base metal ores from the Department of Mines and Department of Science and Technology, New Delhi, HCL, Malanjkhand and HZL, Udaipur is gratefully acknowledged.

REFERENCES

1. Wills, B. A. (Ed.) : Symposium on processing of complex and refractory ores. Special Issue, Minerals Engineering 6(8-10) (1993) 809-1079.
2. Forssberg, K.S.E. (Ed.) : Flotation of sulphide minerals 1990. Special Issue, International J. Mineral Processing 33(1-4) (1991) 1-396.
3. Mavros, P. and Matis, K.A. (Eds.) : Innovations in flotation technology. Dordrecht : Kluwer Academic Pub. 1992.
4. Dobby, G.S. and Rao, S.R. (Eds.) : Processing of complex ores. New York : Pergamon Press 1989.
5. Laskowski, J.S. (Ed.) : Frothing in flotation. New York : Gordon and Breach Science Pub. 1989.
6. Wills, B.A. (Ed.) : Reagents in Minerals Engineering. Special Issue, Minerals Engineering 5(3-5) (1992) 279-567.
7. Gray, P.M.J., Bowyer, G.J., Castle, J.F., Vaughn, D.J. and Warner, N.A. (Eds.) : Sulphide deposits - their origin and processing. London : The Institution of Mining & Metallurgy Pub. 1990.
8. Barbery, G. : In Wills, B.A. and Barley, R.W. (Eds.) : Mineral processing at a crossroads - problems and prospects. Dordrecht : Martinus Nijhoff Pub. 1986, pp 157-194.
9. Jones, M.J. (Ed.) : Complex sulphide ores. London : Institution of Mining & Metallurgy Pub. 1980.
10. Forssberg, K.S.E. (Ed.) : Flotation of sulphide minerals. Amsterdam : Elsevier Pub. 1985.
11. Deng, T. and Chen, J. : Mineral Processing and Extractive Metallurgy Review 7 (1991) 175-207.

12. Jones, M.J. and Oblatt, R. (Eds.) : Reagents in the minerals industry. London : Institution of Mining & Metallurgy Pub. 1984.
13. Somasundaran, P. and Moudgil, B.M. (Eds.) : Reagents in mineral technology. New York : Marcel Dekker Pub. 1988.
14. Pradip : In Prasad, M. (Ed.) : Reagents for better metallurgy. Colorado, USA : SME-AIME Pub. 1994, pp 245-252.
15. Petruk, W. : In Mavros, P. and Matis, K.A. (Eds.) : Innovations in flotation technology. Dordrecht : Kluwer Academic Pub. 1992, pp 125-148.
16. Frew, J.A. and Davey, K.J. : In Proceedings XVIII International Mineral Processing Congress. Sydney : The Australasian Institute of Mining and Metallurgy 1993, pp 905-912.
17. Griffin, L.K., Hart, S., Espinosa-Gomez, R. and Johnson, N.W. : In Proceedings XVIII International Mineral Processing Congress. Sydney : The Australasian Institute of Mining and Metallurgy 1993, pp. 913-922.
18. Pradip : Minerals and Metallurgical Processing 5 (3) (1988) 114-119.
19. Pradip : Metals, Materials and Processes 3(1) (1991) 15-36.
20. Pradip : Minerals and Metallurgical Processing 5(2) (1988) 80-89.
21. Pradip : Current Science 63(4) (1992) 180-186.
22. Fuerstenau, D.W. and Urbina, R.H. : In Chander, S. and Klimpel, R.R. (Eds.) : Advances in coal and mineral processing using flotation. SME-AIME 1989, pp 1-18.
23. Pradip : Mineral Processing and Extractive Metallurgy Review 10 (1992) 121-137.
24. Mingione, P. : SME-AIME Preprint Paper 91-54 (1991) 17 pp.
25. Nagaraj, D.R., Basilio, C. and Yoon, R.H. : In Dobby, G.S. and Rao, S.R. (Eds.) : Processing of complex ores. Pergamon Press 1989, pp 157-166.
26. Nagaraj, D.R., Brinen, J. Farinato, R. and Lee, J. : SME-AIME Preprint Paper 91-181 (1991) 26 pp.
27. Ackerman, P.K., Harris, G.H., Klimpel, R. and Aplan, F.F. : International J. Mineral Processing 21 (1987) 105-156.
28. Marabini, A.M., Barbaro, M. and Alesse, V., International J. Mineral Processing 33 (1991) 291-306.
29. Marabini, A.M., Alesse, V. and Belardi, G. : In Proceedings XVIII International Mineral Processing Congress. Sydney : The Australasian Institute of Mining and Metallurgy 1993, pp 561-67.
30. Pradip : Trans. Indian Institute of Metals 40(4) (1987) 287.
31. Fuerstenau, D.W. and Pradip : In Jones, M.J. and Oblatt, R. (Eds.) : Reagents in the minerals industry. London : Institution of Mining and Metallurgy Pub..1984, pp 161-168.
32. Pradip, Singh, R., Das, K.K., Sankar, T.A.P., Sunilkumar, T.S., Narsimhan, D. and Rao, N.K. : In Proceedings XVII International Mineral Processing Congress. Dresden, Germany IV 1991, pp 151-162.
33. Klimpel, R.R. and Fee, B.S. : In Proceedings XVII International Mineral Processing Congress. Sydney : The Australasian Institute of Mining and Metallurgy 1993, pp 569-576.

34. Nagaraj, D.R. : In Somasundaran, P. and Moudgil, B.M. (Eds.) : Reagents in mineral technology. Marcel Dekker 1987, pp 257-334.
35. Gornostal, A., Solozhenkin, P., Avrakhov, A. and Abdurazakov, A. : In Proceedings XVIII International Mineral Processing Congress. Sydney : The Australasian Institute of Mining and Metallurgy 1993, pp 549-555.
36. Yu, X., Li, F. and Huang, K. : In Gray, P.M.J. (Ed.) : Sulphide deposits - their origin and processing. London : Institution of Mining & Metallurgy 1990, pp 165-175.
37. Pugh, R. J. : International J. Mineral Processing 25 (1989) 101-125.
38. Laskowski, J.S., Subramanian and Nyamekye, G.A. : In Proceedings XVIII International Mineral Processing Congress. Sydney : The Australasian Institute of Mining and Metallurgy 1993, pp 593-600.
39. Liu, Q. and Laskowski, J.S., International J. Mineral Processing 27 (1989) 147-155.
40. Bolin, N.J. and Laskowski, J.S. : International J. Mineral Processing 33 (1991) 235-241.
41. Houot, R. and Duhamet, D. : Minerals Engineering 5(3-5) (1992) 343-355.
42. Wellham, E.J., Elber, L. and Yan, D.S. : Minerals Engineering 5(3-5) (1992) 381-395.
43. Reddy, G.S., Reddy, C.K., Sekhar, D.M.R., Ravindranath, K. and Chulet, M.R. : Minerals Engineering 4(2) (1991) 151-160.
44. Chander, S. and Zhou, R. : In Proceedings XVII International Mineral Processing Congress. Dresden IV 1991, pp 45-57.
45. Zhou, R. and Chander, S. : International J. Mineral Processing 37 (1993) 257-272.
46. Zhang, W. : In Proceedings XVIII International Mineral Processing Congress. Sydney : The Australasian Institute of Mining and Metallurgy 1993, pp 619-624.
47. Zhang, W. and Poling, G.W. : CIM Bulletin 82 (926) (1989) 35-59.
48. Arnold, B.J. and Aplan, F.F. : In Chander, S. and Klimpel, R.R. (Eds.) : Advances in coal & mineral processing using flotation. Colorado : SME-AIME 1989, pp 130-144.
49. Kalapudas, R. and Kalloinen, J. : In Forssberg, K.S.E. (Ed.), Proceedings XVI International Mineral Processing Congress. Elsevier 1988, pp 469-480.
50. Hall, S. : Mining Magazine, June (1991) 379-383.
51. Brewis, T. : Mining Magazine, June (1991) 383-393.
52. Peterson, M.R., Duchene, L.J. and Shirts, M.B. : USBM Report of investigation RI 8309, 1990, 22 pp.
53. Yoon, R.H., Adel, G.T., Luttrell, G.H., Mankosa, M.J. and Weber, A.T. : In Attia, Y.A., Moudgil, B.M. and Chander, S. (Eds.) : Interfacial phenomena in biotechnology and materials processing. Elsevier Pub. 1988, pp 363-374.
54. Yang, D.C. and Holysh, M. : In Dobby, G.S. and Rao, S.R. (Eds.) : Processing of complex ores. Pergamon Press 1989, pp 387-395.
55. Information Brochure from AFT Inc., Cleveland, Ohio, USA.
56. Gopalkrishnan, S., Ye, Y. and Miller, J.D. : Coal Preparation 9 (1991) 169-184.

57. McCulloch, Jr. W.E. : Minerals and Metallurgical Processing, August (1990) 144-148.
58. Lahiri, D. and Ravindranath K. : In Jones, M. J. (Ed.); Silver - Exploration, Mining and Treatment, Institute of Mining and Metallurgy Pub. 1988, pp 193-204.
59. Kapur, P.C. and Mehrotra, S.P. : In Sastry, K.V.S. and Fuerstenau, M.C. (Eds.): Challenges in mineral processing. SME-AIME 1989, Chapt 18.
60. Kapur, P.C., Dey, A. and Mehrotra, S.P. : International J. Mineral Processing 31 (1991) 11-35.
61. Sabne, M.P., Pradip, Kapur, P.C., Simha, S.S., Golani, M.V. and Achariya, S.K.: Computer-aided design of Malanjkhand copper flotation circuit : a case study. Presented at 46th Annual Technical Meeting, Indian Institute of Metals, Udaipur 1992.
62. Wang, X.H., and Yen, X.: Mineral Processing and Extractive Metallurgy Review 7 (1990) 49-79.
63. Yelloji Rao, M.K. and Natarajan, K.A.: Mineral Processing and Extractive Metallurgy Review 7 (1991) 137-173.
64. Rao, S.R., Moon, K.S. and Leja, J.: In Fuerstenau, M. C. (Ed.): Flotation - A. M. Gaudin memorial volume. New York : SME-AIME 1976, pp 509-527.
65. Pozzo, R.L. and Iwasaki, I.: Corrosion 4 (3) (1987) 159-164.
66. Pozzo, R.L. and Iwasaki, I.: J. Electrochem. Soc. 136 (1989) 1734-1740.
67. Yelloji Rao, M.K. : Electrochemical aspects of interactions among ball materials and minerals with reference to grinding and flotation of complex copper-lead-zinc sulphides. Ph.D Thesis. Indian Institute of Science, Bangalore, India 1990.
68. Cheng, X. and Iwasaki, I. : Mineral Processing and Extractive Metallurgy Review 11 (1992) 187-210.
69. Natarajan, K.A. : Private Communication 1993.
70. Berglund, G. and Forssberg, E. : In Richardson, P.E. and Woods, R. (Eds.) : Electrochemistry in Mineral and Metal Processing. 1988, pp 183-197.
71. Majima, H. : Can Met. Quarterly 8 (1969) 269-73.
72. Rey, M. and Frormanek, V. : Proceedings 5th International Mineral Processing Congress, London : IMM 1960, pp 343-353.
73. Ahn, J.H. and Gebhardt, J.E. : International Journal of Mineral Processing 33 (1991) 243-262.
74. Forssberg, K.S.E. and Subrahmanyam, T.V. : In Proceedings XVIII International Mineral Processing Congress. Sydney : The Australasian Institute of Mining and Metallurgy 1993, p 1-6.
75. Development of reagent schemes for beneficiation of mixed sulfide-oxide ore from Malanjkhand, India. Project report submitted to Deptt. of Mines, New Delhi and Hindustan Copper Ltd., India 1991.
76. Smith, P.G. and Warren, L.J. : Mineral Processing and Extractive Metallurgy Review 5 (1989) 123-145.
77. Trahar, W.J. : International J. Mineral Processing 8 (1981) 289-299.
78. Sastry, K.V.S. (Ed.) : Column flotation '88. SME-AIME 1988, pp 320.
79. Finch, J.A. and Dobby, G.S. : Column flotation. Pergamon Press 1990, pp 180.

80. Agar, G. E., Huls, B. J., Hyma, D. B. : Column '91 - Proceedings of International Conference on Column Flotation. Canadian Institute of Mining and Metallurgy, Canada 1991.
81. Kaya, M., Laplante, A. Blades, A., Tan, C., Huls, B., Lacombe, P., Kemp, R., Ayotte, N.: In Onal, G. (Ed.) : Proceedings Third International Mineral Processing Symposium. Istanbul, Turkey 1990, pp 557-567.
82. Crozier, R.D. and Klimpel, R.R. : Mineral Processing and Extractive Metallurgy Review 5 (1989) 257-279.
83. Investigations to improve the performance of lead-zinc ore beneficiation plant at Rajpura-Dariba (HZL). A project report submitted to Hindustan Zinc Ltd., India 1994.
84. Development of reagent schemes for beneficiation of a mixed sulfide-oxide ore from Rampura-Agucha (HZL). A project report submitted to Deptt of Mines, New Delhi and Hindustan Copper Ltd., India 1991.
85. Malghan, S. G. : Mining Engineering 38(9) (1986) 905-910.
86. Wakamatsu, T. and Shibata, J. : In Sastry, K.V.S. and Fuerstenau, M.C. (Eds.) : Challenges in mineral processing. Littleton, Colorado : SME-AIME 1989, pp 130-141.
87. Das, K.K., Pradip and Suresh, B.: In Proceedings, XIX International Mineral Processing Congress. SME-AIME 1995.

Flotation of Malachite Sulfidized with Sodium Tetrasulfide

R. ZHOU and S. CHANDER
Mineral Processing Section, Mineral Engineering Department,
The Pennsylvania State University, University Park, PA 16802

ABSTRACT

A substantial improvement in flotation was observed for malachite sulfidized with sodium tetrasulfide solutions. The effect of tetrasulfide ion concentration, pH, and secondary products of reaction on flotation recovery was investigated. The detrimental effect of sulfidization residue was ascribed to the presence of copper species and thiosulfate ions, which consumed a part of xanthate thereby decreasing the amount of available collector. The sulfidization behavior of sodium tetrasulfide was compared with sodium sulfide to demonstrate the advantages of the former. The tetrasulfide solution, synthesized by mixing stoichiometric amounts of sodium sulfide and elemental sulfur was as effective as the reagent grade tetrasulfide.

INTRODUCTION

Recovery of valuable minerals from mixed sulfide-oxide base metal ores remains a problem of major technological importance in spite of many significant advances in recovering sulfide minerals by flotation and oxidized minerals by leaching. Such a condition exists because oxidized minerals are normally not recovered by thiol collectors used for recovery of sulfide minerals and sulfide minerals are not recovered by leaching under ambient temperature and pressure. In a detailed study of several copper mill tailings from Western U.S. plants, White and Rule [1] showed that they contain, on an average 0.161% Cu (range 0.047-0.384%), of which nearly half, was oxide copper. It is evident that the flotation plants have done a poor job on oxide copper recovery [2].

One of the methods employed for the treatment of oxidized copper ores is sulfidization followed by flotation with xanthate type of collectors [3]. A survey of the literature on the laboratory investigations and plant practice shows that the most commonly used sulfidizing agents are alkali metal sulfides [4]. Bustamante and Castro [5] suggested the following reaction between malachite and sodium sulfide:

$$CuCO_3 \cdot Cu\,(OH)_2 + HS^- + OH^- = Cu\,(OH)_2 \cdot CuS + CO_3^{2-} + H_2O \quad (1)$$

It was suggested that this reaction continues beyond the surface to form multi-layers of copper sulfide. In a similar study of the sulfidization of tenorite, Castro *et al.* [6] observed the amount of sulfide reacted to exceed that for a monolayer formation. They identified the reaction product as cupric sulfide by x-ray diffraction. A copper sulfide coating was also detected in sulfidization of chrysocolla by Raghavan *et al.*[7]. Several authors have suggested that the role of sulfidizing agent is to stabilize the collector coating through formation of a metal sulfide layer on the mineral [5, 8, 9]. Apparently, a bulk precipitate, loosely adherent to the surface in some cases, forms when collector is added without first sulfidizing the mineral. In reactive systems, the formation of a bulk precipitate was predicted by Chander and Fuerstenau [10] using a kinetic model for interaction between a surface active agent and a dissolving solid. In a recent study, Zhou and Chander [11] suggested the formation of Cu^{2+} as an intermediate species in Reaction 1, which was postulated to have a profound effect on properties of the sulfidization layer. Additional results are presented in this article.

In addition to the primary reaction stated above, several secondary reactions could occur. The presence of sulfidization residue was observed to be detrimental to subsequent flotation with xanthate [12]. Successful flotation was obtained only after the sulfidizing residue was removed.

Sulfidization is also known to promote the interaction of the mineral with xanthate collector. Prasad and Rao [13] studied the xanthate/dixanthogen reaction on the surface of sulfidized and non-sulfidized cerussite and anglesite. The extent of xanthate oxidation was less for oxidized minerals whereas it was more after sulfidization. These investigators suggested that sulfidization induces semi-conductor properties which promote oxidation of the collector.

In this article, the superior performance of sodium tetrasulfide as a sulfidizing agent in malachite flotation is demonstrated. A comparison between sodium tetrasulfide and sodium hydrosulfide is made with regard to their effect on the malachite flotation with xanthate. To determine the effect of sulfidization residue, the rate of consumption of xanthate was measured as a function of the amount of thiosulfate and copper in solution. A comparison between sodium sulfide and sodium tetrasulfide was made with respect to the amounts of thiosulfate, sulfate and copper species produced and their effects on the rate of xanthate consumption. The sodium tetrasulfide was observed to be a much better sulfidizing agent than sodium sulfide possibly due to: (1) the sulfidized coating is more uniform and adherent, and (2) the amounts of thiosulfate and solution copper species generated are smaller.

EXPERIMENTAL METHODS AND MATERIALS

Methods

Hallimond Tube Flotation

Flotation tests were conducted with a modified Hallimond tube. Sodium tetrasulfide and sodium sulfide solutions were used as sulfidizing agents and amyl xanthate as the collector. The following procedure was used: (1) a one-gram malachite sample of 65 × 200 mesh was conditioned with 100 ml of a sulfidizing solution for 10 minutes. The sulfidization was carried out, unless specified otherwise, at pH 10 for sodium tetrasulfide and pH 9 for sodium sulfide (a somewhat higher pH was selected for tetrasulfide to prevent reagent decomposition); (2) the sulfidized samples were washed with distilled water, except for those tests in which the effect of sulfidization residue was to be determined; (3) the washed samples were repulped with 130 ml of amyl xanthate solution (16 mg/l) and conditioned at pH 8.5 for another 10 minutes (for the experiments where the sulfidized samples were not washed, the stock solution of amyl xanthate was added directly to the pulp); (4) the conditioned sample was floated for three minutes using purified nitrogen at a flow rate of 55 ml/min. No frother was used in these tests.

Mechanical Cell Flotation

A few flotation tests were conducted in a 1.5-liter Wemco laboratory flotation cell with a rubberized rotor-stator assembly. The following procedure was adopted: (1) a two-gram sample of 65 × 200 mesh malachite was conditioned with 100 ml of a sulfidizing solution in a glass beaker for 10 minutes. The sulfidization pH was 10 for sodium tetrasulfide and 9 for sodium hydrosulfide; (2) the sulfidized samples were washed to remove the sulfidizing residue and placed in the mechanical cell together with 100 grams of – 65 mesh quartz particles, and the cell was filled with tap water to 1.5 liters; (3) the slurry was agitated at a rotor speed of 1100 rpm for various time intervals to test the tenacity of the sulfidized coating on the malachite surface; (4) after conditioning for the desired period, an aliquot of the stock solution of amyl xanthate was added and the pulp was conditioned for an additional minute, (5) after addition of MIBC (methyl isobutyl carbinol) as frother flotation was conducted for two minutes and the products were collected, dried, and analyzed.

Sulfidization Residue Composition

Secondary reaction products such as thiosulfate, sulfate, and copper species in the sulfidization residue were determined after 10 minutes of sulfidization. The amount of thiosulfate and sulfate ions was determined by an HPLC technique, in which a Waters Model 430 conductivity detector and an IC-PAK anion column were employed. The eluent consisted of 0.8 mM trimesic acid and 20% acetonitrile with pH 8.5 in deionized distilled water. The amount of copper species in the sulfidization residue was detected by a Perkin Elmer Model 703 atomic absorption spectrophotometer.

Redox Potential

A platinum wire electrode was used to measure the pulp potential. It was measured using an Orion Specific Ion Analyzer, Model EA 940 with respect to a saturated calomel electrode connected through a salt bridge. The salt bridge was used to prevent damage of the calomel electrode by sulfide ions. To ensure accurate measurement of potential, the reference electrode was periodically checked with respect to another reference electrode that was never exposed to sulfide ions. The measured potentials were converted to the standard hydrogen scale by adding 0.245 V to them. The platinum electrode was cleaned between successive measurements using 600-grit silicon carbide paper.

Surface Morphology

The malachite samples for this study were wet polished using 600-grit silicon carbide paper followed by 0.05 μm alumina suspension to produce a mirror surface. After polishing, the samples were treated with the desired solution for various time periods, dried in nitrogen atmosphere and coated with gold. The surface morphology was investigated using a scanning electron microscope.

Materials

Reagents

The tetrasulfide stock solution was prepared by dissolving sodium tetrasulfide of 96% purity (Alfa Products, Danvers, MA) in de-oxygenated distilled water of pH 10.3. Previous studies show that tetrasulfide ion is the predominant polysulfide species in solution [14]. A slightly higher pH than that normally practiced for sulfidization was used to prevent the decomposition of tetrasulfide ions to elemental sulfur. The sodium sulfide stock solution was prepared by adding $Na_2S{\bullet}9H_2O$ into de-oxygenated distilled water. The concentration of sulfide ions in the stock solution was determined by titration with a standard solution of lead perchlorate [15]. Potassium amyl xanthate collector (Aero 350) was obtained from American Cyanamid Company (Wayne, NJ) and was purified by triple recrystallization from acetone.

Malachite

Lumps of malachite, obtained from Gila Co., Arizona, were crushed and concentrated by hand-sorting. The selected sample was dry-ground in a porcelain mill to minus 65 mesh and the 65 × 200 mesh size fraction was separated for flotation tests by screening. The x-ray diffraction analysis indicated that the sample was essentially pure contaminated with a small quantity of quartz. Chemical analysis of the sample showed that the sample contained approximately 3% quartz.

RESULTS AND DISCUSSION

Effect of Concentration of Sulfidizing Species

The flotation results of malachite, pre-conditioned with various amounts of sodium tetrasulfide, are given in Figure 1 for both washed and non-washed samples. The results with sodium sulfide as the sulfidizing agent are also included for the sake of comparison. It is clear that the flotation recoveries were

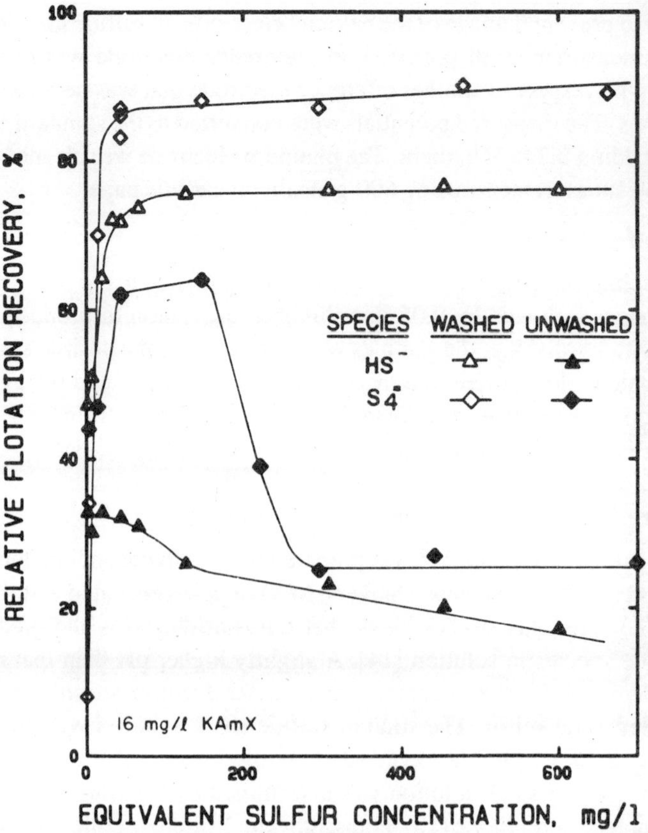

Fig. 1. Effect of the initial equivalent sulfur concentration on malachite flotation with Na_2S_4 and Na_2S as the sulfidizing agent. Sulfidization pH: 10 for Na_2S_4 and 9 for Na_2S; flotation pH 8.5

higher when sodium tetrasulfide was used. The results also confirmed previous findings that the flotation of malachite sulfidized with sodium sulfide improved significantly after removal of the sulfidization residue [12]. Similar benefits were obtained for sodium tetrasulfide also. For the washed samples, the flotation recovery reached a plateau when the equivalent sulfur concentration was greater than about 40 mg/l. The presence of a large excess of tetrasulfide during sulfidization did not decrease the flotation recoveries for the washed samples but it had an adverse effect when they were not washed.

Surface morphology studies revealed an important difference between sodium sulfide and tetrasulfide as the sulfidizing agents. The electron microphotographs of malachite treated with sodium tetrasulfide and sulfide are given in Figure 2. When the malachite was treated with sodium sulfide, crystals of copper sulfide formed on the surface as discrete units, as can be seen in Figure 2b. The nature of the growth was such that the crystals might detach from the surface under conditions of intense agitation in a flotation cell. The detachment

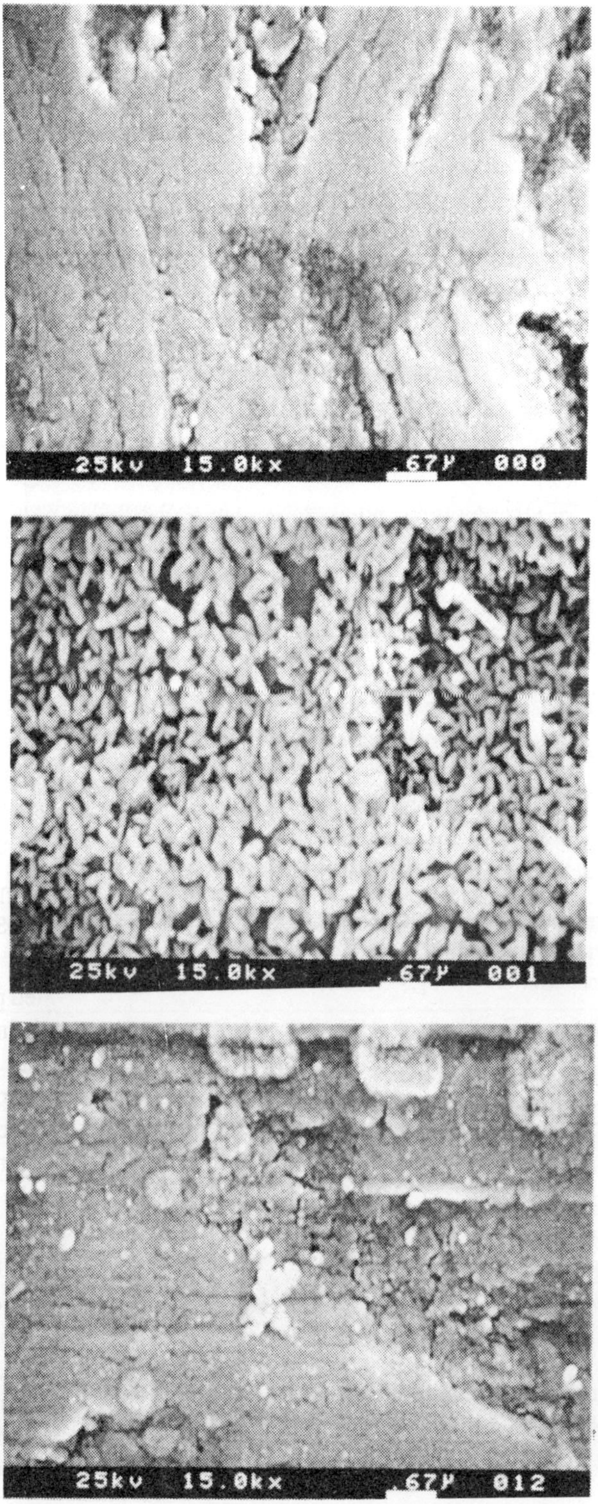

Fig. 2. Scanning electron micrographs of the malachite treated with (a) distill water, (b) sodium sulfide and (c) sodium tetrasulfide solutions (4.2×10^{-2} M)

of sulfidized coatings was suggested by Fuerstenau and Raghavan [4], and confirmed in this study. In comparison, only a few crystals of a new phase were observed on the surface of malachite treated with sodium tetrasulfide, (Figure 2c), for which the majority of surface was free of crystals. Apparently, the surface coating, believed to be that of CuS_4, was not detected by this method. But, the improvement in flotation does suggest that the malachite surface was covered with a sulfidized layer which became hydrophobic in the presence of a collector. The thickness of the sulfidized layer is much too small to be detected by a simple technique but it was postulated based on kinetics of sulfidization [11].

The results of Hallimond tube tests demonstrated that sodium tetrasulfide could be considered as a more effective sulfidizing agent than sodium sulfide in the flotation of malachite. Since tenacity of the surface coating is essential to the flotation conducted in industrial flotation cells, selected tests were conducted in a Wemco laboratory flotation cell. The experiments were carried out for various agitation times and the results are presented in Table 1. At short agitation times, the flotation recovery for the sample pretreated with sodium tetrasulfide was about 65% whereas it was less than 40% with sodium sulfide. As the agitation time increased, the flotation recovery decreased, which might be ascribed to detachment of the sulfidized coatings from the mineral surface. A comparison of the flotation results obtained from the mechanical cell for the two sulfidizing agents confirms that the sulfidized coatings adhere to the mineral surface more strongly when pretreated with tetrasulfide than with sulfide. The use of sodium tetrasulfide as a sulfidizing agent can improve the flotation of malachite in the mechanical cell to a large extent. The benefits of tetrasulfide could be even more if flotation was performed in a column cell.

Table 1: Flotation recovery in the mechanical cell of malachite sulfidized with sodium tetrasulfide and hydrosulfide. Equivalent sulfur concentration: 100 mg/l; xanthate concentration: 10 mg/l; MIBC: 5 mg/l; rotor speed: 1100 rpm

Agitation Time, minutes	Flotation Recovery, %	
	Hydrosulfide	Tetrasulfide
0.2	26, 40	64, 67
5	25	61
10	36	59
15	32	53

Effect of Sulfidization Residue

To determine composition of the sulfidization residue, the solution species were analyzed using UV-VIS and atomic absorption spectroscopies. From the absorption in a wide range of wavelengths in the UV-VIS region it is suggested that colloidal particles of copper compounds were present. The copper compounds might be produced during the sulfidization by the reaction:

$$CuCO_3\,Cu\,(OH)_2 + HS^- = CuS + Cu^{2+} + 2OH^- + HCO_3^- \qquad (2)$$

The dissolved copper ions could precipitate as copper hydroxide or copper sulfide depending upon solution pH and sulfide ion concentration. The precipitation of copper compounds (both hydroxide or sulfide) when occurring in bulk solution is referred to as one of the secondary reactions in this article because such particles might not be recovered in flotation. Other secondary reactions are those which involve oxidation of sulfur species to thiosulfate, sulfate, or elemental sulfur and could be described by the following equations [16]:

$$2HS^- + 5/2O_2 = S^0 + SO_4^{2-} + H_2O \qquad (3)$$

$$2HS^- + 2O_2 = S_2O_3^{2-} + H_2O \qquad (4)$$

In Reaction 3, elemental sulfur is considered to form as a bulk precipitate rather than surface product and therefore does not promote flotation. The sulfidization of malachite with sodium tetrasulfide could be represented by a reaction similar to Equation (2):

$$CuCO_3 \cdot Cu\,(OH)_2 + S_4^{2-} = CuS_4 + Cu^{2+} + 2\,OH^- + CO_3^{2-} \qquad (5)$$

It is demonstrated later in this article that the formation of a sulfidized coating on the malachite surface promotes the flotation of malachite by xanthate type collectors.

Analogous to the malachite-sulfide system, both primary and secondary reactions occur in the malachite-tetrasulfide system. The secondary reactions include formation of colloidal particles of copper precipitates and oxysulfide species in the solution. The oxidation of tetrasulfide ions in bulk solution is known to form S^0, $S_2O_3^{2-}$, and SO_4^{2-} ions [17, 18]. The presence of these species in solution equilibrated with malachite was confirmed in the present study by atomic absorption and UV spectroscopies and HPLC techniques. The fraction of the sulfidizing agent converted to thiosulfate and sulfate depended on the reagent concentration and pH.

In the presence of the sulfidization residue the flotation recovery was considerably less as shown in Fig. 1. For the tetrasulfide ion concentration range of 40 to 150 mg/l, the recovery was about 62%. At higher concentrations, the floatability was substantially decreased. When sodium sulfide was used as a sulfidizing agent, the decrease in recovery occurred in the entire range of concentration investigated. These results showed that sulfidization residue had a detrimental effect on malachite flotation.

Presence of thiosulfate, sulfate, and copper species in the residue was confirmed by HPLC and atomic absorption spectrophotometer, the results of which are given in Table 2. The amounts of these species were determined for both sulfide and tetrasulfide ions. The amounts were much less for sodium tetrasulfide than those for sodium sulfide. The copper species detected in the sulfidizing residue might be present in the form of colloidal particles of CuS, CuS_4 and $Cu(OH)_2$ under the given experimental conditions. It was observed in the present study that the colloids of CuS and CuS_4 were stable in the presence of excess of sulfidizing species. Both thiosulfate ion and the copper species might react with xanthate, decreasing the amount of xanthate available for malachite particles. Castro *et al.* [12] demonstrated that high concentrations of thiosulfate ions were detrimental to malachite flotation with xanthate when sodium sulfide was used as a sulfidizing agent. In the present study, however, the depressant effect of sulfidization residue was also observed under conditions where thiosulfate concentrations were low. Based on these results we conclude that copper colloidal particles also play a significant role in governing the flotation response. At high dosage of the sulfidizing agents where large quantities of thiosulfate appear in the liquid phase, thiosulfate ions also affect flotation adversely as suggested by Castro *et al.*[12].

Table 2: Composition of sulfidizing residues[a] after treatment of malachite with sodium tetrasulfide and sulfide ions.

Initial equivalent sulfur conc. (mg/l)	Solution species (mg/l)[b]					
	Tetrasulfide			Sulfide		
	Cu	$S_2O_3^{2-}$	SO_4^{2-}	Cu	$S_2O_3^{2-}$	SO_4^{2-}
14.7	0.1	1.0	0.5	1.2	3.3	6.6
29.4	0.3	2.1	1.4	3.2	7.2	13.8
44.1	0.7	4.2	3.7	6.3	11.3	21.3

a: Filtrates for analysis were obtained using a 0.2 mm filter.
b: $S_2O_3^{2-}$ and SO_4^{2-} are in equivalent sulfur concentration.

Effect of Sulfidization Time

The effect of sulfidization time on malachite flotation is shown in Figure 3 together with the change in pulp potential. The time required to reach a flotation plateau was shorter for sodium tetrasulfide than that for sodium sulfide. The pulp potential did not correlate with flotation, however. In a previous study, it was demonstrated that the kinetics of sulfidization of malachite is complex and involves a two-stage process [11]. In the first stage, primary sulfidized layer is produced which is necessary for adequate flotation. Secondary reactions occurred during second stage which resulted in products that were harmful to

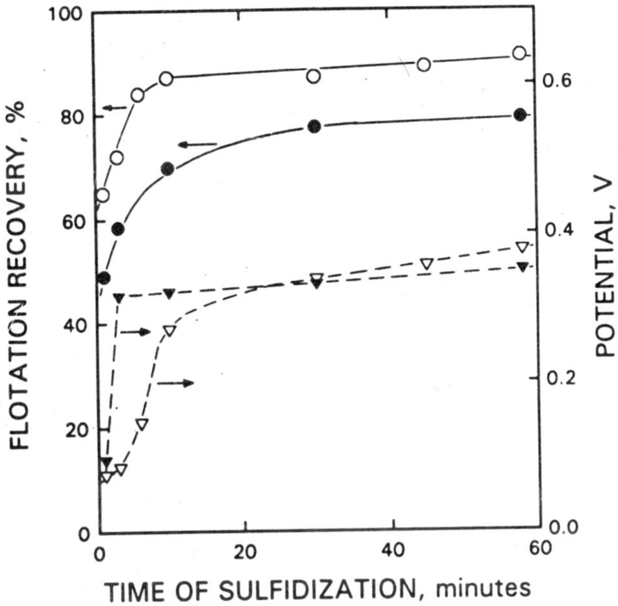

Fig. 3. Effect of the time of sulfidization on the flotation and pulp potential of malachite with Na_2S_4 or Na_2S as the sulfidizing agent. The flotation results are for samples washed after sulfidization. Equivalent sulfur conc. 44 mg/l. o, ∇: $S_4^{=}$; •, ▼: HS^-.

flotation. As a result the amount of sulfidizing agent consumed is not a good measure of the flotation recovery.

To determine the effect of oxidation on sulfidization-flotation of malachite, samples of malachite that had been sulfidized for 10 minutes in various concentrations of tetrasulfide solutions were aerated with nitrogen or oxygen gas for different time periods prior to flotation. The results are given in Fig. 4. At low concentrations of tetrasulfide ion, (44 mg/l), extra conditioning with oxygen decreased the flotation recovery. However, when nitrogen was used additional conditioning did not affect flotation. It is postulated that, at low sulfidization concentrations, most of the reagent reacted and excess tetrasulfide ions were not present in the solution. As a result, less amounts of thiosulfate ions were produced and aeration of the pulp with oxygen led to oxidation of the sulfidized layer on the mineral surface. Soto and Laskowski [19] also observed that if the aeration of pulp was not stopped after oxidation of the hydrosulfide ion in the pulp had been completed, the oxidation of the sulfide coating on the particles began, thus depressing flotation.

At high concentrations of tetrasulfide ion (440 mg/l), additional conditioning improved flotation, in the presence of both oxygen and nitrogen. At this high concentration, there was an excess of tetrasulfide ions in solutions after the normal conditioning time which should hinder the reaction of xanthate with

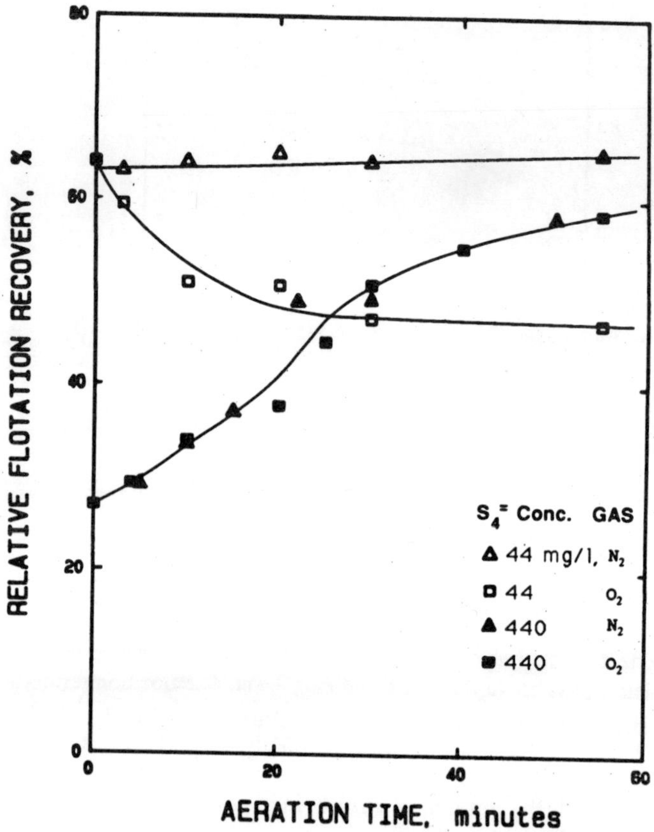

Fig. 4. Effect of aeration of the sulfidizing suspensions on malachite flotation. Gas was introduced after 10 minutes of sulfidization with Na_2S_4 at a flow rate of 120 ml/min.

malachite particles. The excess of polysulfide ions was confirmed by UV spectroscopy and Eh measurements, (the latter gave a reducing environment and a potential of – 0.05 V). Thus, at a higher dosage of tetrasulfide ions, additional conditioning might improve the flotation by removing the excess tetrasulfide ions. This phenomenon could, in fact, be exploited in industrial flotation by aerating the sulfidized pulp.

Effect of Sulfidization pH

The effect of sulfidization pH on flotation recovery is given in Fig. 5. The maximum recovery was obtained in the pH range of 9 to 11. In this set of experiments, the pulp potential was also recorded at the end of sulfidization, and the data are included in Fig. 5. At pH's below 9, colloidal elemental sulfur formed white clouds in the bulk solution and the electrode potential was high. Such a precipitation reaction depleted the sulfidizing agent from solution thereby decreasing flotation. When the pH was increased from 11.2 to 11.8, the

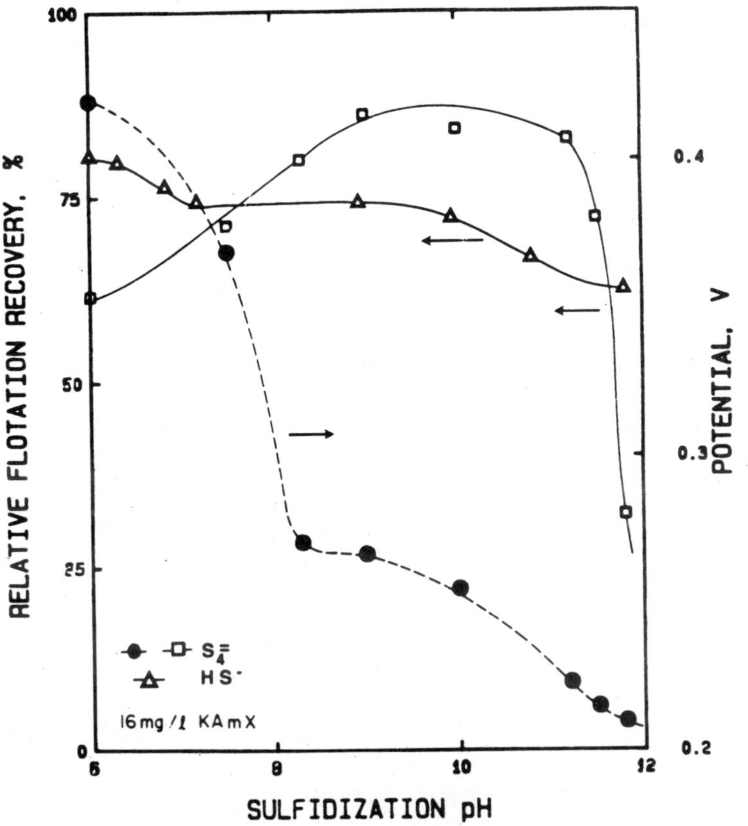

Fig. 5. Effect of sulfidization pH on malachite flotation with Na_2S_4 as sulfidizing agent. Equivalent sulfur conc. 44 mg/l

flotation recovery decreased from 83 to 32%. Such a sharp decrease in the recovery is probably due to the oxidation of tetrasulfide to thiosulfate and sulfate ions and the competitive adsorption of hydroxyl ion on the surface. In a separate study it was observed, for example, that, at pH 11.2 about 30% of the tetrasulfide ions converted to thiosulfate and sulfate ions, and at pH 11.8 the percent conversion was 70.

Flotation of a Mixture of Malachite and Quartz

A few tests were performed to determine the effectiveness of the process using malachite-quartz mixture. Although the separation was good, some activation of quartz was observed as can be seen in Fig. 6. It is also interesting to note that the flotation of malachite improved in the presence of quartz. Some of the secondary reaction products are scavenged by quartz which improves flotation. These results imply that the harmful effect of secondary reaction products would be less while floating an ore. Such a hypothesis needs to be verified, however.

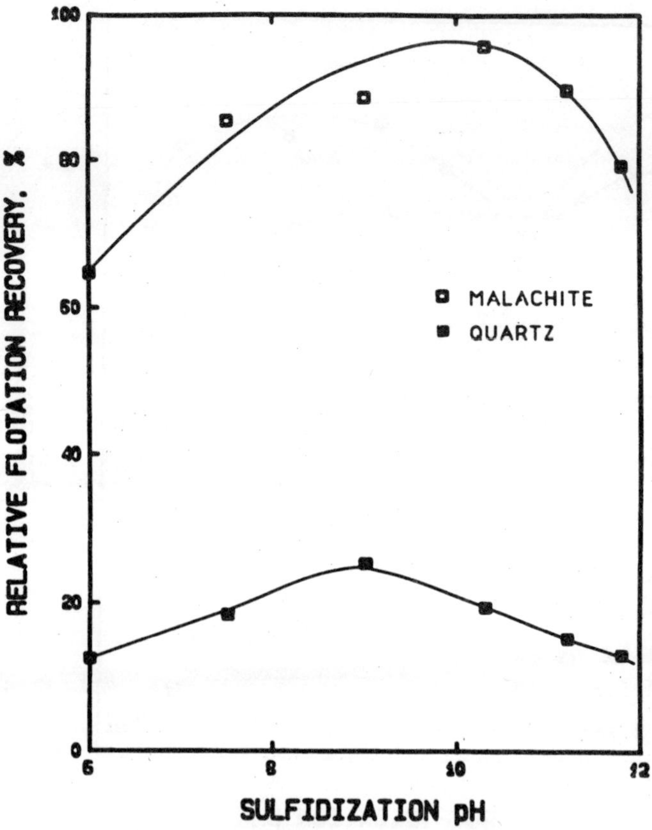

Fig. 6. Flotation recovery of malachite and quartz from a 50% mixture sulfidized with Na_2S_4. The samples were washed after sulfidization. Reagent concentration: S_4^{--} : 44 mg/l; KAmX: 16 mg/l

The role of quartz in the galena flotation was studied by Plaksin and Eliseev [20]. These investigators observed that the sulfide species did not oxidize in the presence of quartz alone; however, when introducing both quartz and galena particles to a hydrosulfide solution, the rate of consumption of hydrosulfide increased considerably compared to that with galena alone. As a result, the flotation of galena was restored quickly in the presence of quartz. It is likely that the lead hydroxyl ions produced via the dissolution of galena adsorbed on quartz, which accelerated the rate of oxidation of hydrosulfide ions as suggested by Chen and Morris [21]. A similar mechanism could occur in malachite-quartz system. In the previous paragraphs, it was discussed that several competing reactions occur involving oxidation of tetrasulfide to elemental sulfur and thiosulfate and sulfate ions. The relative extent of these reactions is altered by quartz activated by metal ions. This behavior was substantiated by analyzing sulfidizing residue, in which 70 percent of tetrasulfide ion was converted to thiosulfate and sulfate ions at pH 11.8.

The effect of quartz on the malachite flotation was further investigated for both sodium tetrasulfide and sodium sulfide and the results are given in Fig. 7. For washed samples, addition of quartz particles improved the flotation recovery of malachite for the conditions investigated. For the non-washed samples, quartz improved flotation for malachite-sodium tetrasulfide system but not for malachite-sodium sulfide system. A further investigation of the role of quartz in malachite flotation is needed to delineate the mechanism.

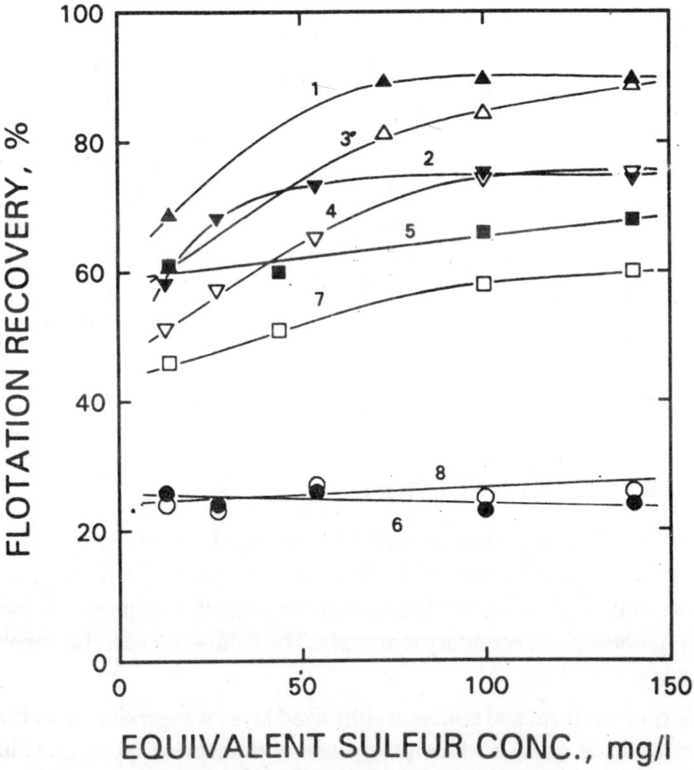

Fig. 7. Effect of quartz on malachite flotation with Na_2S_4 and Na_2S as sulfidizing agents

1 (S_4^{--}), 3 (S^{--}) - washed samples with quartz;

2 (S_4^{--}), 4 (S^{--}) - washed samples without quartz;

5 (S_4^{--}), 7 (S^{--}) - unwashed samples with quartz;

A Low Cost Method of Producing Tetrasulfide Solution

For industrial application, reagent cost could be an important criteria. The reagent cost for the commercial sodium tetrasulfide (40 grams for about $30) is much higher than that for sodium sulfide (2.5 kg for about $50). In order to explore a low cost method for reagent preparation, attempts were made in this study to synthesize tetrasulfide by mixing stoichiometric amounts of elemental sulfur in sodium sulfide solution. The molar ratio of elemental sulfur to sulfide in the mixture was three to one. The effectiveness of the synthesized polysulfide was tested in the mechanical flotation cell. The results are given in Table 3. The flotation recoveries obtained by using the laboratory synthesized polysulfide were close to those obtained by the reagent grade sodium tetrasulfide. Therefore, the procedure described above was considered satisfactory for production of reagent for sulfidization. Alternate schemes for producing the reagent at low cost can not be ruled out.

Table 3: Flotation of malachite in the mechanical cell with sodium tetrasulfide and synthetic polysulfide, equivalent sulfur concentration: 100 mg/l; xanthate concentration: 10 mg/l; MIBC: 5 mg/l; rotor speed: 1100 rpm

Agitation time minutes	Flotation recovery, %	
	Synthetic Polysulfide	Commercial Tetrasulfide
0.2	63, 60	64, 67
5	58	61
10	55	59
15	50	53

SUMMARY AND CONCLUSION

Flotation recovery of malachite could be increased substantially when it was sulfidized with sodium tetrasulfide. The main reasons were the formation of a tenacious sulfidized layer and decrease in the amount of depressants produced through inhibition of secondary reactions. The following specific conclusions are based on this study:

1. A more uniform and coherent sulfidized layer was produced with sodium tetrasulfide and a surface precipitate with discrete crystallites was observed with sodium sulfide. The crystallites have a higher probability of detachment in a flotation cell.
2. The better flotation response in the sodium tetrasulfide system was ascribed also to smaller amounts of colloidal copper substances and thiosulfate ions produced in the sulfidization residue. The colloidal particles consume large amounts of xanthate and decrease the recovery of flotation.
3. The adverse effect of sulfidization residue, which was less with sodium tetrasulfide as compared to sodium sulfide, could be eliminated by washing the sample.
4. The effect of pulp aeration on flotation recovery was a function of reagent type and concentration. At low tetrasulfide ion concentration, excess oxidation destroyed the sulfidized layer and decreased flotation recovery.

Wheras at high tetrasulfide ion concentrations, additional aeration converted excess tetrasulfide ions to sulfidized layer and improved flotation.

5. Effect of sulfidization pH on the flotation was significant for sodium tetrasulfide. Optimum flotation recovery was obtained in the pH range of 9 to 11. At lower sulfidization pH, tetrasulfide ions decompose to form colloidal elemental sulfur in solution. At high pH, values (> 11.5) competitive adsorption of hydroxyl ions on the surface and rapid oxidation of tetrasulfide ions to thiosulfate and sulfate ions decreased recovery.
6. Malachite could be successfully separated from quartz using tetrasulfide ions as the sulfidizing agent. The presence of quartz in the sulfidization process favored flotation by scavenging harmful substances from solution.
7. The polysulfide synthesized by dissolving elemental sulfur in sodium sulfide was as effective as the commercial sodium tetrasulfide.
8. The total amount of sulfur consumed could be decreased when sodium tetrasulfide is used in place of sodium sulfide as the sulfidizing agent.

REFERENCES

1. White, J.C., and Rule, A.R.: Distribution of Sulfide and Oxide Copper in Copper Mill Tailings, USBM RI 7498, 1971.
2. Aplan, F.F. : In Reagents for Better Metallurgy, Littleton, CO : SME 1994, Chapter 12, pp. 103-115.
3. Aplan, F.F. and Fuerstenau, D.W. : Int. J. Miner. Process. 13 (1984) 105-115.
4. Fuerstenau, D. W., and Raghavan, S. ; In Somasundaran, P. (Ed.) : Advances in Mineral Processing, Arbiter Symposium, Littleton, CO : The Society of Mining Engineers, Inc, 1986.
5. Bustamante, H., and Castro, S.: Trans. I.M.M. 84, (1975) C167-C171.
6. Castro, S., Goldfarb, J., and Laskowski, J. : Int. J. Miner. Process. 1, (1974) 141-149.
7. Raghavan, S., Adamec, E., and Lee, L. : Int. J. Miner. Process. 12, (1984) 173-191.
8. Abramov, A.A. Izv. Vyssh. Ucheb. Zaved., Tsvetn. Metall. 12 (5) (1969) 7-13; Chemical Abstract, Vol. 72, 1970, 46299u.
9. Castro, S., Gaytan, H. and Goldfarb, J. : Int. J. Miner. Process. 3, (1976) 71-82.
10. Chander, S. and Fuerstenau, D.W. : Proceedings of the IX International Mineral Processing Congress, Cagliari, Italy, 1975.
11. Zhou, R. and Chander, S. : Int. J. Miner. Process. 37 (1993) 257-272.
12. Castro, S., Soto, H., Goldfarb, J. and Laskowski, J. : Int. J. Miner. Process. 1 (1974) 151-161.
13. Prasad, M.S. and Rao, S.R. : Proceedings of the IX Int. Mineral Processing Congress, Prague, Vol. 1, pp. 157-165, 1970.
14. Giggenbach, W. : Inorganic Chemistry 11 (1972) 1201-1207.
15. Bauman, E. W. : Analytical Chemistry 46 (9) (1974) 1345-1347.

16. Avrahami, M., and Golding, R. M. : J. Chem. Soc. (A) (1968) 647-651.
17. Valensi, G., Muylder, J. V., and Pourbaix, M. : In Pourbaix, M. (Ed.) : Atlas of Electrochemical Equilibria in Aqueous Solutions, Oxford, New York : Pengamon Press 1966, pp. 545-553.
18. Teder, A. : Sevensk Papperstidning 72 (1969) 294-303.
19. Soto, H. and Laskowski, J. : Trans. IMM 82 (1973) C153.
20. Plaksin, I.N. and Eliseev, N.I. : Tsvetnye Metally, N.Y. 8, (1967) 35-37.
21. Chen, K.Y. and Morris, J.C. : Environmental Science & Technology 3 (9) (1972) 838-843.

Optimal Design and Synthesis of Flotation Circuits

S.P. MEHROTRA
Department of Materials and Metallurgical Engineering,
Indian Institute of Technology, Kanpur - 208 016

ABSTRACT

The design and synthesis of flotation circuits is conventionally based on rules of thumb derived from industrial practice. This heuristic approach is dependent on past experience and engineering judgement of experts. With the advent of relatively inexpensive computing power and easy availability of powerful optimization techniques, it is now possible to synthesize optimal multicomponent, multistage flotation separation circuits based on rigorous quantitative analysis of alternate possibilities. Recent advances in this field are reviewed in this paper.

INTRODUCTION

Complete separation of valuable minerals from gangue in a single pass of conventional froth flotation is seldom possible. To achieve effective separation, several passes through a number of interconnected flotation cells/banks are therefore required. This interconnection of flotation cells/banks is referred to as a flotation circuit. The performance of a circuit depends both on the configuration of the network as well as the operating conditions. The latter include: (1) the chemical conditioning of the pulp, which in turn depends on reagents' concentrations, pH, etc., and (2) the physical conditions prevailing in the cell, e.g. feed rate, degree of agitation, air supply and distribution. The optimal performance of a flotation circuit thus requires that the circuit configuration designed at the time of installation of the plant be optimal, and that this circuit is operated under optimal operating conditions at all times. Of course, the optimal circuit configuration should be flexible enough to take into consideration the possible variations in the feed materials on a month to month or a week to week or even a day to day basis.

A typical flotation circuit treats several thousand tons of ore per year. Even a marginal improvement in process efficiency can have a significant economic impact, more so, as mineral rich ores are being depleted and necessitating treatment of progressively lower quality ore with attendant increase in the processing cost. For this reason there have been significant efforts made in the recent past to optimize the performance of flotation circuits.

The conventional flotation circuits are often discussed in terms of rougher, cleaner and scavenger unit operations. Some typical circuits used for sulphide ores are shown in Fig. 1 [1]. While designing a circuit for processing a new ore, the common tendency is to select one of the conventional circuits or their variations on the basis of one's experience. It will certainly be more desirable if the selection and design of the new circuit is based on some quantitative design criteria rather than employing a trial and error approach. In many instances, none of the conventional circuits may give the best performance.

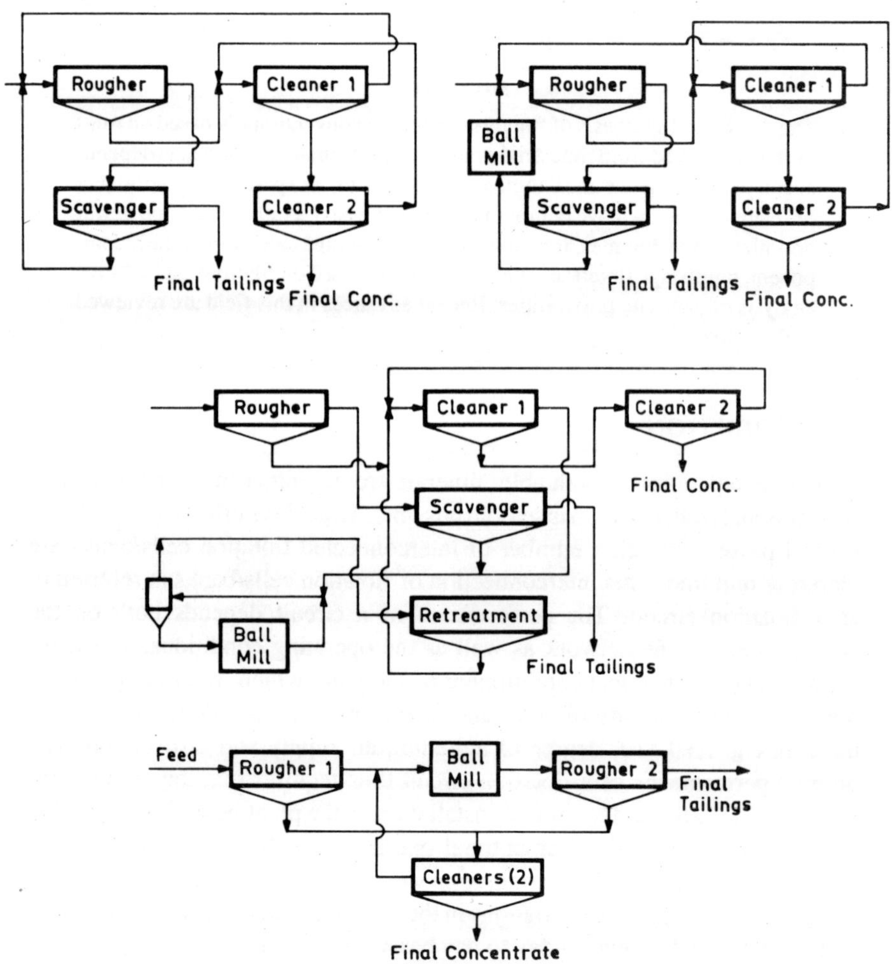

Fig. 1. Some of the conventional flotation circuits used for sulphide ores (after Ref. [1])

The principal aim of this paper is to present and discuss in a consistent form the various strategies and approaches which have already been used and/or have potential for applications in optimal design, synthesis and operation of flotation circuits. Although all meaningful circuit design formulations require, in some form or the other, an understanding of flotation kinetics and circuit simulation, it is beyond the scope of this paper to go into the details of these topics. The readers are referred to recent review papers by Mehrotra and coworkers [2, 3] and Kapur and Mehrotra [4].

OPTIMAL CIRCUIT DESIGN

The design and synthesis of flotation circuits, hitherto, have primarily been dependent on heuristic approach wherein rules-of-thumb based on past experience and engineering judgement are employed. Some examples of these, as compiled by Dey *et al.* [5], are as follows:

1. Feed or recycle streams should join the circuit with minimum possible changes in grade on mixing [6].
2. The greater the number of stages and the higher the circulating loads, the better is the separation [7, 8].
3. Excessive residence time in cleaners leads to deterioration in concentrate grade. For best results, in some circuits at least, the mean residence times in rougher and cleaners should be comparable [9].
4. In a rougher-cleaner- scavenger circuit, the best piece of equipment (i.e. for maximum separation) should be in the rougher position. The feedback loop should step back one stage at a time [10].

The advantage with heuristics is that these lead to quick identification of a network structure which may at times be close to the optimal structure. The drawback however is that there is no reasonable way to establish the quality of networks and often these may even be far from the optimal design [11].

Truly speaking, the synthesis and design of an optimal circuit require detailed consideration of several interacting factors which may be broadly classified as follows [5]:

1. Mineralogy, size distribution, liberation characteristics, etc. of the ore feed.
2. Flotation chemistry which determines the distribution of apparent flotation rate constant of the feed under consideration.
3. Physical parameters and process variables, like aeration rate, pulp density, residence time, froth volume, etc., which affect the flotation behaviour in individual cells [12].
4. Structural variables or design parameters of the circuit including number and size of cells per bank and interconnections between the banks with or without split streams.
5. Objective or profit function which may range from simple aim of maximizing the recovery to highly complex and nonlinear functions of

recovery and grade, equipment cost, operating expenses, etc., subject to appropriate constraints on process variables (e.g. pulp density, flow rates) and structural variables (e.g. cell size, number of split streams).

It is generally assumed that each individual cell is suitably designed, maintained and operated to give the required physical conditioning for flotation. Similarly, chemical conditioning of the pulp, which is of paramount importance, is assumed to be optimally done.

It is only in the recent past that the problem of synthesis and design of an optimal circuit has been formulated using an algorithmic approach. In this approach the problem of synthesizing a circuit is formulated as an optimization problem and rigorous mathematical and/or numerical analysis techniques are employed to obtain the optimal solution; the main advantage being that it provides a more systematic and logical framework for handling a variety of process synthesis problems. The disadvantage, however, is that the computational efforts and cost can be substantial, specially when the problem is not properly formulated.

Prior work on optimization of flotation circuits can be broadly divided into two categories : (1) optimization of conventional circuits - here a fixed circuit configuration, which is generally arrived at using the heuristic approach and is considered optimal, is assumed and only the operating parameters are optimized, and (2) optimization of both circuit configuration as well as the operating parameters - here using an integral approach, the circuit configuration and its structural parameters are extracted from a generalized master circuit or a super-structure and the operating parameters are optimized simultaneously.

The first category is a particular case of trial and error approach where instead of enumerating all possible configurations some conventional configuration consisting of roughing, cleaning and scavenging stages is assumed to be the best configuration. Optimal design here means optimum distribution of total cell volume between stages. In other words, it determines the relative number of cells in roughing, cleaning and scavenging stages of the circuit along with the amount of recycle from stage to stage. In the integral approach the optimal circuit configuration is supposed to be contained in a generalized master circuit or a super-structure consisting of all possible alternate configurations. An objective function is defined to compare the performance of these various configurations. Using an appropriate optimization technique the values of decision variables, which include both the circuit design and operating variables, are determined. These values of decision variables ensure the best performance of the circuit in terms of defined an objective function.

OPTIMIZATION OF CONVENTIONAL CIRCUITS

In this section, the strategies used to optimize the operating parameters of a pre-specified circuit configuration are discussed. From technical considerations alone it can be shown that the performance of a flotation circuit improves as more cells are added to it [7, 8] or in other words more volume is made available within the circuit. An increased number of cells, however, increases both the initial investment as well as the operating cost. It is therefore more logical to impose the restriction of fixed total cell volume for a given feed [9]. This total volume is to be distributed amongst the various banks/cells in the circuit so as to give the best separation. This restriction on volume for a given feed rate implies a constant mean circuit residence time. Thus, the problem of circuit optimization is reduced to finding the optimum distribution of residence time for banks in the circuits, given a fixed residence time based on feed.

Jowett and Ghosh [13] were among the first investigators to attempt optimum design of flotation cells in series. Considering maximum profit as the criterion, optimum operating conditions with respect to feed rate, pulp concentration and number of cells in series were determined. Interestingly their profit function included the cost of operating the cells as well. It was shown that the number of cells in series increased both with increasing feed rate and pulp density. However, it may be noted that cells arranged in series with no circulation only describe a bank of cells and not a circuit. Quantitative analysis using simple flotation models indicates that for a given bank residence time and flotation conditioning, a long bank of small cells is superior to that of a short bank of large cells. An increased number of cells in a circuit, however, increases the initial investment and operating cost. Calculations by Jowett and Sutherland [14] show a strong economic incentive for using a small number of big cells.

Sutherland [9] imposed the restriction of fixed total cell volume for a given feed and optimized the partitioning of this volume into rougher, scavenger and cleaner functions. One of the fixed circuit arrangement considered by him is shown in Fig. 2. Some of his main assumptions were : (1) the flotation of any component can be described by a first order flotation constant (K_j), (2) for any component, K_j is the same in each of the three stages, i.e. roughing, scavenging and cleaning, (3) each stage is a series arrangement of perfectly mixed cells, and (4) the solids hold up per unit cell volume is the same in each cell in all three stages. Considering the configuration with 48 identical flotation cells,

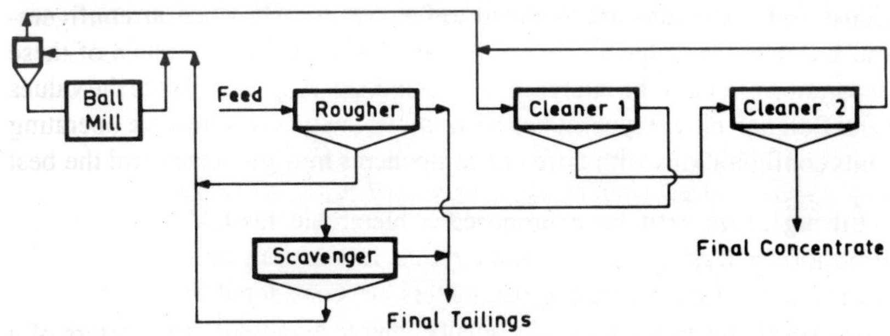

Fig. 2. Schematic representation of a flotation circuit studied by Sutherland (after Ref. [9])

each of $2m^3$, solids hold up of 0.5 t/m^3 and total mass flow rate of 1.2 t/min composed of 12% value and 88% gangue, recovery and grade for a variety of cell arrangements were calculated. His results showed that the optimum circuit arrangements yielding maximum concentrate grade at fixed recovery and plant size, or maximum recovery for a specified grade were such that the residence times in cleaners and scavengers were approximately equal. It was also noted that there were several circuit arrangements which gave almost near optimal performance suggesting that the flotation system was rather insensitive to circuit configuration. This observation, however, should be qualified by his restrictions of a circuit configuration with no split streams and a relatively elementary profit function based on grade only. Jowett and Sutherland [15] subsequently recognized that circuit configuration or interconnections, especially the pattern of recycle streams, was a central issue in design and operation of complex circuits.

Petho *et al.* [16] used an approach based on the principle of combinatorics [17]. Considering the geometric distribution as the discrete analogue of continuous exponential distribution, the recovery from the *r* cell in a bank consisting of *N* flotation cell in series is given as

$$R_r = q^r p \qquad r = 1, 2, ..., N \tag{1}$$

where $q = (1 - p)$, and p is the parameter of geometric distribution and defines cell output - it is related to the flotation rate constant. Considering several configurations and also assuming that the fractional recovery from each cell within a battery is the same, expressions for concentrate, middlings and tailings for minerals of identical as well as different floatabilities have been obtained. The use of geometric distribution allows an optimization to be performed. It is set as an objective to supply products of fixed quality with the highest possible mass recovery, or to fulfil some agreement between the supplier and the user with the highest possible economic gains at minimum investment.

Meloy [10] has employed the circuit analysis approach developed by William and Meloy [18] to optimize flotation circuits. In this approach individual unit operations are assumed to behave linearly when arranged into circuits. Generalized algebraic equations are developed which enable any type of separation circuit to be analyzed and optimized. Applying this technique to several flotation circuit types it has been shown analytically how these various circuits configurations with different components may yield identical outputs.

Bilsing *et al.* [19] have proposed a hierarchic method for modeling, optimization and design of flotation circuits. Interactions between the circuit elements and various operating parameters are considered and coordinated. Starting points for process improvements emerge essentially from the proper adaption of single system parameters which are then optimally adjusted.

OPTIMIZATION OF CIRCUIT CONFIGURATION AND OPERATING PARAMETERS

Various investigations which fall in this category have two features in common [20]. One is that in order to simulate the circuit, instead of using the individual cell models, bank models of unit operations are used. The residence times and pulp environments are assumed to be identical for cells within the bank.

The second common feature is that an integral or structural parameter approach, originally proposed by Ichikawa and coworkers [21, 22, 23] for synthesis of optimal processing systems is employed. A master network comprising of N banks of flotation cell is conceived – N, of course, is specified. One such network with four banks is shown in Fig. 3. The feed to the circuit is split into N streams and fed into each bank. Similarly, each concentrate and tailings stream is split into $(N + 1)$ streams and connected as feed to every other bank as well as short circuited back to the bank from which it originates. The $(N + 1)^{th}$ split stream leaves the circuit as concentrate or tailings output. Clearly every conceivable circuit configuration is embedded in this global architecture and it can be extracted by assigning appropriate values in the range of zero to one to various input, output and connecting streams. It may be seen that in the generalized structural parameter formulation for flotation net works, the total mass feed rate of kth species $(k = 1, 2, ..., X)$ in the feed entering the ith bank in the generalized circuit comprising N banks in all is

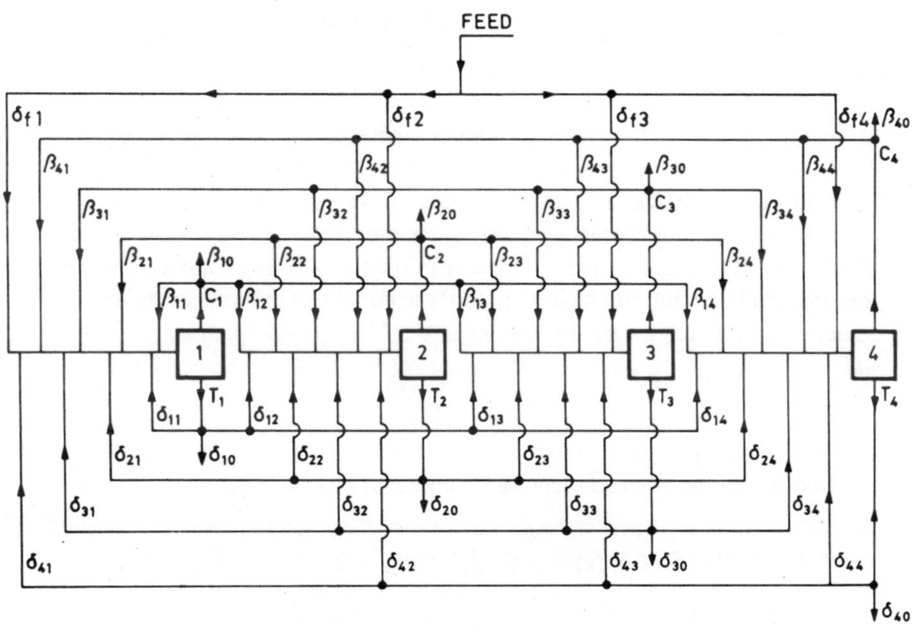

Fig. 3. Generalized master circuit configuration with 4-bank flotation circuit (after Ref. [24])

$$\overline{F}_i(k) = F_i(k) + \sum_{j=1}^{N} C_j(k)\,\beta_{ij} + \sum_{j=1}^{N} T_j(k)\,\delta_{ij}$$

$$= C_i(k) + T_i(k) \qquad \begin{matrix} i = 1, 2, ..., N \\ k = 1, 2, ..., X \end{matrix} \tag{2}$$

where for the species k, $\overline{F}_i$ is the total feed rate to the ith bank comprising new feed and the recycle streams; F_i is new feed to the ith bank; C_j and T_j are total concentrate and tailings flow rates, respectively, of the jth bank and β_{ij} and δ_{ij} are the fractions of concentrate and tailings, respectively, recycled from the jth bank to the ith bank.

Although the various investigations on circuit simulation and optimization are similar in their use of bank models and structural parameter approach, they differ from each other primarily with regards to the application of flotation cell models in the formulation and optimization approach used to solve the problem. Following Yingling's classification the existing literature on circuit synthesis and optimization can be broadly divided into two groups [20]. While the group one investigators have incorporated the flotation models directly in the optimization models, the investigators belonging to the other group have used these models to characterize the constraints on component flows from a flotation bank only. The mass balance equations comprising the optimization models in the latter category are independent of flotation models.

Investigations Incorporating Flotation Models Directly

Studies by Mehrotra and Kapur [24], Dey *et al.* [5], Kapur *et al.* [25] and Yingling [20, 26] are discussed in this section. Considering each individual cell to be a perfect mixer, first order kinetics with distributed rate constant is employed in all these investigations. The mathematical models used by Mehrotra and Kapur and Dey *et al.* are conceptually very similar. While in the case of the former the mass balance equations for the circuit are expressed in terms of the mass flow rates of the kth species in the feed, tailings and concentrate streams, the latter authors have written these equations in terms of the enhancement factor which for the kth species in the jth bank is defined as

$$g_j(k) = \frac{C_j(k)}{T_j(k)} \tag{3}$$

In terms of the enhancement factor eq. (2) is rewritten as

$$F_i(k) = (1 - \delta_{ii} + g_i(k) - g_i(k)\,\beta_{ii})\,T_i(k) - \sum_{\substack{j=1 \\ j \neq i}}^{N} (\delta_{ij} + g_j(k)\,\beta_{ij})\,T_j(k)$$

$$\begin{matrix} i = 1, 2, ..., N \\ k = 1, 2, ..., X \end{matrix} \tag{4}$$

In the matrix notation the circuit equation becomes

$$[F] = [g, \beta, \delta]\,[T] \tag{5}$$

where $[F]$ is $XN \times 1$ feed vector, $[g, \beta, \delta]$ is $XN \times XN$ connection enhancement matrix and T is $XN \times 1$ tailing vector. Conservation of mass requires that:

$$\beta_j^0 + \sum_{i=1}^{N} \beta_{ij} = 1 \tag{6}$$

and

$$\delta_j^0 + \sum_{i=1}^{N} \delta_{ij} = 1 \qquad j = 1, 2, \ldots, N \tag{7}$$

where β_j^0 and δ_j^0 are fractions of concentrate and tailings flow from the jth bank exiting the circuit as outputs. Hence the overall concentrate and tailings outputs of the circuit are

$$C^0 = \sum_{k=1}^{X} \sum_{j=1}^{N} \beta_j^0\, g_j(k)\, T_j(k) \tag{8}$$

and

$$T^0 = \sum_{k=1}^{X} \sum_{j=1}^{N} \delta_j^0\, T_j(k) \tag{9}$$

If $m(k)$ is the mass fraction valuable in the kth species, the valuable flow rates to the circuit in the new feed and the concentrate output are :

$$M_f = \sum_{k=1}^{X} \sum_{j=1}^{N} m(k)\, F_j(k) \tag{10}$$

$$M_c = \sum_{k=1}^{X} \sum_{j=1}^{N} m(k)\, \beta_j^0\, g_j(k)\, T_j(k) \tag{11}$$

Recovery and grade therefore are, respectively,

$$R_c = \frac{M_c}{M_f} \times 100 \tag{12}$$

and

$$G_T = \frac{M_c}{C^0} \times 100 \tag{13}$$

The optimization problem has now the following specifications. Given a feed rate at its entry point into the circuit and range of allowable values of g's, and subject to appropriate constraints on mass flow rate in tailings streams, hold-up, pulp density in cells, grade and/or recovery, etc. determine the optimum values of process variable g's and structure parameters β's and δ's such that on solving eqs. 5, 10 and 11, a specified objective function is maximized. The optimized values of g's, β's and δ's, in turn permit synthesis of the desired circuit. Dey *et al.* have shown that the steady state flotation dynamics can be formulated in terms of g's only, which is related to the mean residence time λ_1 in any cell in the ith bank in the following manner :

$$g_i(k) = [1 + K(k)\,\lambda_1]^z - 1 \tag{14}$$

where $K(k)$ is the specific flotation rate constant of kth species and z is the number of cells in the ith bank. Thus, a complete characterization of the circuit, including sizing of banks, pulp hold-up in a cell, the exit rate, etc. is possible.

It may be noted that the mass balance equations for a generalized network of flotation banks represent a large system of non linear equalities which must be satisfied by specific circuit configuration.

Mehrotra and Kapur [24] used the random search technique and complex method to optimize the objective function. The two main problems encountered in these procedures were: (i) slow rates of convergence, and (ii) convergence to multiple optimal depending on the initial guess. These problems have been eliminated to a large extent by Dey *et al.* [5] who apply the random search technique of Luus and Jaakola [27] after modifying it to incorporate the biased random search numbers. In this modified method the search is deliberately biased towards the parameter bounds depending on drift of the parameter in search iteration. Through several illustrative examples it has been established that this methodology successfully predicts the optimum configuration. It is also observed that several near optimum configurations are relatively simple even when split stream interconnections are not restricted. Considering an illustrative example of a three bank circuit, in which a feed of two species in gangue and mineral is treated, Dey *et al.* [5] have demonstrated that the optimal circuit configuration changes in an expected manner from rougher-cleaner-cleaner to rougher-scavenger-cleaner to rougher-scavenger-scavenger as weightage on grade shifts to weightage on recovery in the objective function. In this example the objective is to maximize the objective function

$$O = W_r R + W_g G \tag{15}$$

where W_g/W_r, the ratio of weights on grade and recovery is varied from 10 : 1 to 1 : 10 in steps. The ratio of flotation rate constants for gangue and mineral species K(Gang.)/K(Min.) is 1/10 with the constraint that maximum feed rate to any cell does not exceed 150% of the feed input rate. Optimal flotation circuits obtained for various W_r/W_g ratios are shown in Fig. 4.

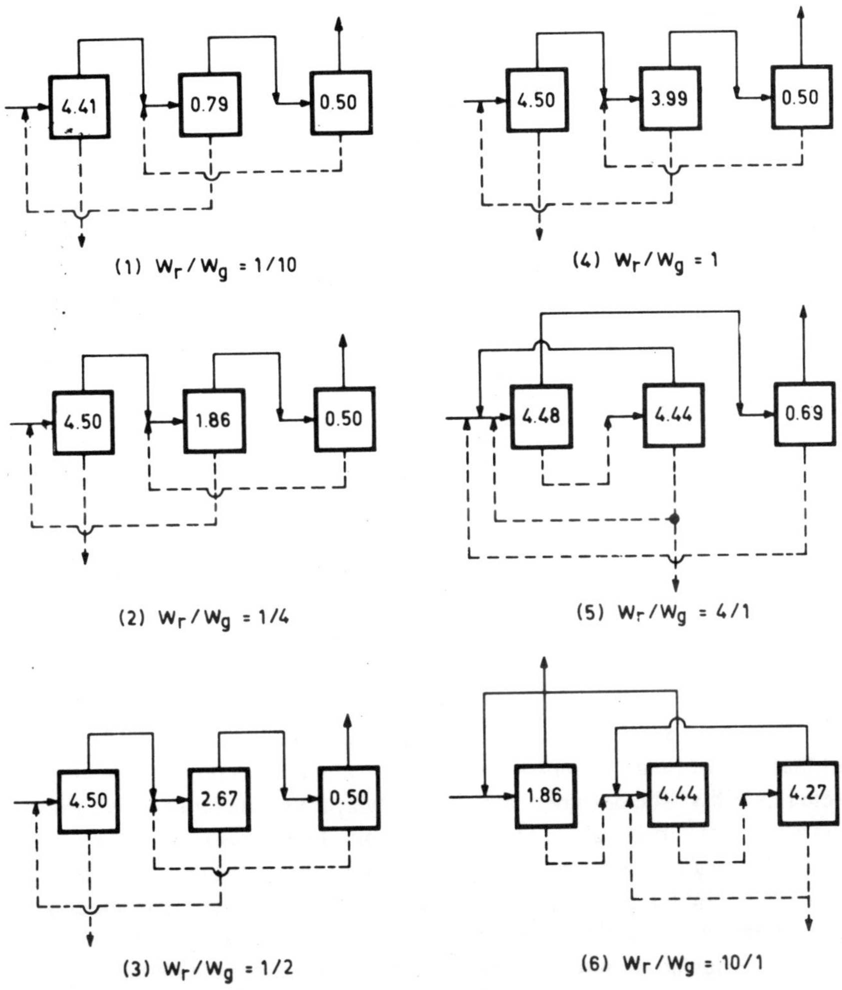

Fig. 4. Optimal circuit configurations for different weights on grade and recovery in the objective function (after Ref. [5])

It must be emphasized that the principal limitations in synthesis of optimum circuit are not in the optimization strategies but rather in our current lack of knowledge of the process variables and of interrelationship between different species with respect to these process variables. Even water balance may be easily incorporated in the formulation provided the relevant parameters are reliably estimated.

In the structural parameter approach described above for an *N* bank circuit, the decision variables associated with each splitting node is an $N-1$ dimension vector of continuous variables. Yingling has discussed in detail the problems that may be associated with these continuous variables [20, 26]. The principal of these is that the objective function of the optimization problem may employ nonlinear concave function which, in turn, may result in numerous local optima that make identification of meaningful solution by the optimization procedure difficult. Extending his flow sheet optimization algorithm [28] he has applied it to flotation circuits for parameter estimation and configuration optimization. The procedure uses individual cell model based on first order distributed rate constant kinetics. A requisite assumption in his formulation is that the unit cell model can be written in the form

$$T(k) = P(k)\, F(k) \qquad k = 1, 2, ..., X \tag{16}$$

where $P(k)$ is the split ratio for particle class k and may be defined as

$$P(k) = \frac{1}{1 + K(k)\,\lambda} \tag{17}$$

where λ, the residence time is defined as the ratio of cell volume to tailing flow rate from that cell. It is recognized that $P(k)$ is a function of vessel's design and control variables but it is independent of changes in the total mass feed rate and the relative distribution of this mass among the various classes, for fixed control settings. Individual cell models are used to optimize the number of cells or stages within a flotation bank. The basis of the mass balance used in this approach is to view the flow of each particle class as the state to state evolution of a Markov chain. For each particle class the mass flow is described by a transition matrix. The algorithm accommodates mixed discrete/continuous formulation of flotation bank interconnection alternatives providing considerable flexibility. The designer may allow single stream interconnection among the banks, or alternatively he may consider split streams interconnections if the performance gains with such a configuration out-weigh the penalties imposed on circuit complexity resulting due to the splitting of streams. This procedure thus optimizes bank interconnection configuration, number of cells per bank and cell design and control parameters. A major improvement in the computational efficiency over direct search techniques is claimed.

Investigations Incorporating Flotation Models Indirectly

The investigations of Green and coworkers [12, 29] and Reuter and Van Devanter [30] fall in this category. The main thrust in these investigations lies in formulating the optimization problem in the linearized form, the optimal solution to which is obtained using linear programming. The fundamental difference between the approaches described in the preceding section and those in this section is that optimization is performed in the latter without the use of flotation models. Once an optimal set of structural parameters and enhancement factors is determined, the model is then used in reverse manner to compute the process variables, e.g. residence time. Flotation models may however be used indirectly to establish the bounds on the enhancement factors required to solve the optimization problem. Thus the separation characteristics of the minerals, and the design and operating features of the flotation cells are included only as bounds on species flows from a particular flotation bank and not as a model included in mass balance equation. In these formulations eq. (2) is rewritten as

$$\overline{F}_i(k) = F_i(k) + \sum_{\substack{j=1 \\ j\neq i}}^{N} T_{ij}(k) + \sum_{\substack{j=1 \\ j\neq i}}^{N} C_{ij}(k) \tag{18}$$

$$T_{ij}(k) = T_j(k)\ \delta_{ij} \tag{19}$$

$$C_{ij}(k) = C_j(k)\ \beta_{ij} \tag{20}$$

$$F_i(k) = F_f(k)\ \delta_{fi} \tag{21}$$

where $T_{ij}(k)$ and $C_{ij}(k)$ are flows of species k in tailings and concentrate streams, respectively, from the jth bank to the ith bank, $F_f(k)$ is fresh feed flow of species k to the flotation circuit and δ_{fi} is proportion of feed stream that is sent to the ith bank. Conservation of mass around any splitting node results in:

$$T_{ij}(k) = \sum_{\substack{j=1 \\ j\neq i}}^{N} T_{ij}(k) \tag{22}$$

$$C_{ij}(k) = \sum_{\substack{j=1 \\ j\neq i}}^{N} C_{ij}(k) \tag{23}$$

Demands imposed on the plant, e.g. grade, production rate and mass flow rate requirements to the other sections of the plant, etc. are incorporated in the form of constraints. An upper and lower bound may be placed on the grade for each species by bounding the enhancement factor for species k at the ith bank, i.e.,

$$[g_i(k)^l]\ T_i(k) \le C_i(k) \le [g_i(k^u)]\ T_i(k) \tag{24}$$

where the superscripts l and u refer to lower and upper bounds, respectively. Upper and lower bounds may also be placed on separation of two species in the ith bank, i.e.

$$S_{kr}^{l} \le \frac{C_i(k)}{C_i(r)} \le S_{kr}^{u} \tag{25}$$

where S_{kr} is the total separation of species k from some reference species r. The enhancement factor for a bank of the circuit is related to the number of cells in the bank, the bank pulp residence time and the flotation constants for various species. A flotation model is used primarily to establish this relationship. Once an optimal set of enhancement factors has been determined, flotation variables are obtained using a flotation model. For instance Green and his coworkers [12,

29, 31] have used gamma function model to obtain various flotation variables and obtained the parameters of the gamma function using the batch test data on a laboratory scale flotation cell.

It may be pointed out that though the mass balance equations in these structural parameter formulation are linear because of the use of variables like $T_{ij}(k)$, $C_{ij}(k)$ and $F_i(k)$, nonlinear constraints of the following form are inevitable:

$$\frac{T_{ij}(k)}{T_j(k)} = \frac{T_{ij}(r)}{T_j(r)} \quad k \neq r \tag{26}$$

and

$$\frac{C_{ij}(k)}{C_j(k)} = \frac{C_{ij}(r)}{C_j(r)} \quad k \neq r \tag{27}$$

These constraints are required for each splitter j and species k, excluding the reference species r, to ensure that the split ratio is equal for all mineral species at the splitting mode j. The number of such nonlinear constraints can be significantly large. The investigations reported in this section differ from each other primarily in the manner in which these nonlinear constraints are dealt with.

Green solved the optimization problem in terms of the flow of a single valuable mineral species using linear programming. Although for a single species problem eqs. (26) and (27) may be omitted, as done by Green [12], Green *et al.* have disregarded these equations even for multiple valuable species formulation resulting in physically unrealistic flows [31]. This discrepancy is recognized by Reuter *et al.* [29] and Reuter and Van Devanter [30] who employ a two stage optimization approach to circumvent the difficulties imposed by eqs. (26) and (27). As summarized by Yingling in his review paper [20], in this approach, beside specifying the bounds on the enhancement factors, the lower and upper bounds on grade of output stream from particular banks are also specified in terms of tailings flow rates, viz.,

$$G_i(k)^l \leq \frac{T_i(k)}{T_i} \leq G_i(k)^u \tag{28}$$

where T_i the total tailings flow rate from the ith bank is

$$T_i = \sum_{k=1}^{X} T_i(k) \tag{29}$$

For the total concentrate flow from the ith bank, the following bounds must hold:

$$\overline{K}_i^l \; T_i \le C_j \le \overline{K}_i^u \; T_i \tag{30}$$

where the lower and upper bounds on the overall separation factor from the ith bank are defined as

$$\overline{K}_i^l = \sum_{k=1}^{X} G_i(k)^l \; g_i(k)^l \tag{31}$$

$$\overline{K}_i^u = \sum_{k=1}^{X} G_i(k)^u \; g_i(k)^u \tag{32}$$

The model used in the first stage of the two stage approach consists of mass balance equations in terms of stream mass flow rates, T_i, C_i and F_{fi} and T_{ij}, C_{ij} and F_i which are defined as

$$T_{ij} = \sum_{k=1}^{X} T_{ij}(k) \tag{33}$$

$$C_{ij} = \sum_{k=1}^{X} C_{ij}(k) \tag{34}$$

and

$$F_i = \sum_{k=1}^{X} F_i(k) \tag{35}$$

Incorporating the constraints given by eq. (30), the linear programming is carried out on these equations. The solution in this first stage gives structure parameters, e.g.

$$\delta_{ij} = \frac{T_{ij}}{T_j} \tag{36}$$

It, however, does not provide any information on product grade as the flows of individual mineral species are not modeled in this stage.

For the second stage of the solution, the structure parameters obtained in the first stage are assumed to be optimal. The second stage takes into consideration the flow rates of individual mineral species and also incorporates the constraints characterized by eqs. (26) and (27). The constraints are linearized by fixing the right hand side at values determined in the first stage. Grade requirements on product stream may also be incorporated in this second stage of solution.

A careful examination of strategies used in this section indicates that in reality the flotation cell models are used directly and indirectly twice - once to specify the lower and upper bounds of enhancement factors and the feasible ranges of operating variables like cell volume, number of cells per bank, pulp densities, residence time, etc, and the second time it is used to determine the value of some of these decision variables from the optimum flow rates of individual mineral species computed in the second stage of the optimization problem. These values of the decision variables, in turn, may be used in the first stage of the computations. As Yingling has rightly pointed out, there are two major drawbacks of such an approach :

1. Within the bounds defined by eq. (24) the enhancement for individual species are assumed to vary independent of each other. It is evident that the separation of various species depends on residence time and number of cells in the bank and these two variables cannot be manipulated to result in arbitrary values of enhancement factors for various mineral species. Furthermore, the flotation rate constants for various species are set almost arbitrarily and independent of each other in the second stage of the optimization problem. This is something highly unlikely to be true in practice.
2. Due to the absence of the flotation model in the system equations, the effect of decision variables on process economics is not taken into consideration. As these variables may have significant economic impact, it would be desirable to incorporate these models directly in the optimization problem itself.

CONCLUDING REMARKS

The paper describes the two broad approaches that have been reported in literature to design and synthesize the flotation circuits :

1. Optimization of conventional circuits
2. Optimization of circuit configuration as well as operating parameters.

The first approach is useful when a flotation circuit already exists, and space, economical and other considerations do not allow changes in the circuit configuration. It is, however, found that often even at the time of installation of a new flotation plant, based on past experiences a conventional circuit configuration is specified a priori completely disregarding the possibility of split streams in the configuration. It is now more or less established that optimal flotation circuits with split stream result in better performance as compared to that of the conventional configuration. Hence, the conventional circuit configuration in most situations may not be the optimal configuration. Optimization of a pre-specified circuit configuration is essentially reduced to optimizing the operating parameters / variables.

Circuit optimization based on the integrated approach is quite novel in the sense that it considers a generalized network of banks of flotation cells which contains in it all possible configurations, including the optimal one. In one strategy within this approach the structural and operating parameters are extracted simultaneously while in an another strategy the optimal configuration is extracted from the generalized network and then the optimal parameters are determined for this optimal configuration. The basic assumption of the latter strategy that the optimal circuit configuration can be determined independent of operating conditions is questionable.

There are two main criticisms that have often been cited by the opponents of the integrated approach: (i) mathematical formulation involves relatively complex equations, and substantial efforts are required to solve them, and (ii) the optimal circuits may contain many split/recycle streams making it too complex to be of much practical utility. With the advent of fast computers and rather easy accessibility to fairly powerful mini/micro computers the first criticism is no longer valid. One can easily develop general computer programmes which accommodate changes in the input data characterizing feed and other constraints that the designers of the circuit may wish to impose, and predict the anticipated. The second criticism is also not serious enough. Illustrative examples of Dey *et al.* [5] and Yingling [26] clearly suggest that even when split streams are not restricted, the optimal circuits obtained in most cases are reasonably simple. Multiplicity of optimal or near optimal circuits has also been demonstrated by Dey *et al.* By imposing realistic penalties for any additional split/recycle stream one can guarantee an optimal or at least near optimal circuit configuration for almost any situation.

As already emphasized, the main problem in optimal design and synthesis of a flotation circuit is not the limitation of the available optimization techniques and algorithm but it is our inability to accurately predict the effect of process variables and parameters on the circuit performance using mathematical models. One of the principal bottlenecks in this direction is the proper characterization of feed material. Identifying various species and/or class of particles constituting the feed and associating a flotation rate constant with each of the species/class, and variation of these rate constants with variations in operating conditions in the circuit still remains very tentative. Kapur *et al.* [25] have recently proposed an inverse modeling approach for characterizing flotation circuit feeds using plant data on total mass and individual mineral flows. Feed is characterized by conjugate distributions of total mass and mineral mass in apparent flotation rate constant as the common attribute. This rate constant is interpreted as lumped kinetic parameter in a reasonably detailed multi-phase and froth stability index. The feed identification procedure involves random search of optimum feed parameters in conjunction with repeated simulation of the circuit in two parallel mathematical streams, one for total mass and the other for the mineral flows only. Validity of this approach of feed identification particularly for the simulation purposes has been demonstrated through an extensive programme of identification of 'synthetic' feeds of various kinds from computer generated flotation circuit data, as well as, actual data from three different circuits reported in literature. Direct application of this feed identification procedure for simulating, in predictive mode, a given flotation plant of arbitrary design and complexity is not yet feasible. Nevertheless, it does show promise and potential for developing a methodology for feed identification required for simulation in the predictive mode.

REFERENCES

1. Lynch, A.J., Johnson, N.W., Manlapig, E.V., Thorne, C.G.: Mineral and Coal flotation circuits. New York: Elsevier Scientific Publ. Co. 1981, pp. 9-12.
2. Mehrotra, S.P.: Minerals & Metall. Processing. 5 (1988) 142-152.
3. Mehrotra, S.P., Padmanabhan, N.P.H.: Trans. Indian Inst. Metals. 43 (1990) 9-21.
4. Kapur, P.C., Mehrotra, S.P.: In Sastry, K.V.S., Fuerstenau, M.C. (Eds.): Modelling of Flotation Kinetics and Design of Optimum Flotation Circuits. Littleton: SME 1989, pp. 300-322.
5. Dey, A., Kapur, P.C., Mehrotra, S.P.: Int. J. Miner. Process. 26 (1989) 73-93.
6. Taggart, A.F.: Handbook of Mineral Dressing, 2nd Ed. New York: Wiley International 1945, p. 103.
7. Imaizumi, T., Inoue, T. : In Robert, A. (Ed.): 6th Int. Miner. Process Congr. - Canes: Pergamon Press 1965. pp.581-593.
8. Loveday, B.K., Marchant, G.R.: J. South African Inst. Mining and Metall. 72 (1972) 288-294.
9. Sutherland, D.N. : Int. J. Miner. Process. 7 (1981) 319-346.
10. Meloy, T.P. : Int. J. Miner. Process. 11 (1983) 89-99.
11. Grossman, I.E.: Comput. Chem. Engg. 9 (1985) 463-482.
12. Green, J.C.A.: Int. J. Miner. Process. 13 (1984) 83-100.
13. Jowett, A., Ghosh, S.K.: In Arbiter, N. (Ed.): 7th Int. Miner. Process. Congr. New York:Gorden Breach 1965, pp.175-184.
14. Jowett A., Sutherland, D.N.: Chemical Engg. (London). 347 (1979) 603-607.
15. Jowett, A., Sutherland, D.N.: Int. J. Miner. Process. 14 (1985) 86-106.
16. Petho, S.Z., Bohm, J., Tompos, E.: Acta Geodaet Geophysica et Montanistica, Academiae Scientiarum Hungaricae. 11 (1976) 61-75.
17. Brualdi, R.A.: Introductory Combinatorics. New York: Elsevier 1977.
18. William, M.C., Meloy, T.P.: Int. J. Miner. Process. 10 (1983) 141-160.
19. Blising, U., Gruner, H., Topfer, E. In Forssberg, E. (Ed.): XVI Int. Miner. Process. Congr. Amsterdum: Elsevier Sci. Publisher 1988. pp. 433-443.
20. Yingling, J.C.: Int. J. Miner. Process. 38 (1993) 21-40.
21. Umeda, T., Ichikawa, A.: Industrial and Engg. Chemistry. 10 (1971) 229-236.
22. Umeda, T., Hirai, A., Ichikawa, A.: Chemical Engg. Sci. 27 (1972) 795-804.
23. Ichikawa, A., Fan, L.T.: Chemical Engg. Sci. 28 (1973) 357-374.
24. Mehrotra, S.P., Kapur, P.C.: Separation Science. 9 (1974) 167-184.
25. Kapur, P.C., Dey, A., Mehrotra, S.P. : Int. J. Miner. Process. 31 (1991) 11-35.
26. Yingling, J.C.: Int. J. Miner. Process. 38 (1993) 41-66.
27. Luus, R., Jaakola, T.H.I.: AIChEJ. 19 (1973) 760-766.

28. Yingling, J.C.: Int. J. Miner. Process. 29 (1990) 149-174.
29. Reuter, M.A., Van Deventer, J.S.J., Green, J.C.A., Sinclair, M.: Chemical Engg. Sci. 43 (1988) 1039-1049.
30. Reuter, M.A., Van Deventer, J.S.J.: Int. J. Miner. Process. 28 (1990) 15-43.
31. Green, J.C.A., Meij, J. T., Du Plooy, A., Van Heerden, P.: Comput. Chem. Engg. 9 (1985) 463-482.

Hydrophobic Force and its Role in Amine Flotation System

S.A. RAVISHANKAR
Center for Coal and Minerals Processing,
Virginia Polytechnic Institute and State University,
Blackburg, Virginia 24061

ABSTRACT

Surface forces play an important role in many applications such as the stability of colloidal suspensions, aggregation, adhesion and phase behaviour of surfactants. The main forces include van der Waals attraction and electrostatic repulsion (known as classical DLVO forces) and structural forces which arises from the fact that the liquid medium is not a structureless continuum (non-classical forces). The recent advances in liquid-state physics and now direct experiments have made it possible to describe and quantify these forces. In this paper, one such force, namely the hydrophobic force, is discussed particularly with respect to its role in flotation systems. The measurements of hydrophobic force between mica surfaces in presence of amine surfactants made with the help of the surface force apparatus are also presented to illustrate the concepts recently developed regarding the origin of hydrophobic force.

INTRODUCTION

The DLVO theory has the ability to predict the stability of colloidal suspensions based on the balance between electrostatic repulsive forces and dispersion attractive forces [1,2]. The theory however fails to explain the stability between very hydrophobic and hydrophilic colloidal suspensions [3-5]. This is due to the influence of some extraneous forces not accounted for by the classical DLVO theory. With the advent of direct force measurement techniques [6-8], these forces have been repeatedly measured and well characterized in the last twenty years [9-12]. It is therefore necessary to extend the classical DLVO theory by including these non-DLVO force components. Yoon and co-workers studied the stability of mineral suspensions [13-17] and bubble-particle interactions [18] using an *extended* DLVO theory by incorporating the structural energy term in the DLVO theory as a third component.

Bubble-particle interaction is the most important subprocess in flotation. For stable adhesion between a bubble and a particle, a net attractive force (or negative disjoining pressure, $-\Pi$) is necessary between the bubble (air/liquid) and the particle (solid/liquid) interfaces. Earlier it was shown that to ensure bubble-particle adhesion, particles have to be hydrophobic in nature suggesting that the major contribution to the net attractive force is the hydrophobic force [19].

Yoon and coworkers [20, 21] modeled the bubble-particle interaction from first principles for amine flotation system and predicted that the driving force for flotation is a long-range hydrophobic force comparable to the values reported by Rabinovich *et al.* [10] for the hydrophobic force measurements between two methylated silica cylinders. However, it is almost an order of magnitude higher than those measured for systems involving water soluble amine and sulfonate solutions [9, 22-25]. Recently Yoon and Ravishankar [17] measured long-range hydrophobic forces in the amine system in the presence of water soluble long-chain alcohols and explained the discrepancy observed between the predicted and experimentally measured hydrophobic forces.

One of the major challenges in flotation research is to develop a comprehensive model based on first principles to control, design and optimize the process. It is therefore important to understand the role of surface forces, particularly the hydrophobic forces, in particle-bubble interactions. It is the purpose of this article to present an overview of hydrophobic force and illustrate its role in amine flotation systems using direct force measurement studies with pure amine and amine/octanol mixture.

HYDROPHOBIC FORCE

Hydrophobicity is a well-known concept that indicates the extent of wetting of water for a solid substrate. Hydrophobicity is commonly determined by measurement of the contact angle subtended by a water droplet on a surface. When two such hydrophobic surfaces are brought together (within 100 nm) in an aqueous media, an attractive force that is sometimes several orders of magnitude higher than the dispersion forces, may originate between them. This attractive non-DLVO force component is considered to be a hydrophobic force. The first direct measurement of the hydrophobic force was made in 1982 by Israelachvili and Pashley between two mica surfaces immersed in a solution of cetyl trimethylammonium bromide at near neutral pH [9]. It was observed to be a short-range force with an exponential decay length of 1-1.5 nm. Later long-range hydrophobic forces were measured between strongly hydrophobized mica and silica surfaces with decay lengths ranging between 9-25 nm [26-36]. Although the existence of the hydrophobic force has been repeatedly confirmed in literature, the origin of this force has not been fully understood. In the following section, an overview of the hydrophobic force is presented with a brief description of different methods of measuring the force, different theories about its origin, and various functional forms of its representation.

Hydrophobic Force Measurements

The most widely used instruments for the purpose of direct force measurements are the Israelachvili type surface force apparatus (SFA) [6, 37] and the Atomic Force Microscope (AFM) [8, 29, 38]. Both of these apparatii are based on Hooke's law where the deflection of a cantilever spring depends on the exerted force.

In the Israelachvili type apparatus, shown in Fig. 1a, two thin (1-4 μm) molecularly smooth mica sheets, each silvered on one side, are glued silver side down on to curved silica disks. One of the disks is mounted on a weak spring while the other is attached to a cylindrical piezo crystal arrangement. The distance between the surfaces is measured using interferometric technique based on the standing wave pattern resulting from reflections between the silvered backsides of the mica sheets [39]. This arrangement enables distance measurements with a precision of 0.1-0.2 nm.

The force is measured by the difference in the deflection of the spring under the desired conditions and the calibrated value at large distances where no force is exerted between the surfaces. From Hooke's law $F = kX$, where k is the spring constant and X is the difference in the deflection of the spring, the magnitude of the force can be calculated. The surfaces "jump" in to contact when the slope of the force $dF/dX > k$. The k can be adjusted in the apparatus to enable the measurement of stronger attractive forces at closer distances.

Another method that became popular in the last 3 to 4 years is the direct force measurement using AFM (see Fig. 1b) [29, 38]. In a conventional AFM, the force is measured between a sample and a pyramidal tip (silicon nitride) as a function of sample displacement. Sample displacement is controlled by a piezoelectric crystal that moves towards the cantilever tip at a desired frequency. The force exerted by the micro-fabricated cantilever tip is calculated from the deflection which is monitored by a laser beam reflected from the back of the cantilever to fall onto a sensitive split photo diode. One can also measure the force using a sphere-plate configuration. For this purpose, a small spherical particle is attached at the end of the cantilever (see Fig. 1b). The force can be converted to energy or disjoining pressure using Derjaguin approximation [40].

Although both the techniques are widely used for force measurements, SFA is preferred over AFM for studies of adhesion, lubrication and wetting. Because of the precision optical method used for the distance measurements in the SFA, it uses transparent samples like mica. The recent modification of SFA which uses piezoelectric-bimorph technology has removed this limitation [44]. Measurements on other surfaces such as silica [41] and alumina [42, 43] have thus been reported. AFM can be used to measure the forces between opaque samples.

Origin of Hydrophobic Force

One of the most intriguing aspects of the hydrophobic force is the lack of understanding about its origin. The explanations put forward by different researchers suggest that long-range hydrophobic forces are due to structural

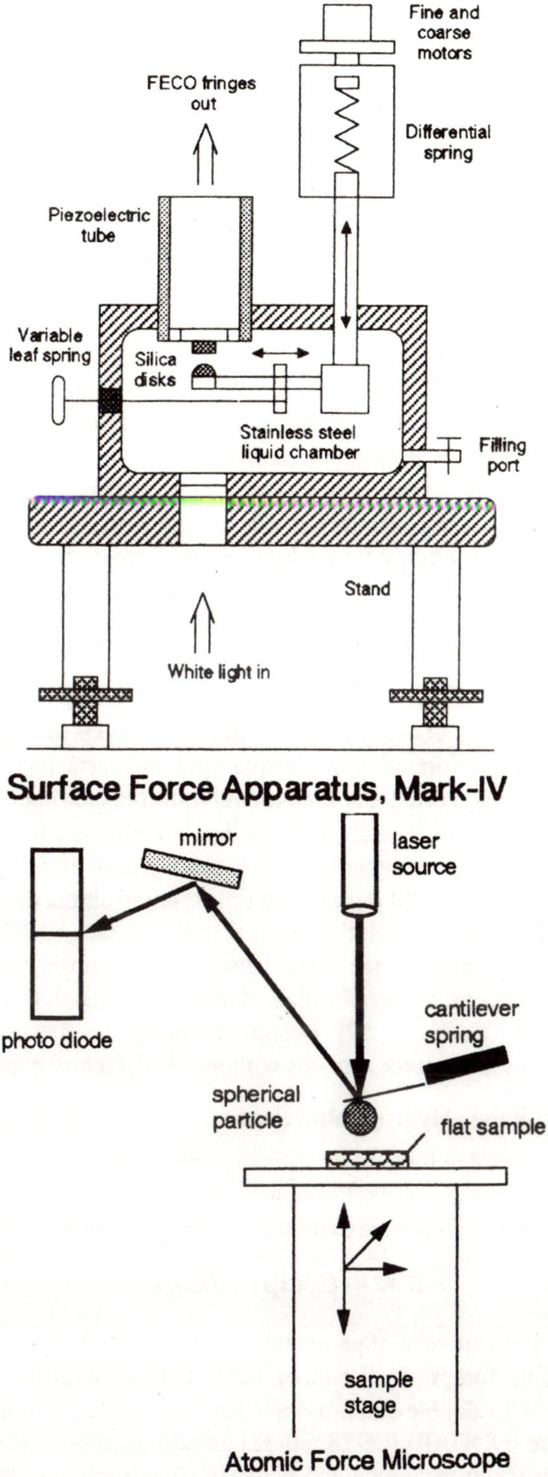

Fig. 1. Schematic diagrams of direct force measuring instruments: A. Surface Force Apparatus (SFA), B. Atomic Force Microscope (AFM).

changes or density fluctuations in the solvent (in the intervening liquid film between two hydrophobic surfaces) or the electrostatic correlation effect similar to dispersion forces described by the Lifshitz theory [45]. In the solvent structure effect, the hydrophobic force was considered to arise from ordering of the liquid layers adjacent to the nonwetting solid surface [46-49]. Yushchenko *et al.* [50] proposed a theory for the hydrophobic force based on cavitation which occurs due to a metastability caused by the local imbalance of the kinetic and cohesive pressures in the liquid film between the hydrophobic surfaces. Ruckenstein and Churaev [51] further stated that the density fluctuations of solvent molecules at the hydrophobic surface/liquid (water) interface may propagate from one interface to another causing long-range hydrophobic forces.

Recently Attard [52] and Podgornik [53] proposed mechanisms based on charge/dipole interactions. Both authors arrive at similar expressions indicating exponential dependence of the force on the distance between particles. Attard assumed anomalous dielectric behaviour of the solvent adjacent to the hydrophobic surface to account for the long-range hydrophobic force. According to Podgornik [53] however, with such an assumption one cannot exceed the limit dictated by the Lifshitz theory [45]. Podgornik instead suggested that the long-range hydrophobic forces are due to fluctuating electric field associated with the lateral mobility of the ions adsorbed at the liquid interlayer boundaries.

On the basis of the above arguments and with the knowledge available about in-plane domain formation at phospholipid and surfactant monolayers [54-56], it has been suggested that the ordered (quasi-crystalline) structure of the hydrocarbon layer can generate in-plane dipole moments (domains) which can induce changes in the structure and in the orientation of water dipoles at the nearby surfaces. The resulting electric fields, in the absence of screening, would penetrate out into the solution normal to the surface on a length scale corresponding to the domain size. Thus between two opposing hydrophobic surfaces, the interactions between in-plane domains may result in a long-range hydrophobic force [30, 31, 36, 57]. Despite the many theories proposed, the origin of the hydrophobic force remains without a satisfactory explanation.

Long and Short-range Hydrophobic Forces

Short-range hydrophobic forces with decay lengths in the range of 1-2 nm are usually observed in presence of self-assembling water soluble surfactants. The measured hydrophobic forces are expressed by an exponential function:

$$F/R = -C_o \exp(-H/D_o), \tag{1}$$

where C_o is the pre-exponential parameter and D_o is the decay length. Long-range hydrophobic forces, on the other hand, are measured between mica surfaces coated with double-chain amines such as diocta decyldimethylammonium bromide (DDOAB) [26-28, 30-32] and fluorocarbons [33-35] using the Langmuir-Blodgett technique and between silica surfaces silanated with trimethylchlorosilane (TMCS) [10, 29] and octadecyltrichlorosilane (OTS) [29]. In each case, the long-range forces are better fitted by a double-exponential force law:

$$F/R = -\ C_1 \exp(-\ H/D_1) - C_2 \exp(-\ H/D_2), \tag{2}$$

where C_1 and C_2 are the pre-exponential parameters, and D_1 values are in the range of 1-2 nm at $H < 10$ nm and D_2 values are in the range of 10-30 nm at longer distances.

It has been shown in the literature that the values of D_2 obtained with double-chain amines (e.g., DDOAB) are higher than the D_o values in the case of single chain water soluble surfactants. This may be attributed to the increased hydrocarbon density and to the molecular ordering of hydrocarbon chain in the monolayer [58]. Recently, Rabinovich *et al.* [57, 59] showed a linear dependency of the order parameters on hydrophobic force constants.

Occasionally, a power law is also used to describe the hydrophobic forces [17, 60]:

$$F/R = -\ K/6\,H^2, \tag{3}$$

in which K is the fitting parameter. Equation (3) is of the same form as the dispersion force. Because of a single fitting parameter, the power law is easy to use for modeling purposes. It is important to note however that the power law, in general, does not accurately describe the measured hydrophobic force data.

Although the theoretical expressions derived for hydrophobic force are similar to the single exponential form [52, 53, 61], they do not provide a rationale for the long-range hydrophobic force. Equations (2) and (3) are therefore still considered empirical in nature.

HYDROPHOBIC FORCES IN THE AMINE FLOTATION SYSTEM

Amine flotation, a process widely used for floating silicate and aluminosilicates minerals, is strongly influenced by pH, concentration of the amine solution [62, 63], as well as the cosurfactant interaction [64]. Although earlier studies on the adsorption of amine collectors at mineral surfaces qualitatively established the conditions for flotation based on the mineral hydrophobicity, those cannot be used directly to model the flotation process from first principles. The model requires the estimates of the energy barrier between the interacting interfaces. With the advent of direct force measurements, it is possible to determine the total force exerted between surfaces and deconvolute the hydrophobic force component using the extended DLVO theory. In the following section, the results of direct force measurements for the adsorption of amine on mica surfaces obtained as a function of concentration, pH and in the presence of a cosurfactant, octanol, have been discussed. The deconvolution procedure based on the extended DLVO theory is outlined for estimating the hydrophobic force parameters.

Single Surfactant Interaction

The force measurements conducted using the SFA apparatus at various concentrations of dodecylamine hydrochloride (DAHCl) solution are described in this section. DAHCl (Eastman Kodak) was recrystallized from ethanol before use, and a stock solution was prepared in nanopure water obtained from a Nanopure II water treatment unit. Fig. 2 shows the results on force vs distance measurements in nanopure water as a function of concentration of DAHCl. The theoretical DLVO curves, based on the constant charge (upper) and the constant potential (lower) model, used to best-fit the experimental data were obtained by the algorithms of Chan *et al.* [65], assuming a suitable ψ_o value and an electrolyte concentration. The nonretarded van der Waal's force was calculated using a Hamaker constant (A_{131}) of 2.2×10^{-20} J. For nanopure water, the ψ_o value of – 170 mV and debye length (κ^{-1}) of 96 nm thus obtained from the DLVO fit, are in close agreement with the literature values [6, 22].

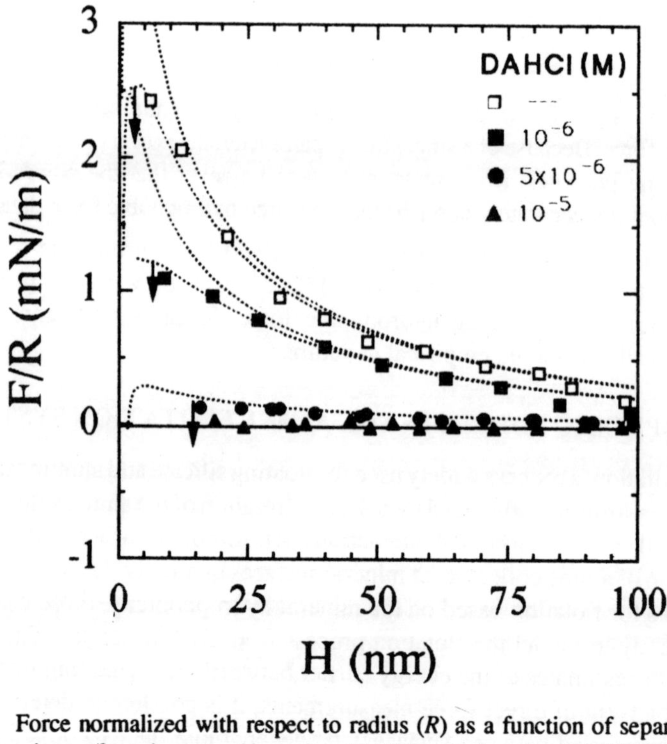

Fig. 2. Force normalized with respect to radius (*R*) as a function of separation (*H*) between mica surfaces immersed in various DAHCl solutions at pH 5.7. The upper and lower dotted lines indicate constant charge and constant potential models respectively of the DLVO theory obtained using $A_{131} = 2.2 \times 10^{-20}$ J, $\psi_o = -170$ mV and $\kappa^{-1} = 96$ nm for nanopure water; and – 125 mV and 76.1 nm at 10^{-6} M DAHCl and – 35 mV and 68.0 nm at 5×10^{-6} M DAHCl. The arrows represent the distance at which the two surfaces jump into contact due to attractive forces. (after Ref. [17, 23])

With the increase of DAHCl concentration in solution, the force-distance curves (measured after 3 hrs of equilibration) indicate reduction in repulsive forces. It may be attributed to the reduction in the double-layer forces as a consequence of charge neutralization of the negatively charged mica surface by the adsorption of positively charged DAH^+ ions. At shorter distances, however, the decrease in the net force may be attributed to the hydrophobic force. The evidence for the hydrophobic force can be seen from the "jump" into contact of the mica surfaces that occurs at a larger distance than what is predicted by the classical DLVO theory. As shown in Table 1, the jump distances, adhesion forces and contact angles increase with increasing amine concentration, indicating the effect of enhanced surface hydrophobicity of mica. At 10^{-5} M DAHCl, the charge on the mica surface is completely neutralized, which is indicated by the absence of double layer forces at all distances as illustrated schematically in Fig. 3a. This finding is consistent with the results reported by Herder [22].

Further increase in amine concentrations above 10^{-5} M indicates an increase in the double-layer repulsive forces which is shown by an increase in ψ_o values and a decrease in the hydrophobic attraction as exhibited by the decrease in the jump distances between the surfaces (see Table 1). This observation may be due to a 'flip-flop' layer (Fig. 3c) or a bilayer formation (Fig. 3e). Also shown in Table 1 are the flotation recoveries of quartz obtained

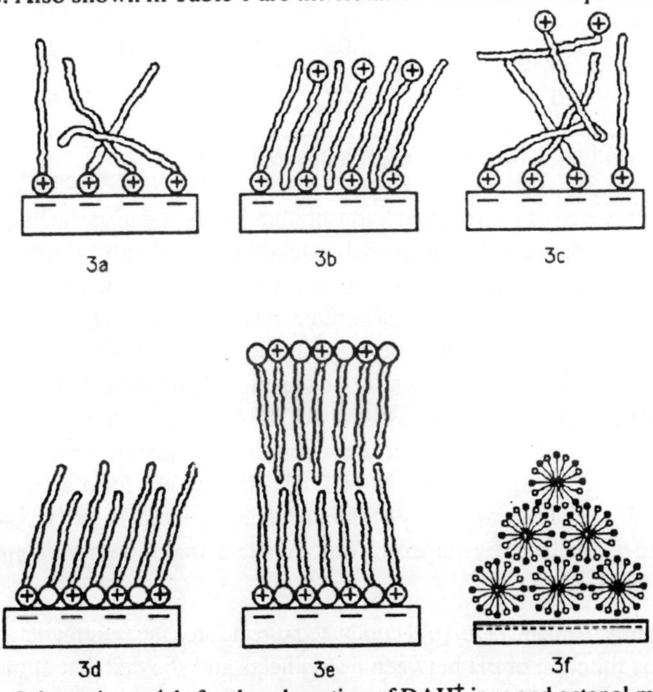

Fig. 3. Schematic models for the adsorption of DAH^+ ions and octanol molecules on mica surface

using amine collector by Yoon and Yordan [66]. Considering the differences in the surface charge density of mica and quartz, it may be said that there is a good correlation between the hydrophobic attraction observed in this study and the flotation response of quartz with amine collector. However, it may be noted that the maximum hydrophobic force measured without electrical double-layer forces (~ 10^{-5} M DAHCl) was still short-ranged with D_o values in the range of 1-1.5 nm [22].

Table 1: Results of surface force, contact angle, and flotation studies with dodecylamine hydrochloride at pH 5.7.

Conc. of Amine (M)	Surface Potential (ψ_o) (mV)	Jump Distance (nm)	Adhesion Force (mN/m) ± 25	Contact Angle (deg) ± 2°			% silica Floated ¥
				θ_a	θ_r	$\theta_a - \theta_r$	
–	– 170	3.0	125	–	–	–	–
1×10^{-6}	– 125	7.0	150	60	35	25	10
5×10^{-6}	– 35	14.0	250	80	60	20	15
1×10^{-5}	0	14.0	250	85	65	20	26
$^*1 \times 10^{-4}$	60	7.0	250	95	65-75	20-30	95
$^*1 \times 10^{-3}$	·150	4.5	300	95	75-80	15-20	90
$^*6 \times 10^{-3}$	220	–	0	65	5-15	50-60	40

*** from Herder [22]; ¥ from Yoon and Yordan [66]**

The presence of short-range hydrophobic forces is due to the hydrocarbon chains in the monolayer being sparsely packed and randomly oriented (see Fig. 3a). Theoretically, each DAH^+ ion occupies the center of the oxygen hexagon vacated by the K^+ ions on the mica surface, resulting in one DAH^+ ion per 0.48 nm^2 of the surface. This is twice as large as the cross-sectional area (0.25 nm^2) of DAH^+ ions. Since the thickness of the monolayer of DAH^+ ions on the surface was measured to be in the range of 0.5-0.7 nm, as compared to 1.8 nm for a fully stretched DAH^+ ion, it is evident that the adsorption of DAH^+ ions at monolayer coverage (at 10^{-5} M of DAHCl) may not form a close-packed layer on mica. The heterogeneity of the surfactant monolayer (at 10^{-5} M) is also evidenced by the large hysteresis of water contact angles observed on the mica surface (see Table 1).

Recently, Rutland *et al.* [67] conducted direct force measurements at 10^{-4} M DAHCl as function of pH between mica sheets and showed that at pH 8-9 the nonionic amines, formed as a hydrolysis product, coadsorb in-between the DAH^+ ions on the mica surface to render the monolayer more compact and hydrophobic in nature. Although these authors claimed that the hydrophobic force had increased under these conditions, it is not unambiguously discernible from their measured force curves, probably due to the charged bilayer formation on the mica surface. When the pH was increased above 9.5, amine precipitated on the mica surface, as evidenced by the appreciable increase in the contact distance. No hydrophobic force was detectable under this solution condition. It has been shown earlier that the precipitation of amine on quartz surface has a detrimental effect on flotation behaviour [68]. It is thus clear from the adsorption and direct force measurements that the key to increasing the hydrophobic force is to enhance the packing density and molecular ordering of the surfactant monolayer on the solid surface [58, 59] without actually forming the bilayer or flip-flop layer.

Co-surfactant Interaction

The co-surfactant interaction is a common phenomenon in the flotation process which could arise either by deliberate addition of organic reagents such as frothers (essentially long-chain alcohols), or by the generation of hydrolysis products as in case of amines and oleates at certain pH, or simply due to the presence of organic impurities in the collector molecules as unreacted starting materials. The bulk solution chemistry of co-surfactant interaction has been studied extensively and, in general, the mixtures of ionic and nonionic surfactants exhibit enhanced activity at the air/liquid interface. There is very little information, however, on the interactions of such mixtures at the solid/liquid interfaces.

Earlier studies on alkyl amine, oleate, and hydroxamate flotation systems have clearly demonstrated the role of neutral molecules in improving flotation perfomance [62-64, 69, 74]. The strong dependence of pH on flotation is well understood and attributed to the coadsorption of hydrolyzed neutral collector molecules in-between the polar groups on the solid surface. As a result of coadsorption, under optimum conditions, the solid surfaces become more hydrophobic and easily floatable. Similar observations were made by Smith [64] who showed that the addition of nonionic alcohol increases the hydrophobicity of amine-coated quartz surfaces by adsorbing between the dodecylammonium ions on the quartz surface. Although the cosurfactant interaction increases the hydrophobicity of the solid surface, it is not clear as to how it influences the driving force (hydrophobic force) for flotation. Yoon and Ravishankar [17] showed, using direct force measurements, that at an optimum combination of amine/octanol mixtures, the presence of octanol actually increased the hydrophobic force from short-range to long-range.

The effect of co-surfactant interaction can be well illustrated based on the direct force measurement work conducted by the author at Virginia Tech [17, 23] at varying compositions of DAHCl and octanol. The stock solution of octanol was prepared in a 1% ethanol solution. After making sure that the octanol alone

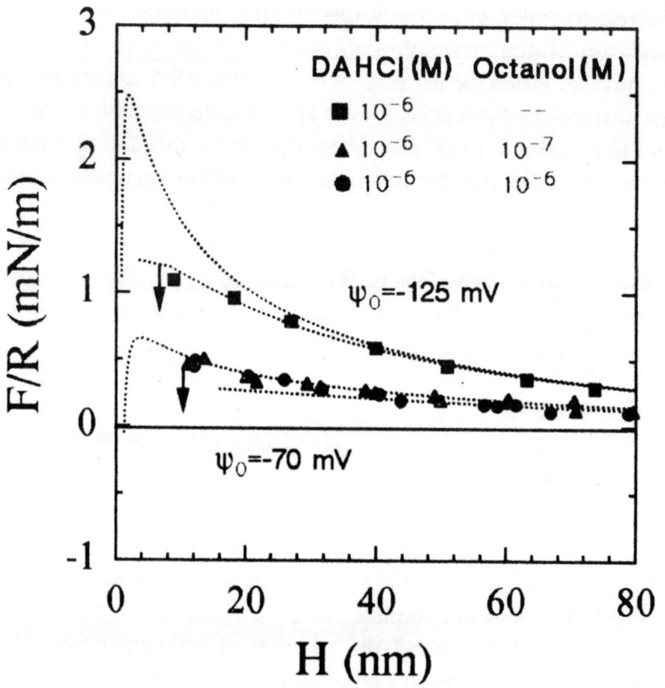

Fig. 4. Force normalized with respect to radius (R) as a function of separation (H) between mica surfaces immersed in 10^{-6} M DAHCl solutions with varying amounts of octanol at pH 5.7. The upper and lower dotted lines indicate constant charge and constant potential models respectively of the DLVO theory obtained using $A_{131} = 2.2 \times 10^{-20}$ J, $\psi_o = -125$ mV and $\kappa^{-1} = 76.1$ nm for 10^{-6} M DAHCl and $\psi_o = -70$ mV and $\kappa^{-1} = 76.1$ with 10^{-7} M octanol; and $\psi_o = -70$ mV and $\kappa^{-1} = 76.1$ with 10^{-6} M octanol. The arrows represent the distance at which the two surfaces jump into contact due to attractive forces. (after Ref. [17, 23]).

does not adsorb onto mica, the force measurements were conducted in presence of DAHCl and octanol mixture. Figure 4 shows the force curves obtained at 10^{-6} M of DAHCl in the presence and absence of 10^{-7} M octanol at pH 5.7. The most important observation to note in these force curves is the reduction of repulsive forces in presence of octanol due to an additional adsorption of DAH^+ ions on the mica surface (as shown by the increase in the ψ_o values from – 125 to – 70 mV). Somasundaran *et al.* [70] made a similar observation while studying the adsorption of allkyl sulfonate on alumina in the presence of a long-chain alcohol.

Another interesting finding is that the coadsorption of octanol also increased the hydrophobic force as evidenced by the increase in the jump distance from 7 to 10 nm. The hydrophobic force however remained short-ranged and

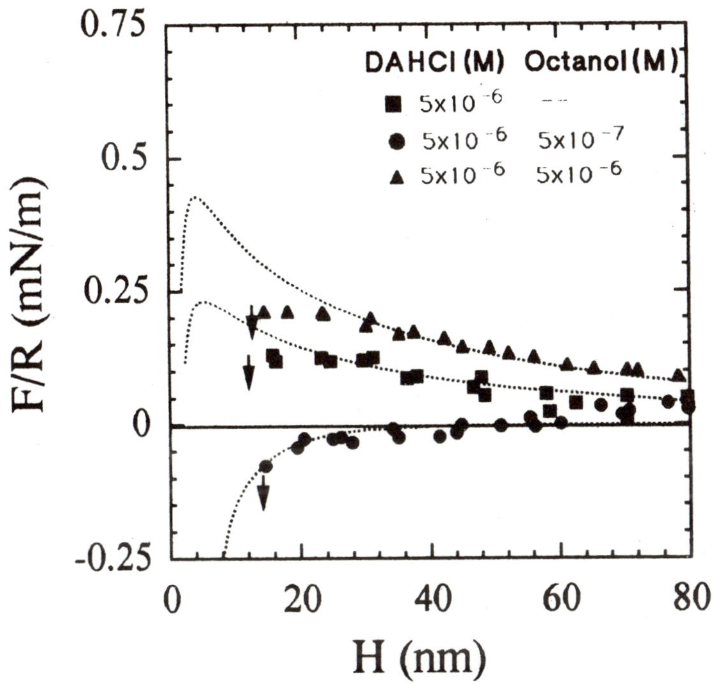

Fig. 5. Force normalized with respect to radius (R) as a function of separation (H) between mica surfaces immersed in 5×10^{-6} M DAHCl solutions with varying amounts of octanol at pH 5.7. The upper and lower dotted lines indicate constant charge and constant potential models, respectively, of the DLVO theory obtained using $A_{131} = 2.2 \times 10^{-20}$ J, $\psi_o = -35$ mV and $\kappa^{-1} = 68.0$ with 5×10^{-6} M octanol. The arrows represent the distance at which tghe two surfaces jump into contact due to attractive forces. (after Ref. 17, 23)

it did not change significantly with further increase in octanol concentration from 10^{-7} to 10^{-6} M.

Figure 5 shows the force vs distance curves at higher concentrations of octanol in the presence of 5×10^{-6} M DAHCl. At 5×10^{-7} M octanol, the double-layer forces were completely suppressed at longer distances, once again indicating the adsorption of DAH^+ ions due to the presence of octanol. But the most striking feature in Fig. 5 is the attractive force observed in the range of 0-45 nm. The measured attractive force, (essentially hydrophobic force) is also manifested by the increase in jump distances from 14 to 16 nm, adhesion forces from 250 to 375 mN/m, and the advancing and receding water contact angles (Refer to Table 2). It is important to note in Table 2 that the increase in receding angles are more than that of advancing angles suggesting a reduced hysteresis behaviour of the contact angles. This indicates better homogeneity of the surfactant coated mica due to the coadsorption of octanol as pictured in Fig. 3d.

Table 2: Results of surface force and contact angle measurements with dodecylamine hydrochloride and octanol at pH 5.7

Concentration Amine	(*M*) Octano	Surface Potential (mV) ψ_o	Jump Distance (nm)	Adhesion Force (mN/m) ± 25	Contact Angle (deg) ± 2°		
					θ_a	θ_r	θ_a-θ_r
1×10^{-6}	1×10^{-7}	− 70.0	10.0	200	63	45	18
1×10^{-6}	1×10^{-6}	− 70.0	10.0	225	64	45	19
5×10^{-6}	5×10^{-7}	0.0	16.0	375	84	72	12
5×10^{-6}	5×10^{-6}	45.0	16.0	350	84	65	19
1×10^{-5}	1×10^{-6}	350.0	–	–	81	48	33

With further increase in octanol to 5×10^{-6} M, however, the repulsive force actually increased, indicating that the adsorption of DAH^+ ion leads to a bilayer (or probably a 'flip-flop' layer) as shown in Fig. 3e. The theoretical DLVO fit to the experimental force curves suggest an increase in ψ_o value from 0 to + 45 mV due to octanol addition.

The charge reversal phenomenon due to the presence of octanol was further studied at higher concentrations of DAHCl (10^{-5} M). As is shown in Fig. 6, the surface charge of mica is completely neutralized at 10^{-5} M of DAHCl which is in close agreement with the results of Herder [22]. However, in the presence of 10^{-6} M octanol, the repulsive force reappeared and is even greater than that measured in nanopure water. The DLVO fit gives ψ_o = + 350 mV and 0.4 nm^2 of surface area per charge (see Table 2). The charge reversal due to the presence of octanol may be attributed, once again, to the formation of bilayer as shown in Fig. 3e.

It is amply clear that the addition of octanol to DAHCl facilitates the adsorption dodecylammonium ions on the mica surfaces. This is initially shown as charge neutralization of mica surface at lower concentrations and charge reversal at higher concentration (Fig. 5). At an optimum combination of amine and octanol, long range hydrophobic forces were observed. These results thus help us understand the interaction of the surfaces coated upto a bilayer of surfactant. It is also important to study the forces between surfaces under the conditions of micelle adsorption and surface precipitation. This situation may arise when the flotation cell is overloaded with collectors or frothers.

In order to understand these phenomena, the force measurements were therefore conducted at higher concentrations of octanol at 5×10^{-5} M of DAHCl. With an increase in octanol concentration from 5×10^{-6} to 5×10^{-5} M, very large repulsive forces (> 40 mN/m) were observed as shown in Fig. 7. The

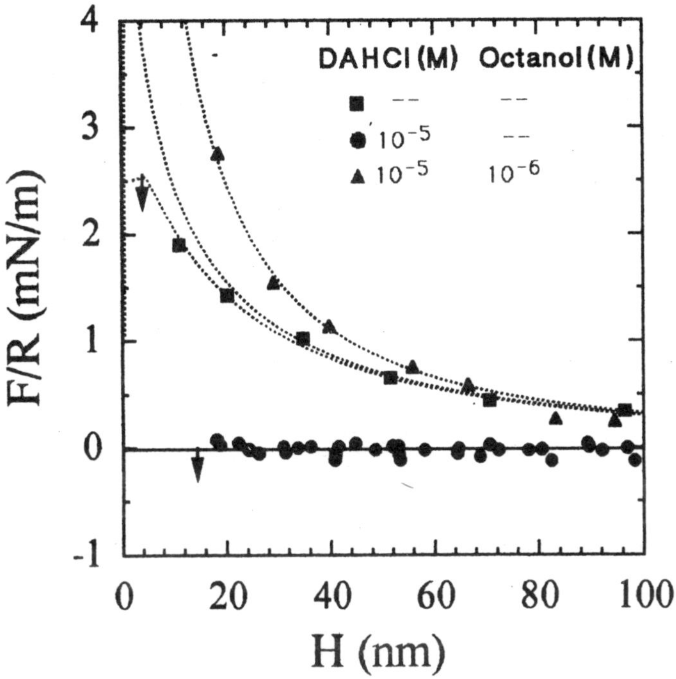

Fig. 6. Force normalized with respect to radius (R) as a function of separation (H) between mica surfaces immersed in 10^{-5} M DAHCl solutions with and without 10^{-6} M octanol at pH 5.7. In the absence of octanol, ψ_o is almost zero, but in the presence of 10^{-6} M octanol large double-layer repulsive force with $\psi_o = +350$ mV and $\kappa^{-1} = 43.0$ nm is observed. This repulsive force indicates charge reversal and is larger than the value measured in nanopure water. The arrows represent the jump distance. (after Ref. [17, 23])

closest contact distance of the surfaces moved out from 0 to 35 nm. Interestingly, the force measurements conducted under these conditions showed hysteresis: the force measurements made while approaching the surfaces are more than those observed while separating them. Similar behaviour was observed with an increase in amine (1×10^{-5} M) and octanol (1×10^{-5} M) concentrations as well as equilibration time (see Fig. 7). The thickness of the adsorbed layer also increased in the process from 35 to 47.5 nm. A similar observation was made by Rutland *et al.* [67] in the presence of pure amine solution (10^{-4} M) at pH above 9.5. These authors attributed the hysteresis to the phase-separated hydrated dodecylamine droplets, which is consistent with the findings of Castro *et al.* [71]. Due to the unlikelihood of any precipitation at pH 5.7 and at low concentration of amine and octanol, as evidenced by the clarity of the solution, an alternate explanation was suggested by Ravishankar and Yoon [23]. They proposed that the mixed micelles formed in the presence of octanol can adsorb at the surface to form multilayers as shown in Fig. 3f.

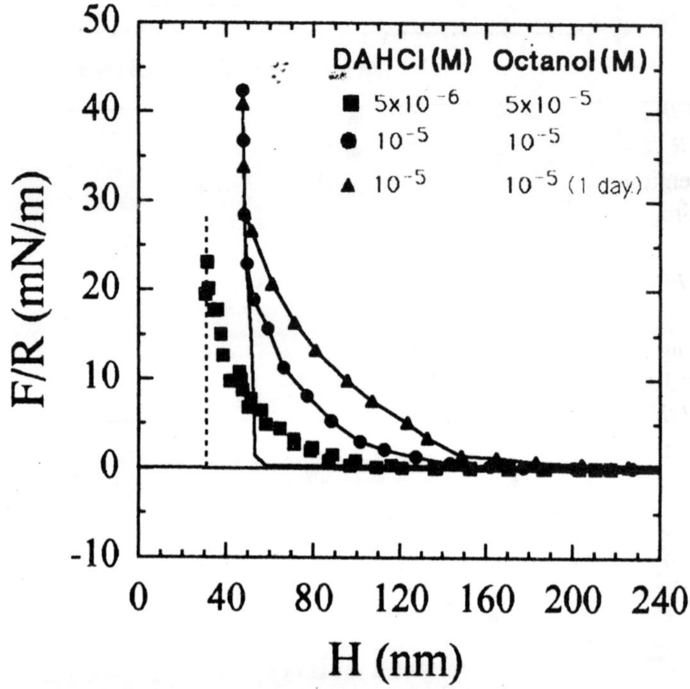

Fig. 7. Force normalized with respect to radius (*R*) as a function of separation (*H*) between mica surfaces immersed in DAHCl solutions with octanol at pH 5.7. At 5×10^{-6} M DAHCl and 5×10^{-5} M octanol, large repulsive forces that cannot be described by DLVO theory, are observed. At higher concentrations of DAHCl and octanol, even larger forces were measured. The force curves also exhibit hysteresis which becomes more pronounced after 24 hrs. (after Ref. [17, 23])

Micelles are known to show oscillatory forces because of their rigid structure [72]; however, the mixed micelles of amine and octanol at low concentrations may not be well ordered and sufficiently rigid structured to show the oscillatory forces. Nevertheless, the large steric repulsive forces can be explained by the adsorption of mixed micelles.

Application of *Extended* DLVO Theory

The experimentally measured attractive force under the optimum combination of amine (5×10^{-6} M) and octanol (5×10^{-7} M) can be analyzed using the extended DLVO theory [15-17] to estimate the hydrophobic force parameters. According to this theory, the total energy V_t can be described as follows:

$$V_t = V_e + V_d \pm V_s \tag{4}$$

where V_e, V_d, and V_s are the electrical, dispersion, and structural components respectively of the total energy V_t between two flat surfaces. One can transform eq. (4) in terms of force that are measured by the direct force measurement technique using the Derjaguin approximation [33] as follows:

$$F/R = 2\pi V_t, \tag{5}$$

where F/R is the normalized force measured between two crossed cylinders of radius R. Since the electrical double-layer forces were absent under the aforementioned condition, the force curves may be fitted to a sum of dispersion and hydrophobic forces using the following equation:

$$F/R = -A_{131}/6H^2 - C_1 \exp(-H/D_1) - C_2 \exp(-H/D_2). \tag{6}$$

Using springs of increasing stiffness, forces at shorter distances were measured with the help of the 'jump' method [6]. The results obtained as $d(F/R)/dH$ vs H values, are shown in Fig. 8. The data points are fitted to the following equation which is obtained by differentiating eq. (6):

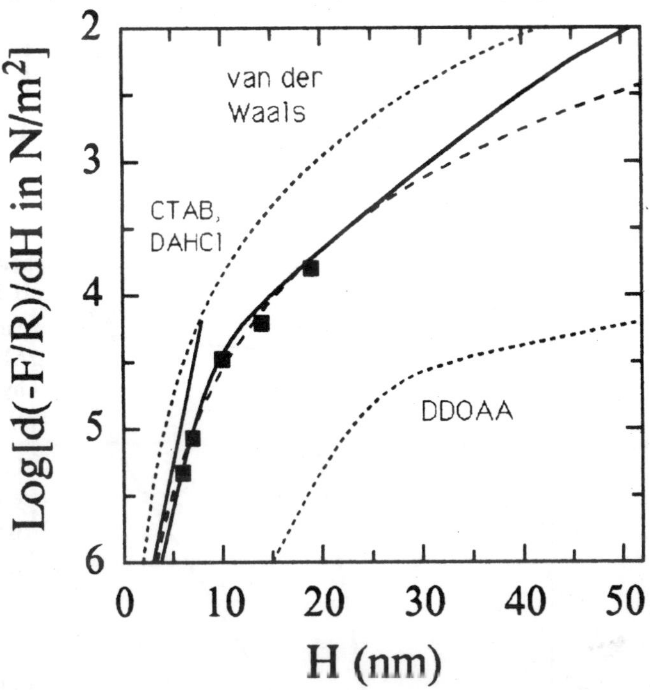

Fig. 8. Normalized force gradient vs H between mica surfaces in 5×10^{-6} M DAHCl and 5×10^{-7} M octanol at pH 5.7. The squares represent the data obtained using the gradient method. The solid line indicates the extended DLVO fit incorporating a double exponential function for the hydrophobic force (eq. 7) where $A_{131} = 2.2 \times 10^{-20}$ J, $C_1 = -40$ mN/m, $C_2 = -0.5$ mN/m, $D_1 = 1.2$ nm, and $D_2 = 6.8$ nm. The dashed line represents the extended DLVO fit using a power law form for the hydrophobic force (eq. 8) $K = -1.0 \times 10^{-19}$ J. Also shown for comparison is the normalized force gradients for nonretarded van der Waals force and for the mica surfaces coated with DAHCl and CTAB and insoluble surfactants such as DDOAB. The soluble surfactants without octanol show only the short-range hydrophobic force.

$$d(F/R)/dH = A_{131}/3H^3 + (C_1/D_1)\exp(-H/D_1) + (C_2/D_2)\exp(-H/D_2), \quad (7)$$

in both eqs. (6) and (7) A_{131} ($= 2.2 \times 10^{-20}$ J) is the nonretarded Hamaker constant and $C_1 = 40$ mN/m, $C_2 = 0.5$ mN/m, $D_1 = 1.2$ nm and $D_2 = 6.8$ nm. An alternative fit to the $d(F/R)/dH$ vs H curve was obtained using the power law expression as indicated below [17]:

$$d(F/R)/dH = A_{131}/3H^3 + K/3H^3, \quad (8)$$

where $K = 1 \times 10^{-19}$ J, which is 4.5 times larger than the Hamaker constant. Also shown in Fig. 8 are the normalized force gradient values for van der Waals force and for soluble surfactants such as DAHCl and CTAB [9, 22, 17] along with those obtained for insoluble double chain amines (e.g., DDOAB) [51]. As seen from Fig. 8, the measured hydrophobic force is not as long-ranged as observed with DDOAB monolayers. It is stronger than that measured with single chain water soluble surfactants. Recently, even stronger hydrophobic forces were measured by introducing dodecanol in amine system [73].

FLOTATION RESULTS

The most significant role of the long-range hydrophobic force as a result of octanol addition is manifested as the enhanced floatation recovery of quartz. The flotation was conducted using standard microflotation cell. In Fig. 9, the

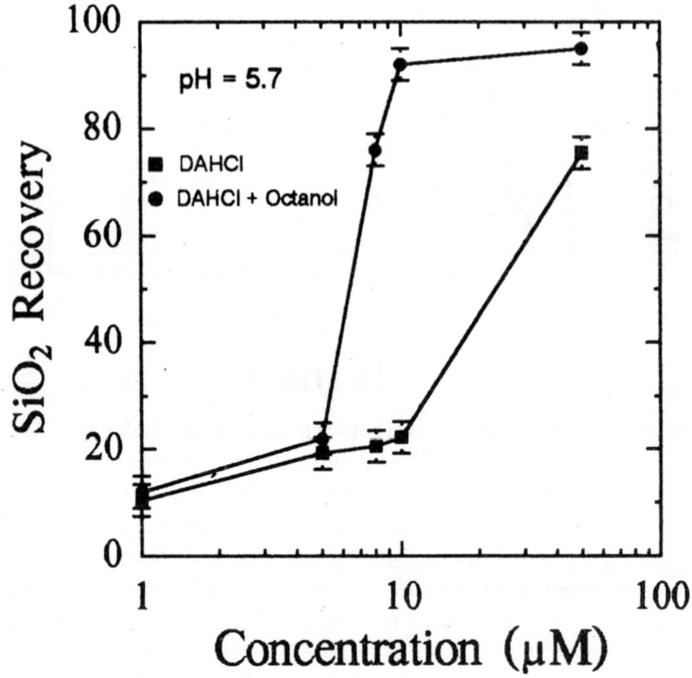

Fig. 9. Floatability of silica in DAHCl and octanol solutions at pH 5.7 (after Ref. [17, 23])

flotabilities of silica using pure DAHCl and a 1 : 1 mixture of DAHCl and octanol were plotted at pH 5.7. The silica recovery remains low up to 10^{-5} M of amine and increases at higher concentrations. However, in the presence of amine/octanol mixture the silica recovery increases sharply at 5×10^{-6} M. Gaudin and Fuerstenau [62] suggested that the floatability of quartz increases rapidly at critical hemi-micelle concentration (c.h.c) of amine that is usually 100 times lower than the critical micelle concentration (c.m.c). It is shown in this study that the interaction of nonionic alcohol at the amine-coated quartz surface appears to reduce the c.h.c, resulting in a more close-packed monolayer on the quartz surface. It is also manifested by an increase in the net attractive force between the surfactant coated surfaces and an increase in the contact angles. The contact angle results are consistent with the earlier findings of Smith [64] who showed that the contact angle of quartz in 4×10^{-4} M dodecylamine is about 30° at pH 5.7, which increases to 50° in the presence of 1.4×10^{-4} M dodecanol. Thus, it is clear from these results that the increase of hydrophobic forces from short to long range is responsible for the higher flotation recovery of silica.

SUMMARY AND CONCLUSIONS

An overview of the hydrophobic force has been presented. The existence of this force has been proved beyond doubt with the help of recently developed direct force measuring techniques. While the mathematical representation of the hydrophobic force in single exponential form is appropriate for short-range forces, in double exponential form it is suitable for describing long-range hydrophobic forces.

Direct force measurement results suggest that DAHCl alone renders short range hydrophobic force and shows poor flotation recovery of quartz at low amine concentration. However, with an optimum combination with nonionic surfactants such as a long chain alcohol, the flotation recovery shows a marked increase due to the presence of long-range hydrophobic forces. At higher concentrations of amine and octanol, the flotation is depressed due to large repulsive forces which may originate from mixed bilayers and mixed micelle adsorption on the surface.

Flotation is one of the most successful fine particle separation processes. Even after eight decades of continuous plant practice, there is no comprehensive model to control and optimize the process due to lack of understanding of the energetics involved in the interaction of different interfaces. In the past two decades, the advent of direct force measurement techniques has improved the understanding of the interplay of the interparticle forces and energies. Currently, direct force measurement techniques are being developed for bubble-particle interaction using AFM. It is believed that a good understanding of these aspects will go a long way toward helping us to understand the phenomena and to develop a comprehensive phenomenological model of the flotation process.

ACKNOWLEDGMENTS

The author is grateful to Prof. R.-H. Yoon for his constant encouragement and financial support. Thanks are also due to Dr. Y. I. Rabinovich, Vivek S, Rajesh. Pazhianur, Darrin H. Flinn and B. Suha Aksoy for their constructive comments on the manuscript.

REFERENCES

1. Derjaguin, B.V., and Landau, L.D. : USSR Acta Physicochim. 14 (1941) 633-652.
2. Verway, E.J., Overbeek, J. Th. G. : Theory of the Stability of Lyophobic Colloids New York : Elsevier 1948.
3. Ninham, B.W. : J. Phys. Chem. 84 (1980) 1423-1430.
4. Claesson, P.M. : Progr. in Colloid & Polymer Science 74 (1987) 48-54.
5. Derjaguin, B.V., and Churaev, N.V. : Colloids and Surfaces 41 (1989) 223-237.
6. Israelachvili, J.N., and Adams, G.E. : J. Chem. Soc. Faraday Trans. 1. 74 (1978) 975-1001.
7. Derjaguin, B.V., and Rabinovich, Ya.I., and Churaev, N.V. : Nature (London). 273 (1978) 313-318.
8. Binnig, G., Quate, C., and Gerber, G. : Phys. Acta. Lett. 56 (1986) 930-933.
9. Israelachvili, J.N., and Pashley, R.M. : Nature 300 (1982) 341-342.
10. Rabinovich, Ya.I., and Derjaguin, B.V. : Colloids Surf. 30 (1988) 243-251.
11. Pashley, R.M. : J. Colloid Interface Sci. 83 (1981) 531-546.
12. Horn, R.G., and Israelachvili, J.N. : J. Chem. Phys. 75 (1981) 1400-1411.
13. Xu, Z., and Yoon, R.-H. : J. Colloid Interface Sci. 132 (1989) 532-541.
14. Xu, Z., and Yoon, R.-H. : J. Colloid Interface Sci. 134 (1990) 427-434.
15. Hiroki, Y., and Yoon, R.-H. : J. Colloid Interface Sci. 157 (1993) 426-433.
16. Hiroki, Y., and Yoon, R.-H. : J. Colloid Interface Sci. 157 (1993) 434-441.
17. Yoon, R.-H., and Ravishankar, S.A. : in press J. Colloid Interface Sci. (1994) .
18. Yoon, R.-H., Xu, Z., and Laskowski, J. : J. Colloid Interface Sci. (1994) submitted for publication.
19. Laskowski, J., and Kitchener, J.A. : J. Colloid Interface Sci. 29 (1969) 670-679.
20. Yoon, R.-H., Plenary paper presented at the XVII International Mineral Processing Congress. September 23-28, Dresden, Germany; Aufberetungs-technik. : 32 (1991) 474-485.
21. Yordan, J. : Studies on the stability of thin films in bubble-particle adhesion. Ph.D Thesis. Virginia Polytechnic Institute and State University 1989.
22. Herder, P.C. : J. Colloid Interface Sci. 134 (1990) 346-356.
23. Ravishankar, S.A., and Yoon, R.-H. : Preprint 93-249, AIME-SME Meeting, Reno, Nevada (1993).

24. Kekicheff, P., Christenson, H.K., and Ninham, B.W. : Colloids and Surfaces 40 (1989) 31-41.
25. Xu. Z. : personal communication.
26. Claesson, P.M., Blom, C.E., Herder, P.C., and Ninham, B.W. : J. Colloid Interface Sci. 114 (1986) 234-242.
27. Pashley, R.M., McGuiggan, P.M., Ninham, B.W., and Evans, D.F. : Science 229 (1985) 1088-1089.
28. Parker, J.L., Cho, D.L., and Claesson, P.M. : J. Phys. Chem. 93 (1989) 6121-6125.
29. Rabinovich, Y.I., and Yoon, R.-H. : submitted to Langmuir (1994).
30. Tsao, Y., Evans, D.F., and Wennerstrom, H. : Science 262 (1993) 547-550
31. Tsao, Y., Yang, S.X., Evans, D.F., and Wennerstrom, H. : Langmuir 7 (1991) 3154-3159.
32. Christenson, H.K., Fang, J., Ninham, B.W., and Parker, J.L. : J. Phys. Chem. 94 (1990) 8004.
33. Claesson, P.M., and Christenson, H.K. : J. Phys. Chem. 92 (1988) 1650-1655.
34. Christenson, H.K., and Claesson, P.M. : Science 239 (1988) 390-392.
35. Christenson, H.K., and Claesson, P.M., Berg, J., Herder, P.M. : J. Phys. Chem. 93 (1989) 1472.
36. Tsao, Y., Evans, D.F., and Wennerstrom, H. : Langmuir 9 (1993) 779-785.
37. Parker, J.L., Christenson, H.K., and Ninham, B.W. : Rev. Sci. Instrum. 60 (10) (1989) 3135-3138.
38. Ducker, A.W., and Senden, J.T. : Langmuir 8 (1992) 1831-1836.
39. Tolansky, S. : An Introduction to Interferometry. London : Longmans 1973.
40. Derjaguin, B.V. : Kolloid. Zh. : 69 (1934) 155.
41. Horn, R.G., Smith, D.T., and Haller, W. : Chem. Phys. Lett. 162 (1989) 404-408.
42. Horn, R.G., Clarke, D.R., and Clarkson, M.T. : J. Mater. Res. 3 (3) (1988) 413-416.
43. Steinberg, S., Ducker, G.V., Hyukjin, C.F., Tseng, M.Z., Clarke, D.R., and Israelachvili, J.N. : Science. 260 (1993) 656-659.
44. Parker, J.L. : Langmuir. 8 (1992) 551-556.
45. Lifshitz, E.M. : Sov. Phys. JETP (Engl. Transl.) 2 (1956) 73-83.
46. Marcelja, S., Mitchell, D.J., Ninham, B.W., and Sculley, M. : J. Chem. Soc. Faraday. Trans. 2. 73 (1977) 630.
47. Luzar, A., Svetina, S., and Zeks, B., Chem. Phys. Lett. 96 (1983) 485.
48. Pratt, L.R., and Chandler, D.J. : J. Chem. Phys. 92 (1977) 3683.
49. Pangali, C., Rao, M., and Berne, B.J. : J. Chem. Phys. 71 (1979) 2975-2981.
50. Yushchenko, V.S., Yaminsky, V.V., and Shchukin, E.D. : J. Colloid Interface Sci. 96 (1983) 307-314.
51. Ruckenstein, E., and Churaev, N.V. : J. Colloid Interface Sci. 147 (1991) 535-538.
52. Attard, P. : J. Phys. Chem. 93 (1989) 6441-6444.
53. Podgornik, V., Rau, D. M., and Parsegian, V.A. : J. Chem. Phys. 91 (9) (1989) 5840-5849.

54. McConnell, M.H., Tamm, K.L., and Weis, M.R. : Proc. Natl. Acad. Sci. USA 81 (1984) 3249-3253.
55. Kenn, R.M., Bohm, C., Bibo, A.M., Peterson, I.R., and Mohwald, H., Als-Nielsen, J., Kjaer, K. : J. Phys. Chem. 95 (1991) 2092-2097.
56. Kjaer, K., Als-Nielsen, J., Helm, C.A., Krayer, T.P., and Mohwald, H. : J. Phys. Chem. 93 (1989) 3200-3206.
57. Rabinovich, Y.I., Guzonas, D.A., and Yoon R.-H. : Langmuir 9 (1993) 1168-1170.
58. Claesson, P.M., Christenson, H.K., Berg, J., and Herder, P.M. : J. Colloid Interface Sci. 136 (1990) 541-551.
59. Rabinovich, Y.I., Guzonas, D.A., and Yoon R.-H. : J. Colloid Interface Sci. 155 (1993) 221-225.
60. Rabinovich, Y.I., and Derjaguin, B.V. : Proc. 5th Hungarian Conference on Colloid Chem. Hungary, Budapest, Lorand Eothos Univ. (1988).
61. Eriksson, J.C., Ljunggren, S., and Claesson, P.M.. : J. Chem. Soc. Faraday. Trans. 2. 85 (1989) 163-176.
62. Gaudin, A.M., and Fuerstenau, D.W. : Trans AIME. 202 (1955) 958.
63. Somasundaran, P., and Ananthapadmanabhan, K.P. : (Ed) Mittal, K.L., Solution Chemistry of Surfactants. New York : Plenum Press 1979, pp. 777-800.
64. Smith, R.W. : Trans AIME. 226 (1963) 427-433.
65. Chan, D.Y.C., Pashley, R.M., and White, L.R. : J. Colloid Interface Sci. 77 (1980) 283-285.
66. Yoon, R.-H., and Yordan, J.L. : J. Colloid Interface Sci. 141 (1991) 374-383.
67. Rutland, M., Waltermo, A., and Claesson , P.M. : Langmuir 8 (1992) 176-183.
68. Laskowski, J.S. : In Sastry, K.V.S., and Fuerstenau, M.C. (Ed.) : Challenges in Mineral Processing. Littleton, CO : Society of Mining Engineers Inc. 1989, pp. 15-34.
69. Natarajan, R., and Fuerstenau, D.W. : Int. J. Min. Process. 11 (1983) 139-153.
70. Somasundaran, P., Fu, E., and Xu, Q. : Langmuir 8 (1992) 1065-1069.
71. Castro, S.H., Vurdela, R.M., and Laskowski, J.N. : Colloids Surf. 21 (1986) 87-100.
72. Bergeron, V., and Radke, C.J. : Langmuir 8 (1992) 3020-3026.
73. Ravishankar, S.A., and Yoon, R.-H. : manuscript in preparation.
74. Fuerstenau, D.W., and Pradip: In Jones, M.J. and Oblatt, R. (Eds.) : Reagents in the Minerals Industry. London : Institute of Mining and Metallurgy 1984, pp. 161-168.

Advances in Beneficiation of Apatite by Selective Flocculation

B.M. MOUDGIL and S. MATHUR
Mineral Resources Research Center,
Dept. of Materials Science and Engineering
University of Florida Gainesville, FL 32611 USA.

ABSTRACT

Selective flocculation has been suggested as a promising technique for beneficiation of dolomitic phosphate rock in Florida which in some cases may require fine grinding to liberate dolomite from apatite. The objective of this paper is to summarize the advances made in the selective flocculation approach for separation of dolomite from apatite. Heteroflocculation, identified to be the major barrier for achieving selectivity was minimized by blocking active sites for flocculant (polyethylene-oxide) adsorption on apatite (the non-flocculating component). This was accomplished by using a site blocking agent (SBA), which is a lower molecular weight fraction of flocculant itself and is incapable of inducing flocculation on its own. The flocculation behaviour of dolomite with PEO was attributed to the presence of clay on its surface. The applicability of selective flocculation process beyond the conventional range of fines (> 50 μm) is also demonstrated. It is further shown that the molecular weight of the polymer and particle size of the mineral affect flocculation and separation efficiency. Also the loss of selectivity in mixed mineral systems due to the formation of cross-coatings during grinding is outlined.

INTRODUCTION

Future reserves of phosphate rock in Florida contain significantly more dolomite (up to 4% MgO) than presently mined deposits. The dolomitic impurities cannot be removed by conventional techniques to yield concentrates less than 1% MgO – a limit set by current fertilizer production technology. Several efforts to eliminate dolomitic impurity from apatite in South Florida rock have been reported, however, economics of most of the suggested strategies is yet to be proved on the commercial scale [1-3].

An appreciable fraction of dolomite is known to exist as inclusions in apatite particles [4]. The particle size of dolomitic inclusions in apatite could vary from relatively coarse to micron size particles, and their liberation can only be achieved by grinding to finer sizes [4, 5]. Beneficiation of finely ground phosphate ore by the conventional froth flotation technique may not be efficient and developing new techniques such as selective flocculation will be necessary to achieve the desired separation.

Selective flocculation technique involves aggregation of one or more fine particulate components from a multicomponent mixture with the aid of polymeric reagents. The aggregates of the polymer coated particles or the "flocs" thus formed are separated from the suspension either by sedimentation/elutriation or floc-flotation. The flocs can be further cleaned, provided they are robust enough, to improve the concentrate grade.

The principles of selective flocculation as well as the underlying surface chemistry have been reviewed earlier by Yarar and Kitchener [6], Friend and Kitchener [7], Attia and Kitchener [8], Attia and Fuerstenau [9], Sresty et al. [10], Somasundaran [11], and more recently by Attia [12, 13] and Moudgil et al. [14].

Selective flocculation process is currently practiced on an industrial scale for the beneficiation of iron ore taconites at Tilden Mine of the Cleveland Cliffs Co. in USA [15] and the processing of potash ore at Cominco Ltd., Canada [16]. Other examples where selective flocculation technique has been attempted on bench-scale include beneficiation of base metal (copper, lead, and zinc) ores, phosphates, tin and tungsten ores, clays, and coal [9, 10, 17-20]. Although primary application of selective flocculation has been in the mineral processing industry but many potential uses exist also in biological and other colloidal systems. These include purification of ceramic powders, separation of hazardous solids from chemical wastes, and removal of deleterious components from paper pulp [21].

Major barriers to the commercial success of the selective flocculation technology are extension of synthetic mixed mineral results to natural ores, and scaling up of bench scale process to plant scale. In general, the selectivity observed in single mineral tests is invariably lost in the mixed mineral or natural ore systems because of one or more of the following reasons: heterocoagulation; charge patch neutralization; dissolved ion interference; slime coating; physical entrapment; entrainment and heteroflocculation [14, 22].

This paper summarizes the efforts, primarily at the Mineral Resources Research Center, University of Florida, devoted towards separation of phosphate from dolomite and silica using the selective flocculation technology. Some of the important developments include the phenomenon of heteroflocculation and its minimization using the concept of site blocking agents (SBA), the effect of particle size and molecular weight on flocculation and separation efficiency, the impact of the source of material on flocculation behaviour, and effects associated with dry or wet grinding of single and mixed mineral systems.

CONTROL OF FLOCCULANT ADSORPTION

The selective flocculation process involves three steps : (1) dispersion of the fine particles, (2) selective adsorption of the polymer on the flocculating component and formation of the flocs, and (3) separation of the flocs from the suspension. Among the three steps, the second step, i.e., selective adsorption of the flocculant is the most difficult one to achieve. In the apatite-dolomite system PEO was identified as the potential flocculant for dolomite after analysis of Infrared spectra of the minerals [22, 24, 25]. Single mineral flocculation tests showed that dolomite flocculated well with PEO of 5 million molecular weight while apatite did not exhibit any significant floc formation [5, 22, 24-28]. However, the single mineral selectivity was not realized in the mixed mineral system primarily due to heteroflocculation.

Heteroflocculation

Heteroflocculation was reported to be the main cause for the loss of selectivity in the apatite-dolomite system [25-28]. Assuming the main mechanism for flocculation to be bridging, it was argued that a small amount of polymer adsorbed on apatite bridges with dolomite surface forming heteroflocs, thereby leading to the observed loss in selectivity.

It is important to distinguish heteroflocculation from the entrainment/entrapment effects discussed in the literature [13, 14]. Entrainment/entrapment occurs due to physical/mechanical factors and a possible remedy is to clean the flocs (provided those are strong enough to withstand shear) by repeated shear and washing to remove the entrained/entrapped gangue. Heteroflocculation, on the other hand, is chemical in nature and cannot be remedied by floc breakage and washing process alone. Efforts to minimize heteroflocculation are outlined below.

Control of Heteroflocculation

A model of the selective flocculation process by Moudgil and co-workers [23, 29], revealed that heteroflocculation is a strong function of the ratio of the flocculant adsorption sites on the various mineral surfaces, and a weak function of the flocculant dosage (surface coverage). These investigations suggested that since only a small amount of adsorption can lead to heteroflocculation, the selectivity in flocculant adsorption can be enhanced by poisoning the active sites or blocking them on apatite, the non-flocculating fraction of the mixture [23, 25]. The blocking of sites on the inert (apatite) particles was achieved by adsorption of the lower molecular weight fraction of the flocculant itself [25, 30]. This is in contrast to the use of specific selective modifiers of mineral surfaces for controlled flocculant adsorption as suggested in the literature [31, 32]. A schematic of the site-blocking process to enhance the selective adsorption of flocculant is illustrated in Fig. 1. The main characteristic of the site blocking agent (SBA) is its capability to block enough sites on the inert particles (non-flocculating or lesser flocculating in single mineral tests) so as to minimize heteroflocculation. The adsorption mechanisms for the flocculant and the SBA are to be similar, if not the same, to ensure blocking of the sites for adsorption of the flocculant. Also, the SBA should not cause flocculation

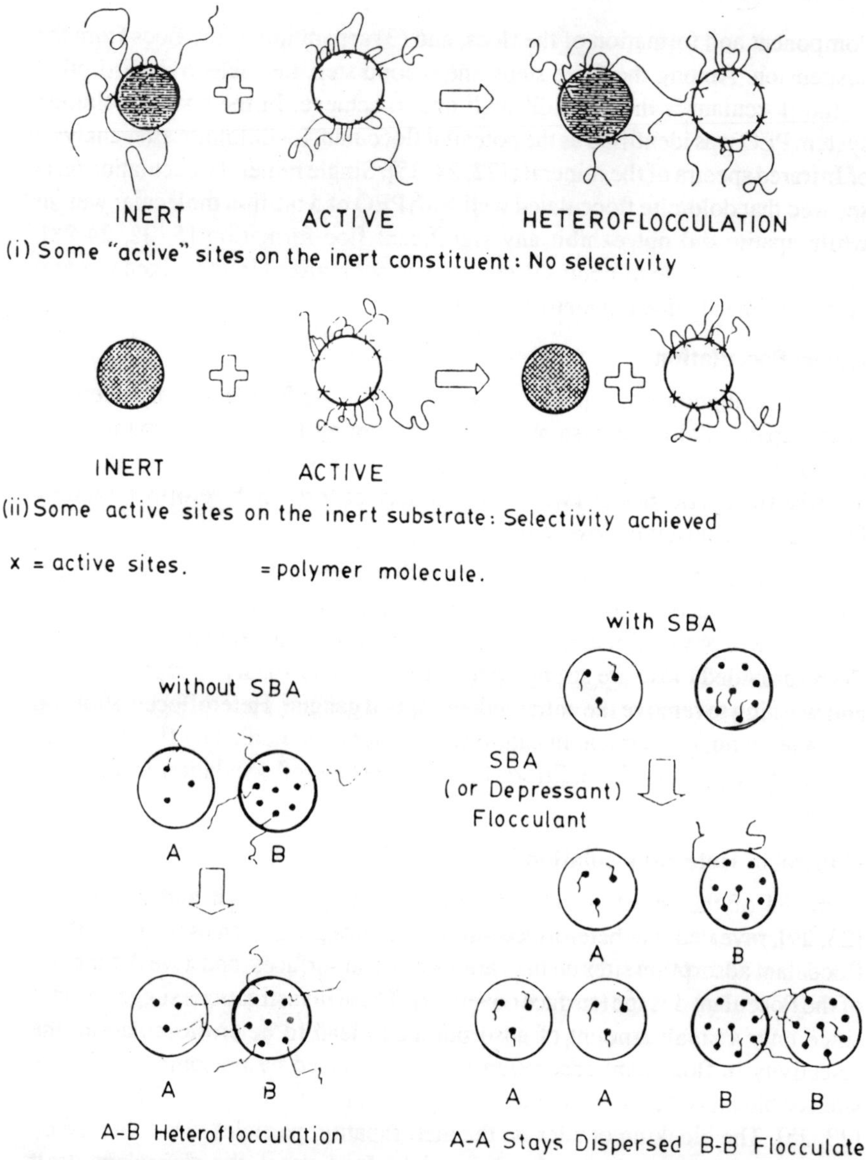

Fig. 1. Schematic diagram illustrating the concepts of SBA and heteroflocculation (after Ref. [23])

by itself. These considerations led to the selection of the lower molecular weight fraction of the flocculant itself as the site blocking agent [23, 25].

Effect of Molecular Weight and Dosage of SBA

The dosage and molecular weight of the polymer used as SBA is critical for obtaining the best selectivity. The most efficient SBA was found to be the highest molecular weight fraction of the flocculant which is not capable of inducing flocculation. For example, the effect of SBA adsorption on flocculation of dolomite, with PEO of 5 million molecular weight as a flocculant was investigated by Behl and Moudgil [25, 30] and their results are presented in Fig. 2. Flocculation is depressed with the lower molecular weight polymers and the extent of depression increases with an increase in the molecular weight till flocculation initiates at molecular weight beyond 6×10^5. This molecular weight fraction, therefore, corresponds to the highest molecular weight polymer which is incapable of inducing flocculation on its own yet causing a maximum decrease in flocculation in the presence of a higher molecular weight flocculant. It is also seen from the figure that adsorption density increases with increase in molecular weight suggesting more effectiveness of SBA molecules in blocking the sites for flocculant adsorption.

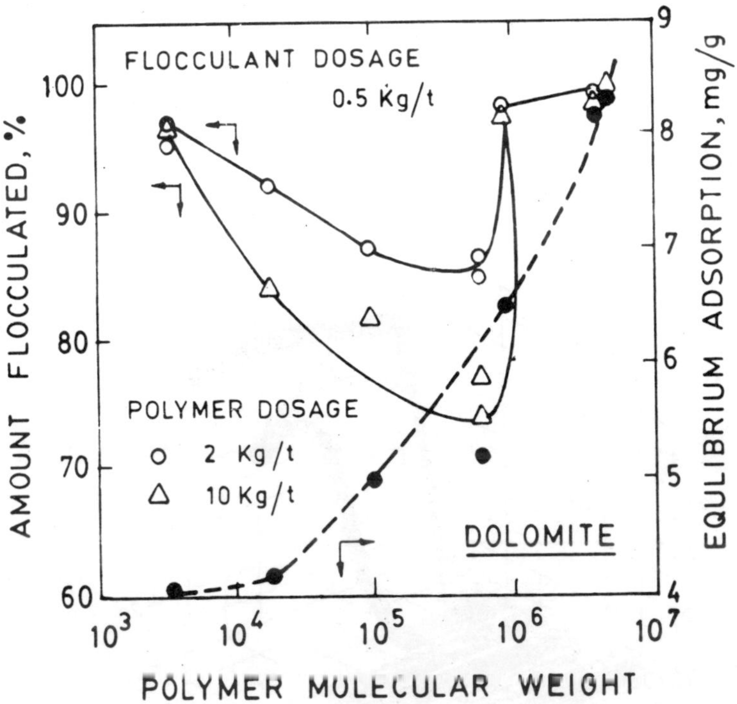

Fig. 2. Effect of SBA molecular weight on dolomite flocculation. The optimum SBA molecular weight is 6×10^5 which is the highest molecular weight fraction of the polymer incapable of inducing flocculation on its own (after Ref. [25])

Effect of SBA on Selectivity

Apatite did not flocculate irrespective of the amount of polymer added. On the other hand, more than 90% dolomite flocculated at a dosage of 0.25 kg/t of PEO. The best separation efficiency for a 80 : 20 mixture of apatite and dolomite was obtained at a SBA dosage of 2 kg/t. The apatite was upgraded from 80% to 95% at 90% recovery level (see Table 1).

Table 1: Effect of SBA on selective flocculation in 20 : 80 dolomite-apatite system. flocculant : PEO (5 million MW); 1 kg/t
Site blocking agent (SBA) : PEO (0.6 million MW)

SBA Dosage, kg/t	Grade, % Apatite	Recovery, % Apatite
0	90.1	26.3
2	95.1	90.6

EFFECT OF PARTICLE SIZE ON SELECTIVE FLOCCULATION

Moudgil et al. [33, 34] have recently discussed the effect of particle size in selective flocculation. A strong correlation between the particle size, molecular weight, and the flocculation efficiency was observed. It was clearly demonstrated that there exists a threshold value of the polymer molecular weight at which particles of certain size can be flocculated. As illustrated in Fig. 3, with increasing particle size a concomitant increase in the molecular weight of the polymer is required for flocculation.

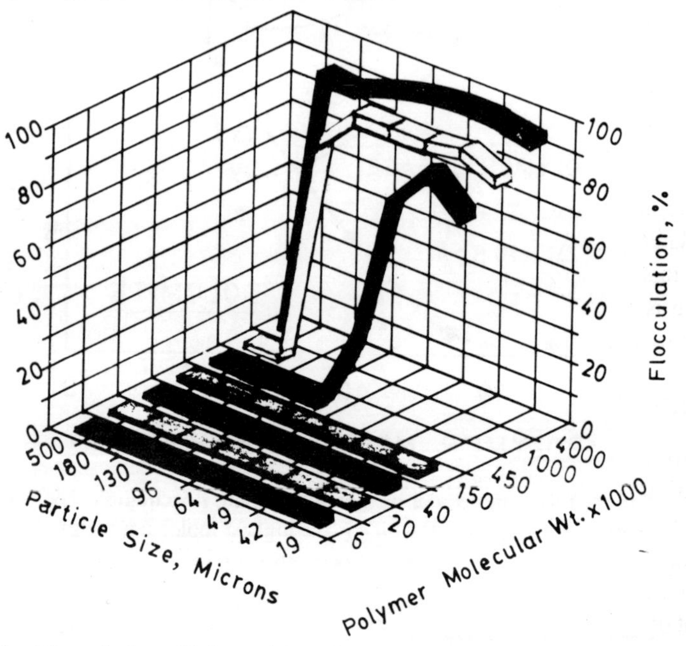

Fig. 3. Flocculation efficiency in apatite-polyacrylic acid (PAA) system as a function of particle size and polymer molecular weight. Coarser particles (as large as 200 μm) can also be flocculated but with higher molecular weight polymers. Similarly, very fine particles also require high molecular weight polymers (after Ref. [33])

The effect of particle size on selective flocculation efficiency was systematically examined in a 50 : 50 synthetic mixture of apatite and silica [34]. It can be seen from Fig. 4 that higher selectivity is observed for coarser size fractions. At a polymer dosage of 1 kg/t, for particle size coarser than 120 μm, a concentrate consisting of more than 80% apatite was obtained. The improved selectivity was attributed to (i) less physical entrapment of coarser size particles, (ii) relatively higher surface coverage of coarse particles affecting the floc characteristic, e.g., higher floc density, and (iii) less adsorption of flocculant on coarser silica particles, hence less heteroflocculation.

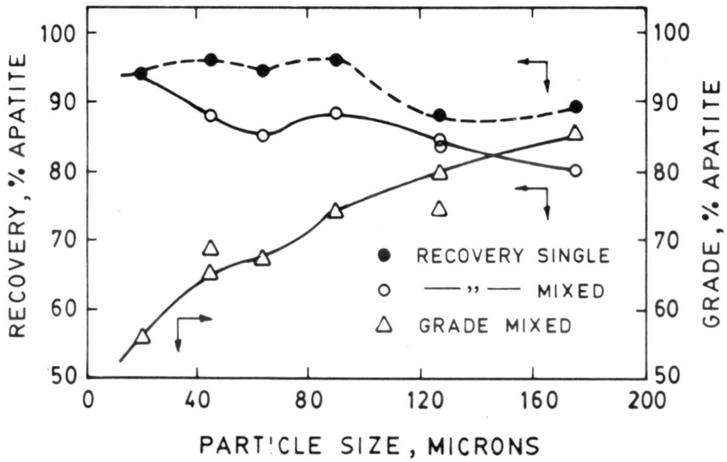

Fig. 4. Recovery and grade of apatite as a function of particle size (PAA mol. wt. = 1 million) (after Ref. [34]).

The above results demonstrated that not only coarser particles can be flocculated, they may lead to higher selectivity. Further, this work also suggested that if the relative particle size distribution of the two components in a mixture is different, the molecular weight can be optimized to enhance the selectivity in the system.

SOURCE OF THE MINERAL CONSTITUENTS

Recently it was shown that the source of the mineral strongly influences its interaction with a given reagent. Dolomite samples from Florida were determined to exhibit flocculation with PEO, whereas those from Perry Mines, Georgia and New Jersey did not flocculate [35, 36]. Detailed characterization of dolomite samples revealed that the flocculating dolomite contained a minor amount of palygorskite, a Mg-rich clay, whereas no such clay was observed on other dolomites [36]. The clay coated particle surfaces exhibited isolated – OH groups which are reported to hydrogen-bond with the ether oxygen of PEO [37, 38] and leading to the flocculation of the particles. The presence of the clay particles on dolomite surface was indicated by their morphology and Si-Al peaks in SEM-EDAX, the existence of structural hydroxyl peak at 3619 cm^{-1} and Si-O stretching peak at 1055 cm^{-1} in IR spectra (Fig. 5) and XRD studies of acid insolubles. The absence of isolated hydroxyl peak at 3619 cm^{-1} in the IR spectra of specific dolomite samples correlated well with their lack of flocculation with PEO.

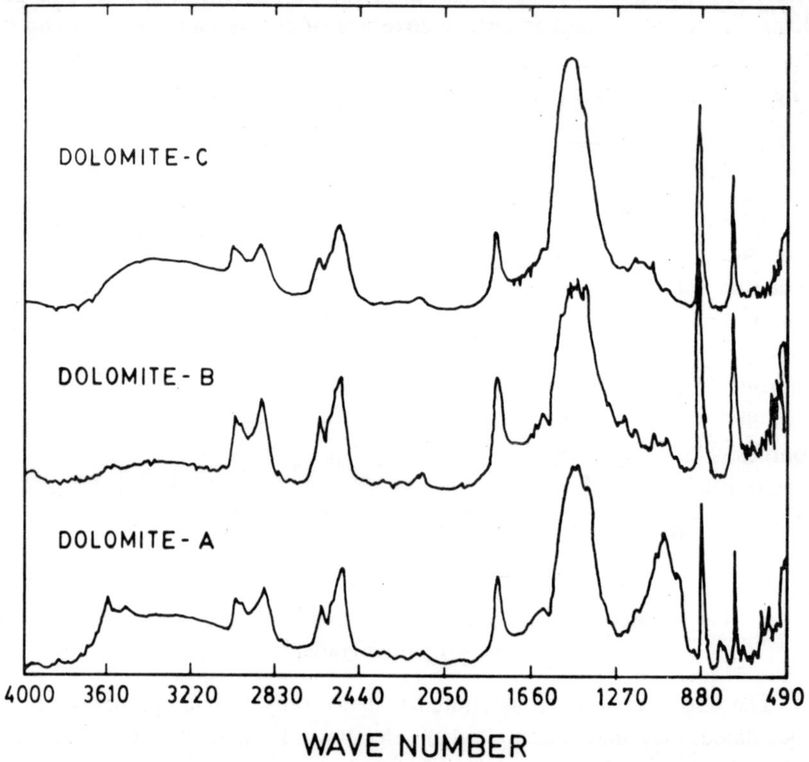

Fig. 5. DRIFT spectra of dolomite samples. Dolomite A (Florida) flocculates and exhibits clay peaks at 3619 cm^{-1} and 1055 cm^{-1}. Dolomite B was obtained from New Jersey and dolomite C procured from Georgia (after Ref. [36]).

In selective flocculation systems fine particles of high surface area are encountered. The presence of even trace level surface-active impurities can remarkably alter the surface characteristics and hence the flocculation behaviour. In order to achieve better control of selectivity, a thorough knowledge of the surfaces of the minerals involved is essential.

GRINDING EFFECTS

The size reduction process has an important bearing on the kind of surfaces present in mixed mineral suspensions. It is, therefore, expected that two mineral mixtures ground under different conditions (wet or dry; in the presence or absence of dispersants) will exhibit differences in surface characteristics and hence in selective flocculation behaviour. Rubio and Marabini [39] for example, observed that selêctivity in ground mixtures varied depending on whether they were wet or dry ground and that selectivity could be improved with the help of dispersants added during wet grinding of these mixtures.

Different flocculation results with dry or wet grinding in the presence of 5 kg/t of sodium silicate dispersant, were observed in a 1 : 1 apatite-silica mixture by Moudgil et al. [40]. Dry or wet grinding resulted in a significant loss of selectivity, as seen from the results in Table 2. Further, the grade of apatite is significantly lower (less than that of the feed) when the synthetic mixture is dry ground. This may be attributed to smearing of the harder silica particles by apatite during dry grinding.

Another effect of the mode of grinding is manifested in the lower recovery upon wet grinding (see Table 2). This implies a decrease in the extent of flocculation when the mineral is wet ground. During a study on flocculation of wet ground apatite, it was observed that only about 66% material flocculated with a dosage of 3 kg/t PAA (molecular weight = 4 million) and only marginal improvement in flocculation was obtained with higher flocculant dosages [41]. Similar flocculation behaviour was observed for wet ground apatite from other sources, whereas more than 90% material flocculated with 1 kg/t PAA if the material was dry ground instead of wet grinding. This was attributed to more efficient reduction in particle size by wet grinding [41]. Materials of similar particle size distribution exhibited identical flocculation behaviour, whether dry ground or wet ground.

Table 2: Effect of grinding on selective flocculation of apatite in 1 : 1 Apatite-Silica system (after Ref. [40])

Flocculant = (PAA molecular weight = 4×10^6)

Flocculant dosage = 1.0 kg/t

pH = 9.5

Type of Feed	Grade, % Apatite	Recovery, % Apatite
Synthetic	90.7	59.2
Dry Ground	41.6	60.0
Wet ground (Na-silicate : 5 kg/t)	78.0	40.0

The effect of grinding of particulate mixtures on the separation efficiency is the least understood aspect of the selective flocculation process. It also holds the key to the success of any selective flocculation separation scheme, particularly in natural ore systems. The loss of selectivity due to grinding can be attributed to slime coating (smearing effect) that is, smaller particles of one constituent coating the other particles and rendering the surfaces indistinguishable from each other after prolonged grinding. Solid coatings may also build up in semi-soluble systems through the dissolution-precipitation type reactions.

The presence of water and dispersants in some cases, does help to prevent coatings yielding better results. There is a need to find alternative methods to alter these surfaces so that the original selectivity is restored.

SUMMARY

Recent advances in the selective flocculation phenomena in mixed mineral systems include the concepts of heteroflocculation and site blocking agents, the role of particle size, and effects of the polymer molecular weight, and the role of trace level surface impurities in determining the selectivity in separation systems. These considerations led to successful separation of apatite from dolomite and silica in synthetic mixed mineral tests. Efforts are underway to further understand the loss of selectivity and to successfully apply the selective flocculation process for beneficiating the dolomitic phosphate ores and other complex low grade ores.

ACKNOWLEDGEMENTS

The authors acknowledge the partial financial support of the work by the National Science Foundation (Grant # CTS 9222854) and the Mineral Resources Research Center, University of Florida, Gainesville, FL.

REFERENCES

1. Lawver, J.E., et al. : Method of Beneficiating Phosphate Ores Containing Dolomite. US Patent 4,372,842, Feb. 8, 1983.
2. Snow, R.E. : Beneficiation of Phosphate Ores. U.S. Patent No. 4,144,969, 1979.
3. Ince, D.E. : Effect of Sodium Chloride on the Selective Flotation of Dolomite From Apatite. Ph.D. Thesis, University of Florida, Gainesville, 1987.
4. Moudgil, B.M. : Separation of Dolomite From the South Florida Phosphate Rock. Publication No. 02-0233-050 and 02-023-051, Florida Institute of Phosphate Research, Bartow, Florida, 1987.
5. Shah, B.D. : Selectivity in Mixed Mineral Flocculation: Apatite-Dolomite System. M.S. Thesis, University of Florida, Gainesville, Florida, 1986.
6. Yarar, B. and Kitchener, J.A. : Transactions of the Institution of Mining and Metallurgy, 79 (1970) C23-C33.
7. Friend, J.P. and Kitchener, J.A. : Chemical Engineering Science 28 (1973) 1071-1080.
8. Attia, Y.A. and Kitchener, J.A. : Development of Complexing Polymers for the Selective Flocculation of Copper Minerals. Proc. 11th International Mineral Processing Congress, Cagliari, Italy, 1975.
9. Attia, Y.A. and Fuerstenau, D.W. : In Li, N.N. (Ed.) : Recent Development in Separation Science, CRC Press, 4, 1978, pp. 93-105.
10. Sresty, G.C., Raja, A., and Somasundaran, P. : In Li, N.N. (Ed.) : Recent Development in Separation Science, CRC Press, 4, 1978, pp. 93-105.

11. Somasundaran, P. : In Somasundaran, P. (Ed.) : Fine Particles Processing. SME-AIME, Vol. 2, 1980, pp. 947-976.
12. Attia, Y.A. : In Attia, Y.A. (Ed.) : Flocculation in Biotechnology and Separation Systems. Amsterdam : Elsevier Science Publishers 1987, pp. 227-246.
13. Attia, Y.A. and Yu, S. : In Attia, Y.A., Moudgil, B.M., and Chander, S. (Eds.) : Interfacial Phenomena in Biotechnology and Materials Processing. Amsterdam : Elsevier Science Publishers 1988, pp. 503-514.
14. Moudgil, B.M., Behl, S., and Pradip : Minerals and Metallurgical Processing 10 (2), (1993) 43-46.
15. Colombo, A.F. : In Somasundaran, P. (Ed.) : Fine Particles Processing. SME-AIME, Vol. 2, 1980, pp. 946-947.
16. Banks, A.F. : In Somasundaran, P. (Ed.) : Fine Particles Processing. SME-AIME, Vol. 2, 1980, pp. 1104-1111.
17. Attia, Y.A., Kogan, R. and Driscoll, K. : In Attia, Y.A (Ed.) : Flocculation in Biotechnology and Separation Systems. Amsterdam : Elsevier Science Publisher 1987, pp. 297-308.
18. Yu, S. and Attia, Y.A. : In Attia, Y.A. (Ed.) : Flocculation in Biotechnology and Separation Systems. Amsterdam : Elsevier Science Publishers 1987, pp. 601-637.
19. Yarar, B. : In Mittal, K.L. and Fendler, E.J. (Eds.) : Solution Behavior of Surfactants: Theoretical and Applied Aspects. New York : Plenum 2, 1982, pp. 1333-1364.
20. Pradip : In Sastry and Fuerstenau, D.W. (Eds.) : Challenges in Mineral Processing. SME-AIME 1989, pp. 409-420.
21. Moudgil, B.M. and Vasudevan, T.V. : In Plumpton, A. (Ed.) : Production and Processing of Fine Particles. Montreal 1988, pp. 249-256, .
22. Behl, S. and Moudgil, B.M. : Minerals and Metallurgical Processing 5 (1992) 92-94.
23. Behl, S., Moudgil, B.M., and Prakash, T.S. : Journal of Colloid and Interface Science 161 (1993) 414-421.
24. Behl, S. and Moudgil, B.M. : Journal of Colloid and Interface Science, 161 (1993) 443-449.
25. Behl, S. and Moudgil, B.M. : Journal of Colloid and Interface Science 161 (1993) 422-429.
26. Moudgil, B. M., Shah, B. D. and Soto, H. S. : Minerals and Metallurgical Processing 4 (2) (1987) 27-31.
27. Moudgil, B. M., Shah, B. D., and Soto, H. S. : Journal of Colloid and Interface Science 119 (2) (1987) 466-473.
28. Moudgil, B. M., and Vasudevan, T. V. : Journal of Colloid and Interface Science, 127 (1) (1989) 239-243.
29. Moudgil, B.M. and Behl, S. : Journal of Colloid and Interface Science 146 (1) (1991) 1-8.
30. Behl, S. : Controlled Polymer Adsorption for Enhanced Selectivity in Flocculation. Ph.D. Thesis, University of Florida, Gainesville, 1993.
31. Attia, Y.A. and Deason, D.M. : Colloids and Surfaces 39 (1989) 227-238.

32. Attia, Y.A., Bavarian, F. and Driscoll, K.H. : Coal Preparation 6 (1988) 35-51.
33. Behl, S., Moudgil, B.M., and Prakash, T.S. : Journal of Colloid and Interface Science, 158 (1993) 511-512.
34. Behl, S., Moudgil, B.M., and Prakash, T.S. : Colloids and Surfaces 74 (1993) 311-312.
35. Moudgil, B.M. and Behl, S. : Flocculation Behavior of Dolomite. Proc. XVII International Mineral Processing Congress. Sydney, Australia : The Australasian Institute of Mining and Metallurgy, Vol.5, 1993, pp. 1309-1314.
36. Moudgil, B.M., Mathur, S. and Behl, S. : Flocculation Behavior of Dolomite With Poly(ethylene-oxide). Presented at the SME Meeting, Albuquerque, New Mexico, Feb. 14-17, 1994.
37. Rubio, J. and Kitchener, J.A. : Journal of Colloid and Interface Science 57 (1) (1976) 132-141.
38. Cheng, Y.C. : The Effect of Surface Hydration on the Adsorption and Flocculation of a Model Silica Suspension Using Polyethylene Oxide. M.S. Thesis, University of Florida, Gainesville, Florida, 1985.
39. Rubio, J. and Marabini, A.M. : International Journal of Mineral Processing 20 (1987) 59-71.
40. Moudgil, B.M., Mathur, S. and Behl, S. : Minerals and Metallurgical Processing, Nov. 1994, 29-34.
41. Mehta, N : Flocculation behavior of Apatite With Polyacrylic Acid. M.S. Thesis, University of Florida, Gainesville, Florida.

Biobeneficiation – Potentials and Prospects

K.A. NATARAJAN
Department of Metallurgy,
Indian Institute of Science, Bangalore - 560012, India

ABSTRACT

The use of microorganisms in the beneficiation of various minerals has been recognised of late. The application of this frontier area of biomineral technology in the field of mineral processing is set to revolutionise the mining and mineral industries during the coming decade. Various types of autotrophic and heterotrophic bacteria as well as fungi are known to interact with sulphide and oxide minerals in such a fashion so as to remove selectively one or more mineral constituents in an ore. In this paper, basic principles and mechanisms governing microbe-mineral interactions are examined from the view point of mineral beneficiation. Typical laboratory studies involving biobeneficiation of lead-zinc sulphides, coal, bauxite, magnesite and other nonferrous sulphides are highlighted to bring out the potential of this technology. Surface chemistry of bacteria as well as of some sulphide minerals is discussed to illustrate the manner in which bacteria could aid in flotation. Future prospects of biobeneficiation is outlined, enumerating the advantages of this process over other conventional techniques.

INTRODUCTION

When microorganisms interact with minerals, several consequences of relevance in mineral processing result, namely,

- adhesion of microorganisms to mineral surfaces
- oxidation/reduction reactions catalysed by organisms
- adsorption/chemical interaction of bacterial metabolic products onto mineral surfaces

The end results of such microbe-mineral interactions manifest as biomineral conversion, bioleaching, biosurface modification, bioaccumulation and biosorption.

Biomineral beneficiation needs to be examined and understood from the following angles:

- Surface chemistry of microorganisms in relation to that of mineral particles
- Mechanisms of bacterial adhesion to mineral surfaces and their role in mineral surface modification

- Selective removal of undesirable constituents leading to enrichment of the desired mineral in an ore body
- Sorption and accumulation of metal constituents by biomass
- Generation of biosurfactants.

Several types of autotrophic and heterotrophic bacteria, fungi, algae and yeasts are implicated in mineral beneficiation processes. The role of such biological agents in bringing about mineral beneficiation may vary; for example, they may serve as flotation aids, flocculants, biosorbents, or as solvents. Microorganisms themselves could be separated from a mixed broth or aqueous medium by flotation, sedimentation or flocculation.

In this paper, microbiological, surface-chemical as well as physico-chemical factors governing biobeneficiation processes are analyzed. The use of different types of microorganisms is considered with pertinent examples. The surface chemistry of *Thiobacillus ferrooxidans* is examined with respect to their ability in conferring either hydrophobicity or hydrophilicity to minerals such as pyrite, chalcopyrite, sphalerite and galena. The use of such autotrophic bacteria in coal desulphurisation is illustrated. Applicability of biobeneficiation using fungi and heterotrophic bacteria is brought out with relevance to biological production of flotation reagents, fine particle flocculation and beneficiation of bauxite, clays, sands, silicates and salt-type and oxide minerals. The commercial potentials and future prospects of this frontier technology are also outlined.

BACTERIAL SURFACE CHARGE AND ADHESION

The surfaces of many microorganisms carry a net negative charge at physiological pH. The cells acquire such charge through the ionization of surface groups such as amino, carboxylate and phosphate which is pH dependent. Competition between electrical and chemical forces governs surface charge neutralisation. Counter ions tend to bind to the charged groups. An electrical double layer is thus established at the interface. The thickness of the double layer depends on the ionic strength of the electrolyte. Similar to solid-solution interfaces, the surface charge properties of microorganisms can be characterized by zeta potential and the isoelectric point (IEP) established by electrophoretic mobility measurements. Surface hydrophobicity is implicated in interfacial interactions of microbial cells with other cells (flocculation), with solids and liquids (adhesion) and with air (flotation). Adhesion of microorganisms to mineral substrates is of relevance in biomineral beneficiation. Bacterial adhesion is initiated with the transport of organisms towards a surface mediated by gravity, diffusion or convection. Once the range of interaction forces is reached, the adhesion process facilitated by different forces commences. Both short range and long range forces may be involved such as Lifshitz-Van der Waals forces, electrostatic interaction, specific and chemical interaction as well as hydrophobic interaction. Contact angles reflect the overall hydrophobic character of an organism. In the adhesion process, new interfaces are created and a free-energy balance of all interfaces involved in adhesion can be used to predict as to whether adhesion is energetically favourable or not [1].

$$\Delta F_{\text{adhesion}} = \gamma_{BM} - \gamma_{ML} - \gamma_{BL} \tag{1}$$

where

BM = bacteria-mineral interface

ML = mineral-liquid interface

BL = bacteria-liquid interface

Adhesion is favourable when the free energy of adhesion is negative. A physico-chemical approach to bacterial adhesion including the influence of electrical charge interactions is based on the DLVO theory which is used primarily to understand the kinetics of adhesion.

CELL SURFACE HYDROPHOBICITY

The role of hydrophobicity in bacterial adhesion to mineral surfaces has been thought to be significant. Cell hydrophobicity is conferred by a combination of certain hydrophobic structures or molecules present on cell surfaces. The role of electrostatic interactions in influencing microbial cell surface hydrophobicity need also to be understood. Since surface charge increases the likelihood of polar interactions with water molecules in the vicinity, the more charged the cell surfaces, the less hydrophobic they become. Hence, any treatment that lowers the cell negative charge, would enhance its hydrophobic character. Cells of higher hydrophobicity and lower electrophoretic mobility would then be more adherent. However, relatively hydrophobic cells also tend to exhibit highly negative electrokinetic potentials. Even in the absence of a net surface charge, the cells may be covered by a large number of positively and negatively charged moieties, all of which tend to lower the hydrophobic effect. Cell appendages such as fimbriae could also simultaneously confer both electronegativity and hydrophobicity [2].

Cell surface hydrophobicity is often governed by the chemical composition and physical arrangement of the cell wall outer layers. Surface proteins are involved in bacterial adhesion. For example Rhamnose-rich polysaccharides are involved in the hydrophobic adhesion of *Rhodococcus sp.* Hydrophobicity-related proteins and polysaccharides may serve merely as a bridge connecting associated fatty acids or lipids to the cell wall. It may also be possible that relatively hydrophobic portions of protein or even hydroxyl-deficient polysaccharide molecules can be so oriented as to be exposed at the cell surfaces while the hydrophobic regions of the molecules are anchored to the cell walls. Such molecules possessing both hydrophobic and hydrophilic parts may confer dual properties to the organisms in that they could migrate through aqueous solutions and colonise on newer solid substrates. There is reason to believe that many organisms exhibit both hydrophilic and hydrophobic surface regions. Cell surface hydrophobicity is not a permanent attribute of the cell envelope, but could be modified from time to time to suit changing environmental conditions and other physiological factors. Interfaces may provide a better environment for survival and growth of organisms than bulk aqueous phase due to accumulation of nutrients at interfaces. Adhesion may also allow better interaction with and utilisation of nutrients available on substrates. Bacterial adhesion is dependent not only on the biochemical and surface chemical properties of the organism but also on the interfacial and surface properties of the mineral in an aqueous system. The polar flagellum of the bacteria acts as a locomotor organ. When grown in a liquid nutrient medium, the cells acquire appropriate surface properties to survive in a liquid medium, while if grown on a solid substrate, would orient themselves accordingly.

SURFACE CHEMISTRY OF THIOBACILLUS FERROOXIDANS

A typical example of such a behaviour pertains to *Thiobacillus ferrooxidans*, the most widely implicated autotroph in bioleaching and biomineral beneficiation. It is a gram-negative chemolithotrophic, mesophilic, monoflagellated acidophile obtaining its energy by the oxidation of ferrous iron, sulphur and reduced valence sulphur compounds. They are rod shaped (0.5 μm wide by 1-1.5 μm long) and occur singly or in pairs as illustrated in Fig. 1. Relative hydrophobicity of the cells of *Thiobacillus ferrooxidans* was studied through hydrophobic interaction chromatography and phase partitioning after growing them in a liquid 9 K medium and on solid mineral substrates such as sulphur, pyrite and chalcopyrite [3]. Over 50% of the cells grown on mineral substrates were observed to be bound to octyl sepharose whereas only less than 18% of the cells grown in the liquid nutrient medium were so bound. Also, cells of the bacteria grown on mineral substrates were observed to be increasingly transferred to an organic phase, hexadecane, unlike those grown in a liquid medium. Mineral-grown cells of *Thiobacillus ferrooxidans* thus exhibit higher hydrophobicity. Common immunoreactivity was observed between the various mineral-grown cells, while liquid-grown cells did not exhibit such a behaviour, indicating that a secretion is released on mineral-grown cells alone for aiding surface adhesion. FTIR analyses of the cell surfaces grown on sulphide mineral substrates readily indicated the presence of NH_3, NH_2, NH, CONH, CO, CH_3, CH_2, CH and COOH functional groups. Such a proteinaceous secretion synthesised by *Thiobacillus ferrooxidans* was completely absent if grown in a liquid nutrient medium, suggesting the unique environmentally necessitated alteration of bacterial cell composition. The bacteria prepare the mineral particle surfaces for adhesion through the secretion of surface proteins and firmly attach thereafter. Such a bacterial adhesion process brings about a change

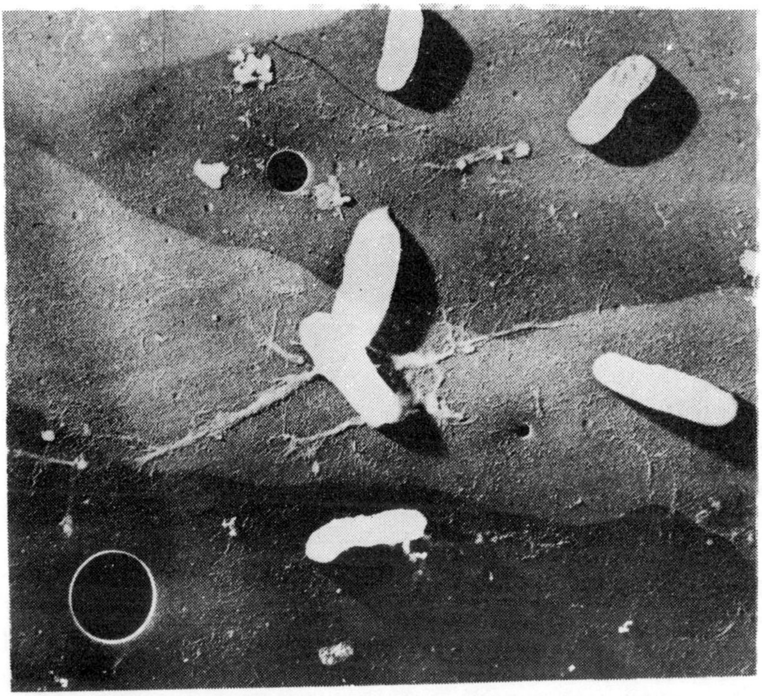

Fig. 1. Electron Micrograph of *Thiobacillus ferrooxidans*

on the surface properties of not only the organism, but also of those of mineral surfaces on which they attach.

MICROBE-MINERAL INTERACTIONS AND SURFACE MODIFICATION

Surface-chemical changes brought about by bacterial adhesion to mineral surfaces can be studied through zeta potential measurements [3]. Figure 2 illustrates the electrophoretic mobility of *Thiobacillus ferrooxidans* grown on different solid substrates such as sulphur, pyrite and chalcopyrite and in a liquid 9K medium containing ferrous iron. It could be seen that the liquid medium grown bacteria exhibit an isoelectric point (IEP) at about pH 2, while those grown on mineral substrates exhibited IEPs corresponding to a pH of about 3.8. The increasingly positive surface charge of the mineral-grown bacteria at acidic pH ranges indicates the presence of cationic amino groups at cell surfaces. Electrostatic interaction between positively charged cells and negatively charged sulphide mineral particles would favour bacterial adhesion. This observation is of interest in the conditioning of sulphide minerals with *Thiobacillus ferrooxidans* at an acidic pH in the range of 2 to 3 before leaching or flotation.

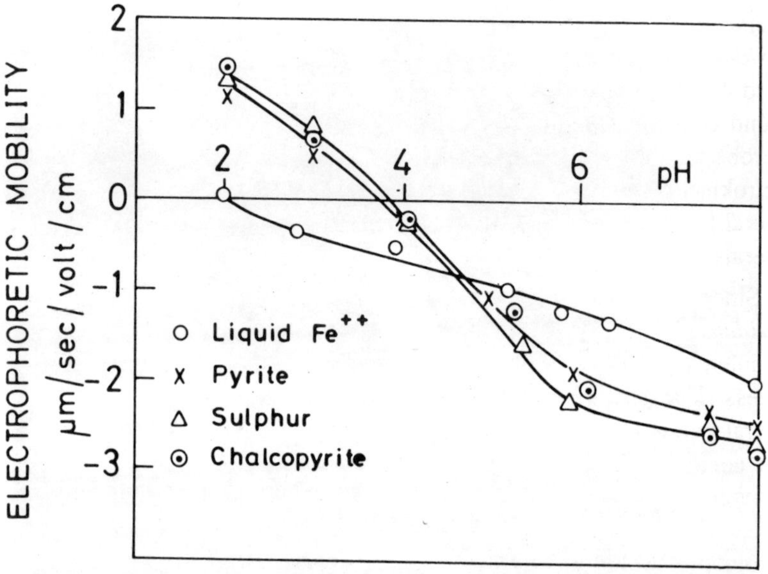

Fig. 2. Electrophoretic mobility of *Thiobacillus ferrooxidans* grown on solid substrates such as sulphur, pyrite and chalcopyrite and in a liquid 9 K medium, as a function of pH (after Ref. [3])

The electrokinetic behaviour of minerals such as sulphur, pyrite and chalcopyrite before and after interaction with *Thiobacillus ferrooxidans* is illustrated in Fig. 3. The IEP of sulphur observed at about pH 2.0 in the absence of bacterial interaction is seen to be shifted to higher values between 3.8 to 4.5 after conditioning with the bacteria. A similar shift in IEP could also be

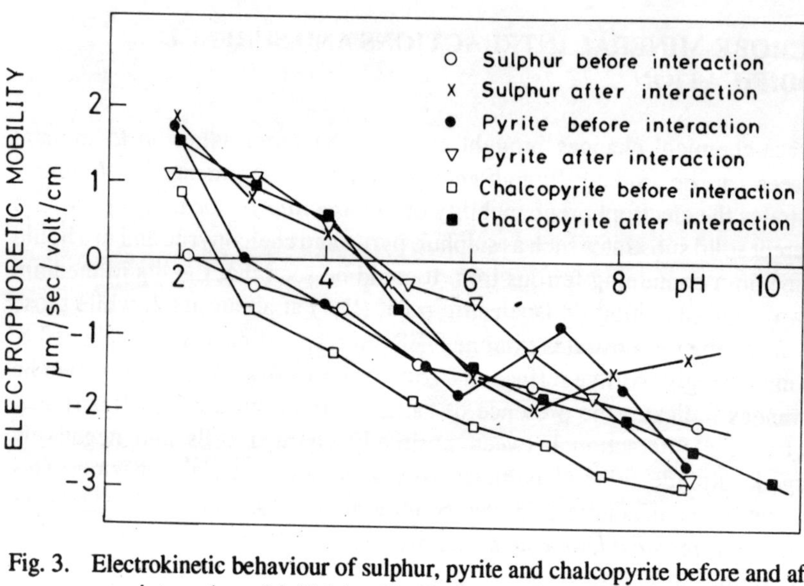

Fig. 3. Electrokinetic behaviour of sulphur, pyrite and chalcopyrite before and after interaction with *Thiobacillus ferrooxidans* (after Ref. [3])

observed with pyrite and chalcopyrite after interaction with *Thiobacillus ferrooxidans*. For example, the IEP of a fresh sample of pyrite shifted from pH 2.9 to 4.7 after conditioning with the bacteria. Similarly, the IEP of freshly ground chalcopyrite shifted from pH 2.6 to 4.4 after bacterial interaction. Microbe-mineral interactions thus bring about significant changes in the electrokinetic behaviour of both the bacterial cells and the minerals. Since, mineral flotation is influenced by the change in the surface properties of the minerals, bacteria could influence their flotation behaviour.

Since the optimum pH for the growth and activity of *Thiobacillus ferrooxidans* is about 2-3, bacterial adhesion to mineral surfaces at this pH range is of interest. As shown above, electrostatic interactions could play a role in this process. Besides, hydrophobic interactions between the bacteria and the mineral surfaces play a very significant role in bacterial adhesion. Hydrophobic cells adhere to hydrophobic mineral surfaces to a greater extent than to relatively hydrophilic surfaces. A combination of high hydrophobicity and low surface charge would favour strong adhesion. Typical scanning electron micrographs illustrating the adhesion of *Thiobacillus ferrooxidans* to sulphide minerals are illustrated in Fig. 4. Surface morphology of the mineral particles

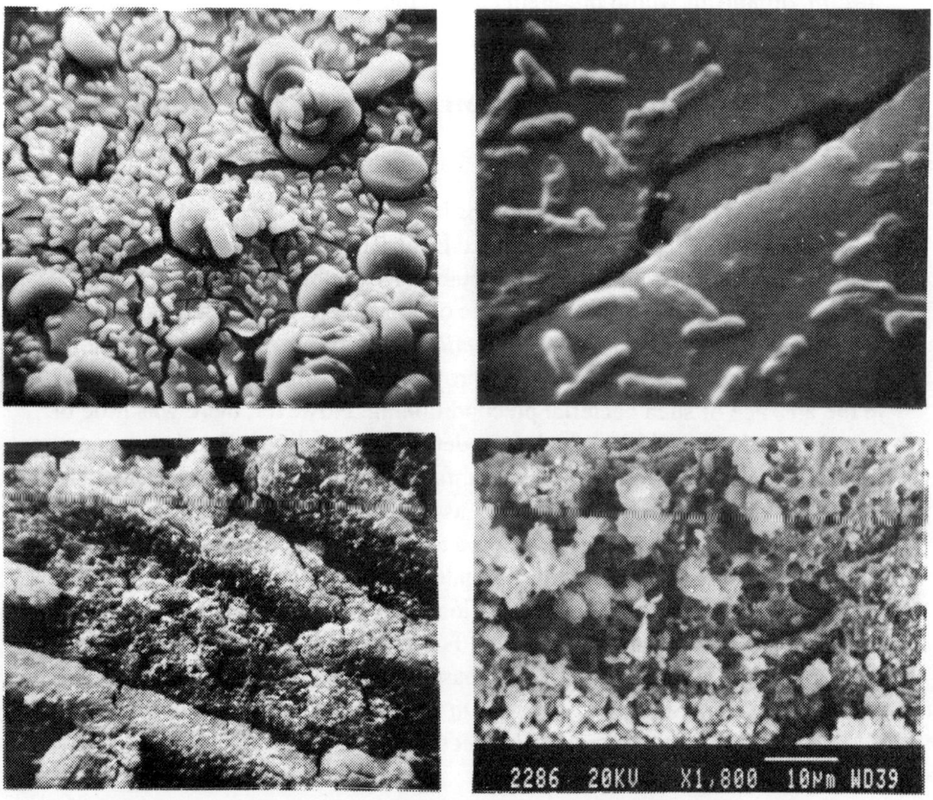

Fig. 4. Scanning Electron Micrographs illustrating adhesion of *Thiobacillus ferrooxidans* on sulphide minerals and surface morphology after bacterial interaction

after bacterial interaction could also be seen. Surface precipitates consisting of metabolic and interaction products cover the reacted mineral surfaces. Although electrostatic and hydrophobic forces play a significant role in bacterial adhesion to mineral surfaces, the exact mechanisms by which bacterial cells adhere to mineral particles could be quite complex. Surfactants secreted by the organisms, bacterial pili and fimbriae, nature of the electrolyte, solution pH as well as the crystal structure and surface morphology of the solid substrates, all influence bacterial adhesion processes. Besides the electrokinetic and hydrophobic properties of the minerals, electrochemical as well as mineralogical characteristics of minerals available in an ore body also influence the extent and intensity of bacterial growth [4]. Bacteria prefer mineral sites that yield nutrients in a complex ore substrate. For example, *Thiobacillus* type of organisms would preferably adhere to sulphide mineral sites rather than to inert quartz or other oxide minerals in an ore matrix. This aspect is of significance in biomineral beneficiation since preferential bioremoval of mineral constituents from an ore can be brought about by appropriate selection of microorganisms which would derive energy by utilisation of the desirable components. Thus, 'silicate' bacteria need be used for removal of silica, while phosphate specific organisms would preferably act on that moiety.

INFLUENCE OF BIOPRETREATMENT ON THE FLOTABILITY OF GALENA AND SPHALERITE

The utility of *Thiobacillus ferrooxidans* in the flotation beneficiation of some sulphide minerals can now be examined [5-6]. The effect of biotreatment with *Thiobacillus ferrooxidans* on the floatability of sphalerite and galena is illustrated in Figs. 5 and 6. In the absence of any added flotation reagents, when sphalerite was pretreated with *Thiobacillus ferrooxidans* in a pH 2 sulphuric acid medium, its flotation recovery improved when compared to the conditions in the absence of such bacterial preconditioning. However, there was little or no effect on the floatability of galena under similar conditions. Further, in many cases, bacterial pretreatment was harmful to the floatability of galena. The availability of initial biomass as well as the duration of conditioning of the mineral with the bacteria were observed to be important variables influencing the floatability of both sphalerite and galena. With increase in cell concentrations and duration of biopretreatment, floatability of both sphalerite and galena was found to be deleteriously affected. For sphalerite, 94% recovery obtained after treatment with 10^8 cells/ml decreased to about 53% when the cell concentration was increased by ten fold. On the other hand, floatability of galena was observed to be more seriously affected by increasing the initial biomass. Typical results are shown in Table 1.

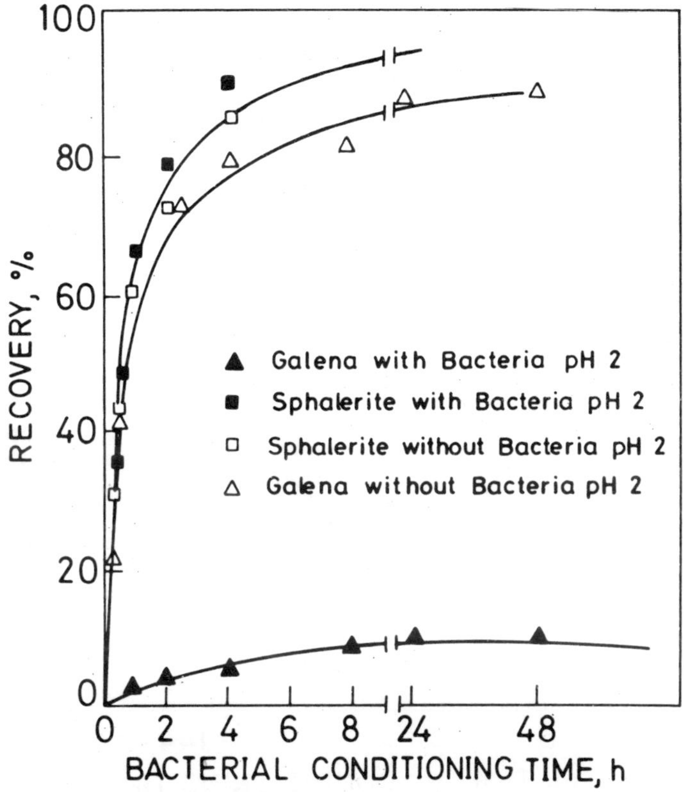

Fig. 5. Effect of conditioning with *Thiobacillus ferrooxidans* (10^8 cells/ml, pH 2) on the floatability of sphalerite and galena (after Ref. [5-6])

Table 1: Effect of initial concentration of *Thiobacillus ferrooxidans* on the floatability of galena and sphalerite (24-h, bacterial conditioning, 20% pulp density; sphalerite floated in the presence of copper sulphate and isopropyl xanthate, and galena with isopropyl xanthate) [5-6].

Initial cell population cells/ml	Flotation recovery, percent	
	Sphalerite	Galena
10^2	95	84
10^4	96	78
10^8	94	48
10^9	53	23

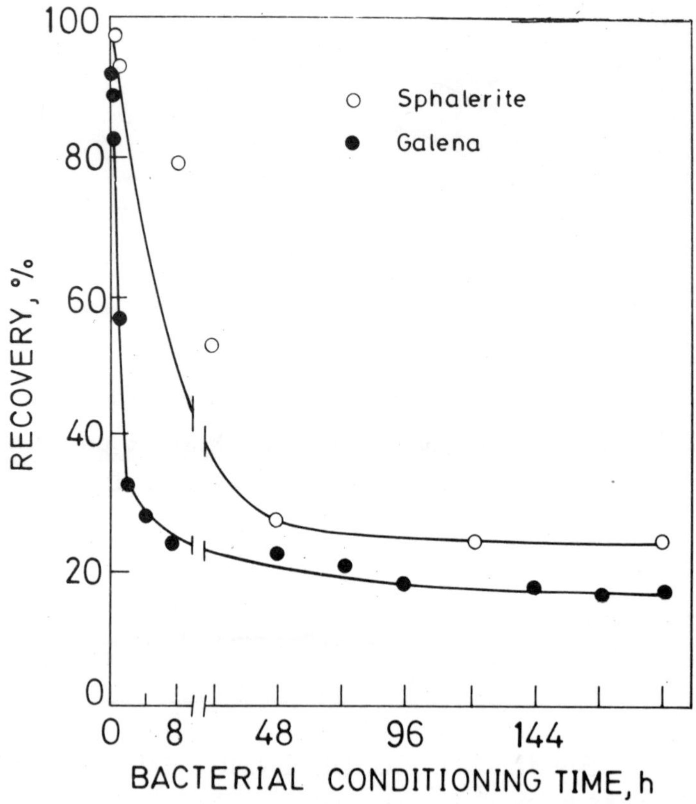

Fig. 6. Effect of conditioning with *Thiobacillus ferrooxidans* at higher cell density (10^9 cells/ml) on the floatability of sphalerite and galena (after Ref. [5, 6])

It would thus be possible to preferentially float sphalerite from a mixed zinc-lead sulphide ore through bacterial pretreatment under carefully controlled conditions.

During biopretreatment, significant number of bacterial cells were found to be tenaciously adhered to the above sulphide mineral surfaces. Cell concentrations in the liquid phase as a function of duration of biotreatment for both sphalerite and galena suspensions are illustrated in Fig. 7. In both cases, about 45% of the cell population were dispersed in the liquid phase after about one minute of bacterial treatment, the rest being adhered to the mineral surfaces. With increase in the duration of bacterial treatment, the cell population in the liquid phase tended to increase in a sphalerite suspension, while it decreased with time in case of galena. The fluctuations in cell numbers are due to continuous attachment and detachment of organisms with respect to the mineral phase. These results indicate that at any point of time, relatively higher number of cells are adhered to galena surfaces than to sphalerite. Unlike sphalerite, galena surfaces can then be expected to be covered with a dense layer of bacterial cells along with their metabolites and reaction products. Such a

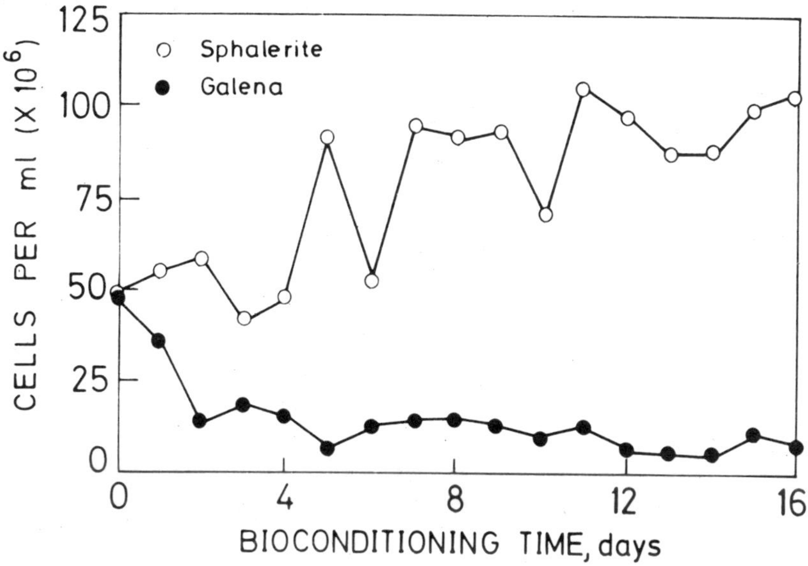

Fig. 7. Population of *Thiobacillus ferrooxidans* in the supernatant solution after conditioning sphalerite and galena for different periods of time (after Ref. [5, 6])

difference on the surfaces of these minerals brought about by bacterial interaction could then explain the enhanced hydrophobicity of sphalerite compared to galena.

An initial product of interaction of *Thiobacillus ferrooxidans* with sulphide minerals at acidic pH is elemental sulphur which tends to coat the mineral surfaces improving their floatability. However, prolonged interaction of the bacteria results in further oxidation of the elemental sulphur to various sulfoxy compounds and ultimately to sulphate. Lead sulphate being insoluble in the medium remains as a adherent film on galena surfaces, while the continuous dissolution of zinc sulphide as its sulphate promotes the generation of fresh mineral surfaces. The presence of oxidised insoluble products on the galena surface hampers its floatability. Bacterial adhesion promotes the above biochemical reactions, the rate of which is dependent on the number of available cells and duration of interaction. Very high cell concentrations and prolonged interaction periods would result in excess and random bacterial coverage of mineral particles leading to an increase in surface hydrophilicity.

PYRITE REMOVAL FROM COALS USING *THIOBACILLUS FERROOXIDANS*

Another important application of *Thiobacillus ferrooxidan* is in the desulphurisation of coals. Pyritic sulphur from coals can be efficiently removed by bacterial conditioning for pyrite depression followed by coal flotation [7-9]. Amount of biomass, bacterial conditioning period, percent solids in the pulp as well as the flotation pH influence pyrite sulphur removal from coals. The effects of conditioning period, percent solids and flotation pH on the removal of pyrite sulphur are illustrated in Tables 2 to 4. Bacterial conditioning was also found to be useful in the lowering of ash content of coals [10].

Table 2: Effect of bacterial conditioning period on pyrite flotation from coals (Particle size – 28 mesh, conditioning pH 2.0, flotation pH 9.0) (after Ref. [10])

Conditioning period (hrs)	Flotation yield %	Percent pyritic sulphur removal
2	74.5	66.3
3	61	74.8
4	78	77.6

Table 3: Effect of percent solids on pyrite flotation from coals (– 28 mesh; Conditioning period 4 hrs at pH 2.0, flotation pH 9) (after Ref. [10])

Pulp density %	Yield %	Percent removal of pyritic sulphur
10	65.6	76.3
15	78.0	77.6
20	63.8	76.5

Table 4: Effect of flotation pH on pyrite removal from coal (Conditioning for 4 hrs. at pH 2, 15% solids) (after Ref. [10])

Flotation pH	Yield %	Percent pyritic sulphur removal
3	80.2	58
5	67.1	78.4
9	78.0	77.6
10	63.0	81.3

Using a novel technique for sulphur reduction involving bacterial pretreatment, 80 to 90% of pyrite was released from – 400 mesh coal during oil agglomeration of the coal. The bacterial conditioning time to achieve the desired result has been reduced to 4 hours from an earlier observation of 1-3 days. The rationale behind such bioconditioning is to modify the surface chemistry of pyrite through bacterial adhesion and interaction, resulting in suppression of pyrite in many cases. Even bacterial liquors from a fermentor can be directly used to suppress the pyrite from coals instead of using a cell biomass. A short period of 2.5 minutes was found to be sufficient for bacterial conditioning [11-12]. The temperature dependence on the process is negligible.

Suppression of pyrite is dependent on the bacterial density available during conditioning as illustrated in Table 5.

Table 5: Effect of Bacterial density on pyrite flotation (– 212 + 106 μm size, pH 2.0) (after Ref. [10])

Cell density cells/ml	Pyrite flotation, %
2×10^9	80–85
8×10^9	40
2×10^{10}	15
2×10^{11}	5–10

Higher number of cells lead to generation of hydrophilic mineral surfaces. From the above results, it becomes evident that biomass density and duration of bacterial conditioning are the important factors influencing the floatability of pyrite. With lower biomass, pyrite flotation instead of suppression may thus be facilitated.

MINERAL BENEFICIATION USING SULPHATE REDUCING BACTERIA

Anaerobic bacteria of the type *Desulfovibrio* reduce sulphate ions to sulphide ions and hydrogen sulphide in the pH range 4-8 at temperatures upto 40°C. These organisms find a variety of applications in mineral beneficiation as detailed below.

Treatment of oxidised lead and antimony minerals with a culture of sulphate reducing bacteria (SRB) significantly enhanced their flotation [13]. Floatability of cerussite pretreated with SRB was found to be enhanced by 20-25%. The duration of bacterial pretreatment in these cases is very important in that shorter periods lead to activation while prolonged treatments result in suppression of the biotreated mineral. These organisms essentially serve as cheaper sulphidising agents for oxidised ores. Through appropriate controls, the bacterial reactions can be so manipulated as to generate a sulphide product (promote flotation) or H_2S gas (depressant). The ability of SRB to generate larger volumes of H_2S can be beneficially used in leaching also; for example, leaching of tin from difficult to beneficiate hydrostannite-type ores. The treatment of poorly soluble sulphates such as those of calcium with SRB results in the conversion of sulphates to H_2S and stimulate their conversion to carbonates.

$$CaSO_4 + 2\,(CH_2O) \xrightarrow{SRB} CaS + 2\,CO_2 + 2\,H_2O \tag{2}$$

$$CaS + H_2O = Ca\,(OH)_2 + H_2S \tag{3}$$

$$CO_2 + H_2O = H_2CO_3 \quad (4)$$

$$Ca(OH)_2 + H_2CO_3 = CaCO_3 + 2\,H_2O \quad (5)$$

These organisms also serve as effective solvents for antimony in an alkaline medium and could extract as much as 96% of antimony within an hour. Another significant application of SRB is in the differential flotation of mixed sulphides. Selective separation of chalcopyrite and molybdenite on the one hand and of galena and sphalerite on the other becomes possible through controlled interactions with SRB. Yet another important use of these anaerobes is in the desorption of xanthate and its oxidised product coatings from sulphide mineral surfaces [14]. This aspect offers greater promise in the bulk and selective flotation of lead-zinc and copper-molybdenum sulphides.

Sulphate reducing organisms can stimulate the conversion of iron oxides to ferrous sulphide, malachite to covellite, silver chloride/carbonate to argentite, and smithsonite ($ZnCO_3$) to sphalerite (ZnS). Thus. they are good sulphidisers of relevance in mineral beneficiation.

BIOCONVERSION OF MINERALS

Bioconversion of minerals from one form, to another could be brought about by bacterial action. For example, reduction of haematite to magnetite, oxidation of hausmanite to pyrolusite, reduction of MnO_2 to lower oxides, conversion of sulphide to sulphate and metal sulphates to sulphides. The bacteria-mineral cycles involved in these bioprocesses are of relevance in mineral beneficiation [15]. In all the above beneficiation processes, various metabolic products secreted by microorganisms play a significant role as lixiviants and chelating agents for specific mineral forms, as illustrated in Table 6 [16].

Table 6: Metabolic products secreted by some microorganisms (after Ref. [16])

Microorganisms	Major metabolic products
Pseudomonas Sp.	Citric, oxalic and gluconic acids
Bacillus polymyxa	Acetic acid
Bacillus circulans	Succinic, formic, fumaric and maleic acids
B. mucilagionosus	Exopolysaccharides
Thiobacillus Sp.	Sulphuric acid, proteins
Aspergillus niger	Citric, oxalic and gluconic acids
Penicillium Sp.	Citric, oxalic acids
Bacillus Sp.	Amino acids
Yeasts and algae	Proteins and nucleic acids

Biodegradation of carbonates through direct and indirect biological activity is of relevance in the removal of calcareous gangue from minerals.

$$CaCO_3 + H^+ = Ca^{++} + HCO_3^- \quad (6)$$

Metabolic acid generation helps the above dissolution process. Besides, mere metabolic generation of CO_2 during bacterial respiration may have this effect.

$$CO_2 + H_2O = H_2CO_3 \quad (7)$$

$$CaCO_3 + H_2CO_3 = Ca^{++} + 2\,HCO_3^- \quad (8)$$

Insoluble carbonates can thus be biologically removed. Organic acids and CO_2 generated from heterotrophic metabolism result in limestone decay. Autotrophic nitrifying and sulphur-oxidising bacteria were also shown to promote limestone degradation. Endolithic cyanobacteria, algae and fungi cause dissolution of calcite.

BIOLOGICAL REMOVAL OF SILICON, IRON, CALCIUM AND PHOSPHOROUS

Silicon compounds in the form of clays stimulate certain microbial metabolism. Clays tend to modify the microbial habitat physicochemically. Some fungi have been known to accumulate silicon. The role of biological factors in the silicon cycle is of great significance. Many autotrophic and heterotrophic bacteria are capable of breaking the siloxane bond, Si–O–Si. At high pH values, the biological destruction of quartz increases sharply. The solubilization may involve the cleavage of Si–O–Si or Si–O–Al bonds or removal of cations from the lattice of silicate, leading to disintegration of the silicate lattice structure. The mode of attack may be through microbially produced ligands, organic or inorganic acids, ammonia or amines or extracellular polysaccharide. For example, *Pseudomonas* produces 2-keto gluconic acid from glucose dissolving silicates of calcium, zinc, magnesium and minerals such as apophyllite, wollastonite and olivine. Fungi such as *Penicillium notatum* and *Aspergillus niger* degrade spodumene. *Sarcina ureae* releases silicon readily from plagioclase and quartz. Ammonia resulting from the hydrolysis of urea creates an alkaline pH. *Bacillus mucilaginosus* excretes polysaccharides that could react with silica and silicates. All the above biological processes are of significance in mineral beneficiation with respect to removal of silica, silicates, and oxides of iron, calcium and aluminium from ores [17-19].

Typical examples with respect to mineral processing are illustrated below.

Biologically mediated silica removal from magnesite ores has been reported using *Bacillus circulans, Bacillus mucilaginosus* and *Bacillus licheniformis*. Successful desiliconisation of bauxite, iron ores, manganese and phosphate ores could be achieved through the choice of appropriate microorganisms and controlling the process parameters. All the above microbial species aid the above process. Another approach in processing low-grade bauxites is selective leaching of aluminium which could be brought about by *A. niger*, *B. circulans*, and *P. aeroginosa*. In the beneficiation of bauxites, the type of ore minerology plays a significant role. *B. mucilaginosus* and *A. niger* were found to be useful for the removal of Al_2O_3, SiO_2 and Fe_2O_3 from different bauxite ores as illustrated in Table 7 [20].

Table 7: Bioremoval of silica, alumina and iron oxide from Bauxite ores using *Aspergillus niger* (after Ref. [20])

Bauxite type	Removal after 35 days with *A.niger* (ppm)		
	SiO_2	Al_2O_3	Fe_2O_3
Hematite-chamosite-diaspore	3390	1070	1152
Hematite-kaolinite-gibbsite	360	345	388
Gibbsite-kaolinite-goethite	495	662	4657
Siderite-kaolinite-gibbsite	475	810	8587
Kaolinite-hematite-boehmite	1850	545	1780

Bacterial removal of iron from kaolins and clays is of industrial significance in the production of commercial grade raw materials. Oxalic and citric acids produced by *A. niger* could effectively remove iron from clays. The metabolites were found to have higher iron removal rates if the process could be carried out at 90°C. The iron content of different kaolins could be brought down to 0.4-0.7% from an initial values as high as 1.5%. Such bioremoval could be efficiently carried out in bioreactors at pulp densities in the range of 20-40% within a period of a few hours [21].

Removal of phosphorous and silicon from manganese ores is yet another example of biomineral beneficiation. *Hypomicrobium Sp.* and *A. niger* are efficient phosphorous removers. Also, *B. mucilaginosus* aids in phosphorous removal along with silica.

BIOFLOCCULATION

Many microorganisms possess flocculant growth habits and produce extracellular polymers such as polysaccharides, polypeptides and polyglycons. Such biologically generated polymers cause flocculation of not only the microorganisms but also of the mineral particles with which they come into contact. The mechanisms by which dispersed cells are repelled and the forces responsible for their aggregation depend on the molecular architecture of the cell surfaces. Interfacial interactions are the primary driving force for flocculation similar to adhesion. Microbial flocculation and aggregate formation are of practical significance in many industrial processes such as biomass production, bioconversion, waste management and mineral separation.

Mycrobacterium phlei, a gram-positive, rod shaped procaryotic cell has been observed to be useful as a flocculant for phosphate slimes, haematite and coal fines [22-23]. The cell wall contains glycolipids, phospholipids and free lipids. Free long chain fatty acids have been isolated from this organism. Surface functional groups such as R-COOH, R-NH_2, R-OH, R-CONH-R, $(RO)_2$HO-P=O and R-C-O-C-R are present in the cells. Accumulation of polar groups at the surface imparts a high negative charge and the large number of nonpolar groups confer hydrophobicity to the surfaces of these organisms. The cells could strongly adhere to mineral surfaces such as haematite which could be effectively floated after bacterial conditioning under controlled conditions. At mildly acidic pH values, the organism flocculates haematite but not quartz. Substantial settling of haematite fines could be achieved within a few minutes after bacterial interaction. Selective flocculation of fine coals with respect to separation of pyritic sulphur could also be achieved in the presence of *M. phlei* as illustrated in Table 8. More than about 85% of pyritic sulphur from the coal could be removed by bioflocculation.

Table 8: Bio-flocculation results on coal (after Ref. [22])

Materials	Sulphur %	Ash %	Yield %
Feed	2.5	12.1	100.0
Floc. concentrate	1.2	5.8	85.2
Unflocculated rejects	9.3	45.3	14.8

Biosurfactants derived from such organisms may be primarily responsible for the flocculant action. The mechanisms involved in bioflocculation may be similar to those encountered in nonbiological systems involving bridging, charge neutralization, hydrophobic bonding and particle to particle contacts.

BIOSORPTION AND BIOACCUMULATION

Microorganisms are capable of concentrating metals from aqueous solutions. The major mechanisms through which microbes immobilize, complex or remove metals include extracellular precipitation, volatilization, extracellular complexation and accumulation, binding to cell surfaces and accumulation in intracellular spaces. These phenomena could be advantageously made use of for the removal of toxic metals such as copper, lead, zinc, mercury and uranium from mineral processing effluents. Both living and dead biomass could be used for the purpose. The use of dead biomass would be more advantageous since

- No nutrients need be supplied
- Easily and cheaply available as a waste product
- Toxicity of metals will not affect the process.

The following factors affect the metal uptake capacity of the biomass

- Nature, type and amount of biomass
- Solution chemistry – pH and types of ions present.

Table 9 illustrates some of the important microorganisms capable of sorption and accumulation of metals [21]. Depending on the type of the effluent appropriate organisms can be chosen. In the choicc for an industrial application, the ability to grow the organisms on a massive scale need be considered.

Table 9: Microorganisms capable of metal accumulation (after Ref. [21])

Organisms	Metal ions
Thiobacillus ferrooxidans	Cd^{+2}, Co^{+2}, Cu^{+2}, Cr^{+3}, Fe^{+3}, Ni^{++}
Cyanobacteria	Au^{+3}
Chlorella pyrenoidosa *Chlorella regularis*	Cd^{+2}, Co^{+2}, Ni^{+2}, Zn^{+2}, UO_2^{+2}
Bacillus subtilis	Cd^{+2}, Co^{+2}, Cu^{+2}, Zn^{+2}, Fe^{+3}, Au^{+3}, $UO_2{}^{+2}$
Pseudomonas Sp.	Cd^{+2}, CO^{2+}, Ni^{+2}, $UO_2{}^{+2}$ radio nucleides
Yeasts	Cd^{+2}, Co^{+2}, Cu^{+2}, Ni^{+2}, Zn^{+2}
Aspergillus niger	Radio nucleides, Co^{+2}, Cu^{+2}, Zn^{+2}

Sulphate reducing bacteria of the type *Desulfovibrio desulfuricans* can precipitate dissolved metal ions as their sulphides. These organisms can easily be grown in the presence of organic substrates in tailing ponds.

A number of technological developments have taken place in bioenvironmental control. Continuous Stirred Tank Reactors of different designs and capacities are available for cultivation of biomass. Procedures for immobilisation of biomass have also been developed.

An aerobic growth biological treatment process has been developed at the Homestake Mining Company in the USA for the degradation and decomposition of cyanide wastes. Algae have been effectively used to treat effluents from lead mining and milling in Missouri, USA. Green algae promoted by nutrients in sewage effluents could enhance biosorption of several metals.

POTENTIALS AND PROSPECTS

The potentials for biomineral beneficiation are enormous with significant commercial implications. Since the useful microorganisms relevant to minerals processing are present ubiquitously in ore deposits and mine waters, harvesting and utilizing them become simpler and practical. The utility of microbes in a variety of beneficiation process such as flotation, selective leaching, flocculation as well as metal sorption and accumulation has been illustrated with examples. The following applications are noteworthy :

- Beneficiation of complex sulphides with special reference to selective flotation
- Desulphurisation and deashing of coals.
- Desulphurisation of oils
- Selective flocculation of mineral fines
- Beneficiation of nonsulphide ores such as bauxite, dolomite, magnesite, phosphates, fluorites, manganese and iron oxides through selective leaching of undesirable constituents
- Generation of surface active reagents as well as flotation modifiers and depressants
- Conversion of one mineral form to another to facilitate adaptability to a chosen process and production of desirable compounds
- Detoxification of solid and liquid wastes.

Recent developments in mineral biotechnology have paved an entirely new path not only for mineral beneficiation and extraction of metals but also for effective and cheap environmental control. Biomineral processing is poised for a big leap forward in the coming decade and may revolutionise future mineral processing operations, the world over. The future scenario in this context is loaded with what may now appear as exotic possibilities.

- Genetically engineered super strains will be developed in laboratories for enhancing microbe-mineral interactions.
- Specially designed bioreactors will be used to cultivate special strains.
- Reagents for minerals processing will be generated through biogenic reactions. Biosorbents, biosurfactants and bioflocculants will be commercially produced and would replace the existing costly, environmentally hazardous chemicals.
- Microbes will be increasingly used to modify surface properties of minerals to facilitate flotation, flocculation and selective leaching.
- Bioenvironmental control would assume greater practical acceptance. Highly toxic solid and liquid wastes would be efficiently biodegraded in an environmentally acceptable manner.

REFERENCES

1. Busscher, H.J., Sjollema, J., Van der Mei, H.C.: In Doyle, R.J. and Rosenberg, M.(Eds.): Microbial Cell Surface Hydrophobicity. Washington, D.C.: American Society for Microbiology 1990, pp. 335-346.
2. Irvin, R.T.: In Doyle, R.J. and Rosenberg, M. (Eds.): Microbial Cell Surface Hydrophobicity. Washington, D.C., American Society for Microbiology, 1990, pp.137-177.

3. Preston Devasia, Natarajan, K.A., Sathyanarayana, D.N. and Ramananda Rao, G.: App. Environ. Microbiology 59 (1993) 4051-4055.

4. Natarajan, K.A.: In Ehrlich, H.L., Brierley, C.L. and Brierley, J.A., (Eds): Microbial Mineral Recovery. New York : McGraw Hill 1990, pp. 79-106.

5. Yelloji Rao, M.K., Natarajan, K.A. and Somasundaran, P. : In Smith, R.W. and Misra, M.(Eds): Mineral Bioprocessing. Warrendale, PA : The Minerals, Metals & Materials Society 1991, pp. 105-120.

6. Yelloji Rao, M.K., Natarajan, K.A. and Somasundaran, P.: Minerals and Metallurgical Processing 9 (1992) 95-100.

7. Dugan, P.R. and Apel, W.A.: In Murr. L.E., Torma, A.E. and Brierley, J.A. (Eds.): Metallurgical Application of Bacterial Leaching and Microbial Phenomena. New York: Academic Press 1978, pp. 223-250.

8. Silverman, M.P., Rogoff, M.H. and Wender, I.: Appl. Microbiol. 9 (1961) 491-496.

9. El Zeky, M.A. and Attia, Y.A.: Coal Preparation, 5 (1987) 15-37.

10. Dogan, M.Z.: In Karavaiko, G.I., Rossi, G. and Avakyan, Z.A.(Eds): Biohydrometallurgy. Moscow: Centre for Int. Projects 1990, pp. 275-281.

11. Capes, C.E., Mcllhinney, A.E., Sirianni, A.F. and Puddington, E.I.: Can. Inst. Min. Met. Bull. 66 (1973) 88- 91.

12. Townsley, C.C., Atkins, A.S. and Davis, A.J.: Biotech. Bioengg. 30 (1987) 1-8.

13. Solozhenkin, P.M. and Lyubavina, L.L.: In Karavaiko, G.I. and Groudev, S.N. (Eds.) : Modern Aspects of Microbiological Hydrometallurgy. Moscow: Centre for Int. Projects 1985, pp. 409-414.

14. Solozhenkin, P.M., Lyubavina, L.L., Sherepova, S.A. and Lyalikova, N.N.: In Rossi, G. and Torma, A.E.: (Eds.) : Recent Progress in Biohydrometallurgy. Roma: Associazione Mineraria Sarda 1983, pp. 705-715.

15. Ehrlich, H.L.: GeoMicrobiology. New York: Marcel Dekker 1990.

16. Groudev, S.N., Groudeva, V.I. and Petrov, E.C. : Travaux ICSOBA 13 (1983) 249-256.

17. Karavaiko, G.I., Belkanova, N.P., Eroshchev-shak, V.A. and Avakyan, Z.A. : Mikrobiologiya 53 (1984) 976-981.

18. Avakyan, Z.A., Belkanova, N.P., Karavaiko, G.I. and Piskunov : Mikrobiologiya 54 (1984) 301-307.

19. Mehta, A.P., Torma, A.E. and Murr, L.E. : Biotech. Bioengg. 21 (1979) 875-885.

20. Karavaiko, G.I., Avakyan, Z.A., Ogurtsova, L.V. and Safonova, O.F. : In Salley, J., McCready, R.G.L. and Wichlacz, P.L. (Eds): Biohydrometallurgy. Ottawa : CANMET 1989, pp. 93-102.

21. Karavaiko, G.I. and Rossi, G. (Eds.): Bio-Geotechnology of Metals. Moscow: Centre for Int. Projects, 1988.

22. Smith, R.W., Misra, M. and Dubel, J: Minerals Engineering 4 (1991) 1127-1141.

23. Misra, M., Smith, R.W., Dubel, J. and Chen, S.: Minerals and Metallurgical Processing 10 (1993) 20-23.

Electroflotation

S. VENKATACHALAM
Department of Metallurgical Engg. & Materials Science,
Indian Institute of Technology, Bombay 400 076 India.

ABSTRACT

Electroflotation is based on the use of electrogenerated hydrogen/oxygen gas bubbles instead of air normally used in the conventional flotation process. Presently, this technique is being used for purification of municipal and industrial waste waters. Considerable amount of work has been done to use the electroflotation process for the beneficiation of low grade, finely disseminated ores. The fundamental and applied aspects of this technique have been critically reviewed. The dynamic process of electrogeneration of gas bubbles as affected by a number of variables and the charge on gas bubbles has been discussed. Application of electroflotation to processing of fine particles in a variety of mineral systems and the advantages and limitations of this technique are presented.

INTRODUCTION

Electroflotation is nothing but flotation using electrolytically generated gases. Electrolysis of an aqueous solution releases hydrogen and oxygen gases which can be utilised either separately or in combination, instead of air in conventional flotation. Is there any specific advantage of H_2 or O_2 in preference to air? How do the different mineral systems respond to hydrogen or oxygen? Apart from the chemical effect of the gas, if any, what is the size distribution of the electrogenerated gas bubbles and what are the charges on these gas bubbles? Is there any other special advantage in employing electroflotation? Mineral beneficiation apart, what are the other areas where electroflotation can be made use of? It is essential to know the answers to these questions before an assessment can be made with regard to the use of electroflotation in preference to conventional flotation. In this review an attempt has been made to consider the fundamental and applied aspects of electroflotation. The physical process of gas evolution as influenced by a number of variables, the influence of size and charge of particles and bubbles in flotation and electroflotation studies on a variety of mineral systems have been critically reviewed.

THE PHYSICAL PROCESS OF GAS EVOLUTION

Electrolytic gas evolution is a dynamic phenomenon. The interaction among the process variables affects it. The electrode, its potential, the electrolyte, small amounts of impurities, current density, electrode morphology and pretreatment affect the dynamics of gas evolution. Since bubble evolution is very fast, special techniques are required for its study. The physical process of gas evolution can be divided into three stages: nucleation, growth and detachment. The first step in the gas evolution is nucleation. Bubbles of a critical radius and larger grow, while, bubbles having a radii less than this diameter tend to decay [1]. With regard to electrogenerated gas bubbles, we have to discuss the influence of current or rather the current density at the electrodes. If the current density is very low, the dissolved gas is removed from the electrode in the direction of the liquid bulk by molecular diffusion and convection and no gas phase is formed. If the dissolved gas reacts homogeneously and quickly like chlorine, this might happen even at high current densities. If the current density is sufficiently high, gas bubbles form at predetermined nucleation sites on the electrode surface. Adjacent to the electrode, sufficient supersaturation of the liquid would have to occur prior to nucleation [2].

Real surfaces contain pits, scratches and grooves of various sizes which act as nucleation sites. Westerheide and Westwater [3] reported that nucleation on their microelectrodes occurred at preferred sites such as pits and scratches. Janssen and Hoogland [4] observed that bubbles nucleated on a rotating platinum wire at specific sites that depended on pretreatment as well as current density.

Bubble growth takes place after nucleation. Bubbles grow initially from the critical radius by expansion due to the high internal pressure and by mass transport of dissolved gas to the gas/liquid interface. Scriven [5] using a similarity transform has provided a general analysis of the diffusion controlled growth of a bubble.

Westerheide and Westwater [3] studied the evolution of electrolytic hydrogen bubbles and quantitatively compared their growth data to the diffusive square root of time growth dependence. They found agreement for a single bubble but multiple bubbles interfered with each other's growth. Their work established the importance of mass transfer of dissolved gas to the bubble surface as the asymptotic mechanism by which bubbles grow before and between coalescence. Glas and Westwater [6] extended this work to include oxygen, chlorine and carbon dioxide bubbles at various electrode materials. Coalescence in gas evolution is also important. The validity of the growth law is restricted to where the adhering bubbles do not mutually interfere at moderate current densities. As soon as singular bubbles touch each other while growing, coalescence may occur. Smaller bubbles are swallowed by larger ones. Another reason for coalescence is the sliding of the big bubbles upward along the electrode while embodying small bubbles in a kind of scavenging effect. Hydrogen bubbles produced in acid medium have been shown to grow by such a mechanism. Coalescence occurs within short times of about 10^{-4} s. Sides and Tobias [7] have proposed a cyclic mechanism of oxygen bubble growth in a basic medium. The period of the cycle at a current density of 100 ma/cm^2 was approximately 0.1 sec. It may be noted here that coalescence seems to be unimportant in hydrogen evolution from basic solutions because the bubbles evolved are quite small. Furthermore, large bubbles formed by coalescence need not detach from the electrode but at least under certain conditions adhere to the electrode while smaller bubbles depart.

The growing bubble remains attached to the surface of the electrode until a sufficient size is attained. Kabanov and Frumkin [8] have found that the buoyant gas bubbles detach when the surface adhesive forces related to bubble contact angles can no longer restrain them. The departure diameter of a given bubble cannot be predicted with our present knowledge. Horizontal or vertical positioning of the electrodes also affects the bubble size.

Size Distribution of Electrogenerated Bubbles

The size of the electrogenerated bubble is dependent on several variables. Fig. 1 shows the various parameters affecting the bubble size. Electrode material influences the bubble size. Venczel [9] has noticed that large bubbles are formed on platinum because the bubbles grew uniformly and coalesced, whereas, small ones formed on copper and iron because the bubbles detached from the electrode surface before reaching the size for coalescence. The influence of cathode material on the size of hydrogen bubbles is extremely pronounced in acidic media while the effect is less marked in alkaline media.

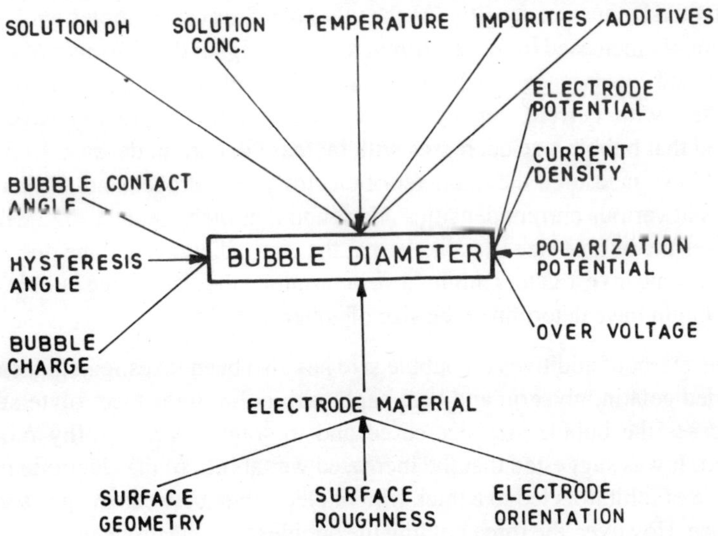

Fig. 1. Parameters affecting bubble size.

In neutral media the size of the hydrogen bubble is independent of the cathode material. The following series shows the change in the bubble size as influenced by the electrode material [10].

H_2 bubble size increases

$$\xrightarrow{\hspace{6cm}}$$

Pb, Sn, Cu, Ag, Fe, Ni, W, Pd, Pt

$$\xleftarrow{\hspace{6cm}}$$

O_2 bubble size increases

Kabanov and Frumkin [8] have found that there was a direct relationship between the bubble size at detachment and the electrode potential and that the connection between the two was the variation of interfacial tension with potential. The size of the electrogenerated gas bubbles has been found to depend on the surface geometry of the electrode also. The bubble size has been found to increase with increase in the thickness of the wire mesh. Experiments with wire electrodes have shown that the bubble diameter depends on the angle of curvature of the electrode.

The influence of current density on the bubble size distribution is quite significant. For the same electrode area when the current is increased the current density and the amount of gas evolved increase. There is some contradiction in the literature regarding the influence of current density on bubble size. Janssen and Hoogland [11] found that with platinum electrode when the current density was increased past 10 ma/cm^2 all the hydrogen bubbles evolved in base increased in size probably due to increased coalescence. Fig. 2 shows that hydrogen bubbles evolved in alkaline medium are the smallest and their size is not affected by current density. On the other hand, oxygen bubbles evolved in base quickly increased in size at current densities higher than 30 ma/cm^2 owing to increased coalescence. On the other hand, Venczel [9] found that bubble size decreased with current density. Several other workers [12-14] have also reported that bubble size decreases with increase in current density. Ketkar et al. [14] have measured the diameter of electrolytic hydrogen and oxygen gas bubbles at various current densities and found that increase in current density decreases the bubble size. It is likely that the contradictions may be due to the fact that some investigators might have determined the size of free bubbles and others might have determined the size of adhering bubbles.

The effect of additives on bubble size has also been investigated. Venczel [9] added gelatin, glycerin and beta naphthachinolin to the electrolyte and in most cases the bubble size decreased and in some cases a frothy mixture resulted. It was suggested that the increased wettability of the electrode in the presence of inhibitors led to a thick film of electrolyte between the gas and the electrode. However, the force holding the bubble to the electrode was probably weakened proportionately with the reduction in both the perimeter of the

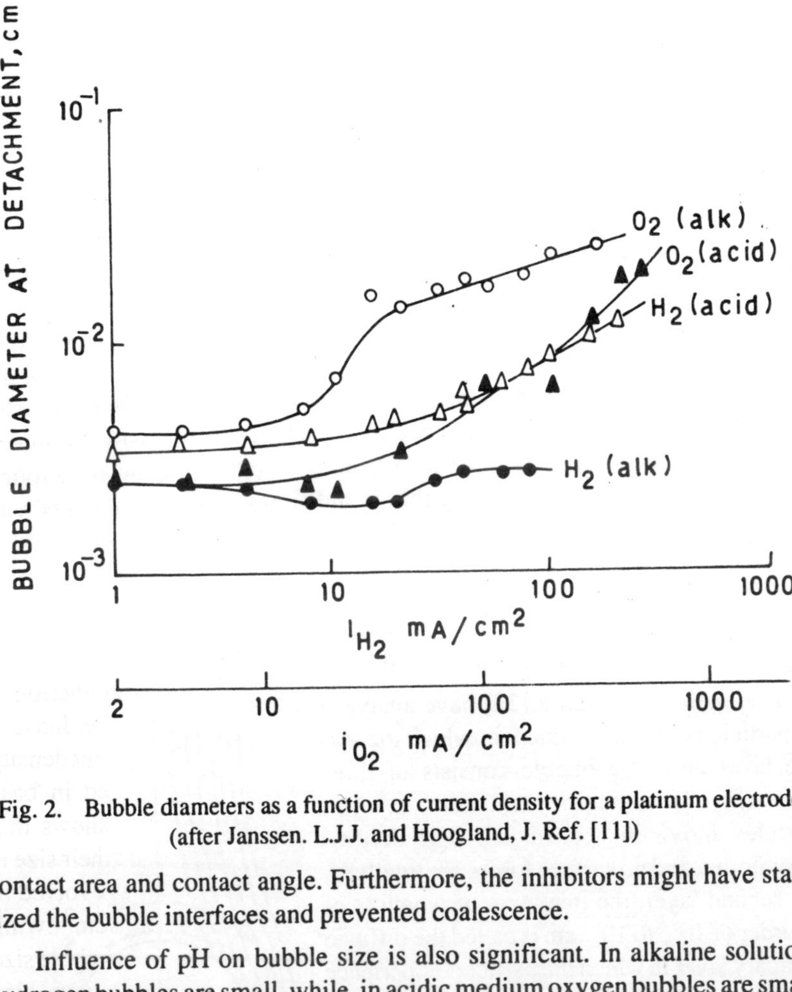

Fig. 2. Bubble diameters as a function of current density for a platinum electrode. (after Janssen. L.J.J. and Hoogland, J. Ref. [11])

contact area and contact angle. Furthermore, the inhibitors might have stabilized the bubble interfaces and prevented coalescence.

Influence of pH on bubble size is also significant. In alkaline solutions hydrogen bubbles are small, while, in acidic medium oxygen bubbles are small. Probably, the phenomenon of coalescence is not important in hydrogen evolution in basic medium which is responsible for the production of small hydrogen bubbles. On the other hand, in alkaline solution coalescence of oxygen bubbles occurs frequently, practically each collision leading to coalescence. The diameter of oxygen gas bubbles evolved in alkaline solution increases suddenly at about 30 ma/cm^2 due to increasing coalescence.

The growth kinetics of hydrogen, oxygen and chlorine gas bubbles electrogenerated at microelectrodes has been studied by Brandon and Kelsall [15]. Three modes of growth of individual bubbles on micro electrodes have been observed during their lifetime: liquid inertia control, then diffusion control followed by 'direct injection' when the bubble to electrode diameter ratio was large. The bubble departure diameter in surfactant free electrolyte was found to be pH dependent. The departure diameters of hydrogen bubbles were large at low pH and much smaller at high pH, the converse of the behaviour of oxygen and chlorine bubbles. These trends confirmed earlier reports although the marked change in the slopes of the departure diameter-pH relationships and their crossing at pH 2-3 has not been reported previously.

TREATMENT OF MINERAL FINES

The unsuitability of conventional flotation techniques for the treatment of mineral fines has been attributed mainly to the large bubble size. The behaviour of fine particles in a flotation cell involves complicated hydrodynamics. The role of particle size in flotation has been reviewed by a number of workers [16-21].

Two stages can be considered in the flotation process. These are the collision of the particle and bubble and the attachment of the particle on the bubble. The attachment of the particle to an air bubble in liquid media depends upon the rupture of the intervening thin liquid film between air and particle surfaces. However, particle bubble attachment in flotation is characterized by the process occurring under dynamic conditions in which not only the surface forces but also hydrodynamic factors should be taken into consideration. The complexity is increased by the interfacial electrical conditions around a moving bubble which differs considerably from those of a moving solid.

Derjaguin and Dukhin [22] have analyzed the particle bubble attachment in which the surface layer around a bubble consists of three layers. (Fig. 3). In the first and outermost layer, particles move in accordance with hydrodynamic law and no surface forces are involved. The second layer, the thickness of which is of the order of 10^{-2} to 10^{-3} cm is called the diffusio boundary layer in which the particles experience diffusiophoretic forces depending on their surface charge and electrical field in this layer. In the third and innermost layer, particle motion is governed by surface forces which consist of double layer interaction forces and molecular interaction forces. It is necessary for particles to come into the third layer to achieve particle-bubble attachment. Relatively large particles move under the effect of inertial force in an almost rectilinear way until it collides with the bubble surface. While the large particles can enter the second layer, the small particles travel along the streamline and cannot enter the second layer from the first layer.

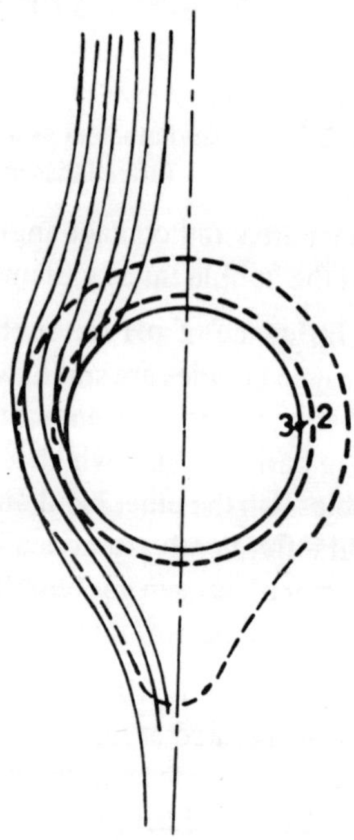

Fig. 3. Schematic representation of a bubble rising in a liquid medium.

Attempts to develop new techniques in the flotation of fine particles can be categorized into two approaches: the processes which are based primarily on increasing the probability of collision between gas bubbles and fine mineral particles and those which are based on more favourable changes in the energetics of bubble – particle contact. The probability of collision between bubbles and mineral particles increases when the particles are agglomerates or attached to hydrophillic particles which act as carriers. Agglomeration of fine particles can be enhanced by chemisorbing collectors.

In column flotation, counter current flow of mineralized bubbles and liquid in the upper zone of the column helps removal of entrained gangue particles from the bubble. Vacuum or pressure release flotation directly nucleates bubbles on surface of hydrophobic minerals thereby eliminating bubble – particle collision.

Advantages of Electroflotation

Production of fine bubbles in the flotation cells can increase the particle – bubble contact. One of the methods employed to produce fine bubbles is electrolysis. Flotation by electrogenerated gas bubbles, called as electroflotation, has been reviewed by a number of workers [23-26]. It has several advantages compared to conventional flotation.

Electroflotation produces extremely finely dispersed bubbles in the range of 10-100 microns depending on the electrolysis conditions as compared to 2 mm bubbles in pneumatic cells and about 0.8-0.9 mm bubbles in conventional mechanical cells. In addition to size, bubble flux and bubble number density are also important. Bubble flux may be defined as the number of bubbles per unit area per unit volume. Bubble number density may be defined as the number of bubbles per unit volume. It is also possible to use hydrogen or oxygen separately or in combination. Air can also be used along with finer bubbles of hydrogen and oxygen so as to ensure that the bigger particles are floated by the bigger air bubbles and the fine ones by finer bubbles. Parameters like current density, pH and electrode materials with different geometries can be suitably chosen to yield the desired bubble size distribution.

By varying current density, it is possible to create the desired gas bubble saturation in the flotation medium as well as in the flotation chamber not usually achieved in conventional cells. During electroflotation, it is also possible to obtain a continuous adjustment of the bubble dimension. Electrogeneration of bubbles can give a narrow size distribution.

Finely dispersed electrolytic gases are not only capable of floating very small particles, but can also activate the flotation of the larger particles when the electrolytic gases are used in combination with air bubbles. The improvement in the adhesion of large bubbles to mineral particles when these are previously adhered to small bubbles is well known in conventional flotation.

The advantage of electroflotation lies not only in the finer bubble sizes produced but also in the ability of hydrogen and oxygen bubbles generated by electrolysis to alter the flotation properties of the mineral to a great extent than the molecular gases. This is because the nascent gases are more effective. The thin film of water between the mineral and the gas can be easily removed by the smaller bubbles because of the high capillary pressure inside them. The electrolytic gas can either activate or passivate the mineral surfaces which can give rise to an improvement or deterioration in the flotation properties of the minerals. The oxidation-reduction changes that can occur on the surfaces of some minerals with electrolytic gases is very striking and these are discussed in a later section.

Technology of Electroflotation

The use of electrogenerated bubble for flotation was suggested by Elmore [27] as early as 1904. However, attempts to use this process in some beneficiation plants in Broken Hill, Australia were not successful. Interest in this method was revived almost 50 years back both in Germany and in the Soviet Union. Later, several other countries also showed interest. A large number of papers are available about the Russian work. Although this technique did not make a breakthrough in mineral beneficiation, in the field of waste water treatment in silk and meat industries, leather industries, oil refining etc. electroflotation has been successful.

Rivkin et al. [28] developed several designs of laboratory flotation cells. The anode was installed at the bottom of the cell and the cathode was placed either in the pulp surface or vertically above the anode. The electrolytic preparation of the slurry was also carried out in the flotation cell simultaneously with flotation. The rate of flotation increased significantly under the influence of gases produced by electrolysis. Electroflotation of chalcopyrite, zinc blende and pyrite were studied by Zozulya [29] who concluded that the flotation rate was increased by a factor of 2 or more. Bergmann [30] reported the separation of limonite and quartz in an electric field at pH 4. Iron was used as the cathode and the copper rod as the anode. The iron ions formed by the dissolution of limonite migrated towards the cathode, whereas, the negatively charged quartz migrated to the anode. Activation of quartz by iron ions was thus prevented. Results of the electroflotation of manganese minerals like pyrolusite and psilomelane was reported by Romanov et al. [31]. They reported that the recovery of manganese with hydrogen bubbles was 92 to 95% and was greater than that achieved in column flotation. The possibility of using aluminium, iron and magnesium as natural supporters during electroflotation of uranium was also studied. It was found that the hydroxides of aluminium and iron acted as carriers of uranium hydroxide. Malco [32] has published nomograms for selection of optimal electroflotation conditions. The effectiveness of electroflotation for the selective separation fine minerals – pyrite, manganese minerals, diamond has been studied by Glembotsky et al. [33]. Nebara et al.

[34] obtained high recoveries of ferrocyanides, oxyquinolates and hydroxides of copper, nickel, cobalt, titanium, zinc and molybdenum by electroflotation using a mixture of hydrogen and oxygen over a wide pH range above 2.5.

Mamakov et al. [35]. studied the electroflotation of cassiterite with hydrogen and oxygen. A diaphragm cell was used. C-7 to C-9 aliphatic acids were used as collectors. Hydrogen was found to be more effective than the mixture of gases. Electroflotation of cassiterite using a mixture of hydrogen and oxygen was also studied by Hogan et al. [36]. Conditions for optimization of the process and overall energy requirements were deduced. Although the upgrading was better than conventional flotation, the actual energy consumption was about 50 times greater than conventional flotation.

In an earlier section, it was pointed out that oxidation-reduction changes may be brought about on the mineral surfaces by the use of electrolytic oxygen and hydrogen respectively. Hydrogen has been found to be better for cassiterite flotation. But, hydrogen and oxygen did not have any significant effect on the surfaces of certain oxygen-salt minerals like barite, calcite and chromite. Their floatability depended mainly on the bubble dimension and pH.

On the other hand, the influence of gases was very significant in the flotation of sulphide minerals. It has been found that electrolytic oxygen gave 98% recovery of pyrite whereas hydrogen achieved practically no flotation [33]. The activating effect of oxygen was so great that it can be successfully used for flotation even in an alkaline medium. Electrolytic oxygen made the surface of pyrite, chalcopyrite and sphalerite particles so strongly hydrophobic that there was no need to use a collector also [33]. It is pertinent to note that the effect of electroflotation differs greatly from one sulphide to another and it is difficult to predict the floatability. Flotation of chalcocite with electrolytic oxygen decreases the recovery [33].

Bhaskar Raju et al. [37] have studied the electro-flotation of chalcopyrite fines with potassium ethyl xanthate as collector. Experiments were carried out either hydrogen and oxygen separately. It was observed that electroflotation with oxygen was very effective. Electroflotation experiments were also done with chalcopyrite fines using sodium diethyl dithio carbamate [38]. The mineral surface products were examined by UV-Visible and IR spectrophotometric methods and indicated the presence of copper diethyl dithio carbamate complex. The qualitative nature of the surface product was found to be unaffected by electroflotation.

Ledesma and Guzman [39] have reported electro-flotation studies on chalcocite. Electrogenerated hydrogen oxygen mixture was used using isopropyl ethyl thiocarbamate as collector. Results were compared with conventional flotation using nitrogen. For the fine particles, electroflotation gave about 30% more recovery than conventional flotation.

Rao et al. [40] have studied electroflotation of fine anglesite in hydrogen and oxygen. Flotinor S-72, an alkyl sulphate of sodium and potassium amyl-xanthate were used as collectors. With the former collector more than 95% recovery could be obtained using oxygen. Potassium amyl xanthate gave a maximum recovery of only 75%. It was presumed that Flotinor S-72 must have a strong adsorption affinity in acid medium towards the anglesite surface.

Electroflotation of copper sulphides has also been attempted from thickener overflow [41]. When calculating thickener size, it is usually assumed that the maximum size of particles carried away with overflow should not exceed 3-5 microns. The thickener overflow normally contains finely dispersed mineral particles the recovery of which in some cases would make it possible to increase the total recovery at the concentrator by 1-3%. Tests have shown that all the sulphide copper lost in the thickener overflow may be recovered. Magnesiun hydroxide particles precipitated in the electroflotation process when the pH increased to 9.5 to 10.5 served as a carrier. Continuous tests in electroflotation of copper on thickener overflow gave a recovery of about 99%.

Ketkar et al. [42] studied the electroflotation of quartz fines of various sizes using electrogenerated hydrogen and oxygen bubbles. The effect of change in the diameter of hydrogen and oxygen gas bubbles on the recovery of quartz fines have been investigated. Changes in the electrode surface geometry and the current density changes the bubble diameter as well as bubble flux influencing the recovery [Fig. 4].

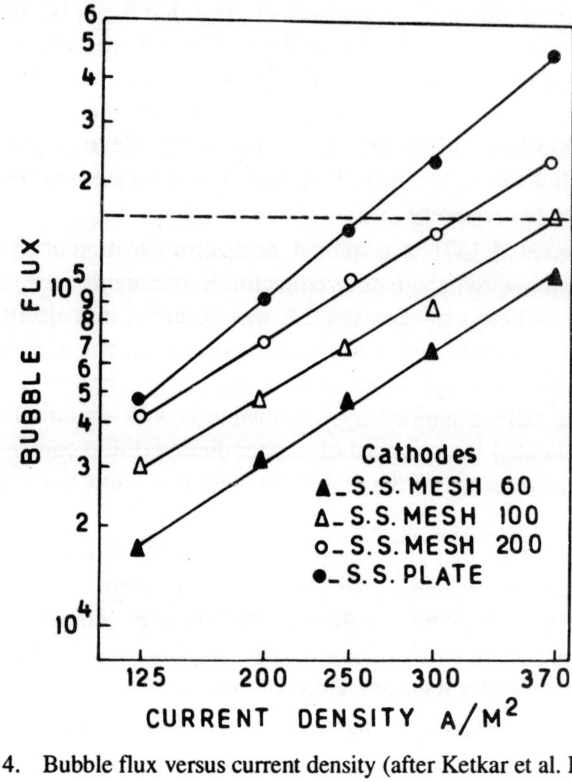

Fig. 4. Bubble flux versus current density (after Ketkar et al. Ref. [42]).

Khosla et al. [43] have studied the effect of interrupted current (pulsed) electrolysis on the generation of gas bubbles. A microcomputer controlled current source designed to generate the required pulses has been described along with the typical results obtained using the system. It has been shown that a decrease in the duty cycle at a given pH and average current density causes an increase in fine sized bubbles and consequent increase in bubble flux.

Flotation of alumina particles has been carried out by Khosla et al. [44] employing electrogenerated hydrogen and oxygen bubbles. The effect of current density was studied over a range of 0.1 to 1.6 kA/m^2. The results suggested an optimum current density for a given particle size distribution. At a given gas evolution rate, oxygen bubbles resulted in higher flotation recovery as compared to hydrogen. For a given particle size distribution, the rate flotation of alumina particles increased with increasing current density. Although flotation with electrogenerated gas at low current densities was kinetically poor, utilization of the gas was better at lower current densities as compared to electroflotation at higher current densities.

Electroflotation of quartz fines using dodecyl amine as a collector has been studied by Setty [45]. Hydrogen and oxygen bubbles generated from a stainless steel plate cathode and platinum anode under various experimental conditions were investigated. Flotation recovery of quartz was found to increase with current density. For all the particle sizes and all current densities investigated, flotation recovery increased with finer mesh electrodes. Electrolytic and conventional flotation tests were compared under two conditions. In one case the total bubble surface area and in the other case the total number of bubbles were kept identical. When the total surface area of bubbles was kept identical electroflotation with hydrogen gave considerably higher recovery followed by gases obtained through glass frits G2 and G1 revealing the importance of the bubble size, bubble flux and bubble number density. On the other hand, flotation recoveries obtained using gas bubbles produced through frits G4 and G5 were slightly higher than those obtained using electrogenerated hydrogen, when the total number of bubbles were kept identical.

CHARGE ON GAS BUBBLES

Electrokinetic phenomena at liquid-gas interfaces have not been fully explored as those at solid-liquid or liquid-liquid surfaces, mainly due to the difficulties in experimentation. Gas bubbles are reported to be generally negatively charged in pure water and in solutions of inorganic electrolytes. This charge can play an important role in flotation. Hence, the determination of the charge on the surface of a gas bubble is of interest. Over the past 100 years a number of attempts have been made to measure the electrophoretic mobility of gas bubbles. All these methods had drawbacks. Collins et al. [46] have used a Rank Brothers particle microelectrophoresis apparatus to measure the charge on small gas bubbles. The measured value of the bubble zeta potentials showed them to be in the range of 50-64 mV. The charges were of the same sign and magnitude as those of polystyrene latex particles with the same electrolyte and surfactant concentration.

Usui and Sasaki [47] measured the Dorn effect or the sedimentation potential created when finely divided argon gas bubbles are allowed to rise through aqueous solution of surfactants such as sodium hexadecyl sulphate, polyoxyethylene dodecyl ether and 1-butanol. The zeta potential of bubbles in distilled water showed a mean value of – 15 mV, probably due to preferential adsorption of OH^- ions. With increasing SHS concentration, negative zeta potential increased. With HTAB solution, the zeta potential increases its positive value with surfactant concentration. With non-ionic surfactant solutions, the zeta potential of bubbles remained nearly constant. Zeta potential in mixed solutions of ionic and non-ionic surfactants have also been carried out. Higher zeta potentials in the presence of 1-butanol seem to indicate an enhanced adsorption of ionic surfactant through the coadsorption of 1-butanol at a gas-solution interface.

Usui et al. [48] have also studied the Dorn effect caused by the rising of finely divided argon gas bubbles of varying sizes in sodium hexadecyl sulphate, butanol aqueous solution and distilled water. It was noticed that the negative zeta potential increased with decreasing bubble size in each system.

No information so far was available as to how the charge of the bubble depends on inorganic salts. Recently Li and Somasundaran [49] have studied the electrokinetic behaviour of nitrogen gas bubbles in aqueous electrolyte solutions. It was found that bubbles were negatively charged in NaCl solutions, whereas, under certain conditions, they became positively charged in $MgCl_2$ solutions. The charge reversal in the pH range 9-11 was attributed to the precipitation of $Mg(OH)_{2\,(s)}$ at the gas liquid interphase, while, the reduction in the magnitude of the zeta potential of the bubbles observed in the acidic pH range was considered to be due to the specific adsorption of Mg^{2+}.

The effect of trivalent aluminium ions on the zeta potential of bubbles has also been studied [50]. The bubbles were found to exhibit unusual positive surface charge characteristics in Al^{3+} solutions. The reversal of bubble charge has been attributed to specific adsorption of Al^{3+} and its hydroxo complexes at the gas-liquid interface in the low pH range and to precipitation of $Al(OH)_3$ in intermediate pH range.

WASTE WATER TREATMENT

The removal of suspended substances from effluents is important in waste water treatment from the environmental view point. Electroflotation is ideal in waste water treatment compared to conventional treatment. In the small and compact electroflotation apparatus, the amount of electrogenerated gas can be controlled, corresponding to the amount of flocs. Since fine bubbles can be generated, even small flocs can be collected and floated. It is also important to note that the temperature of waste water does not affect the generation of gas bubbles. Further, since oxidation occurs due to electrolysis, in addition to gas bubble generation, it can reduce BOD and COD.

Although electroflotation has been applied to the treatment of domestic sewage in 1911 in US, there were problems because electrodes tended to scum and the scale on the electrode surface prevented the passage of electric current. In Japan, a few companies are reported to have solved these problems by using an alloy of aluminium as electrodes. Fukui and Yuu [51] using polystyrene lattices of mean diameter 0.6 micron have studied the effect of charge on both particles and bubbles on the total collection efficiency during electroflotation. The results clearly demonstrated that the collection efficiency depends on the charge on both the particle and bubble. The theoretical total collection efficiency has been found to be in close agreement with the experimentally determined values. This work was subsequently extended [52] where the surface properties were kept unchangeable by conducting the experiments with cationic surfactant CTAB whose concentration was maintained at 90% of the initial concentration during the run. Hence the Hamaker constant and zeta potential of the particle and bubble were considered to be constant. Fukui and Yuu [53] have also developed a new apparatus for electroflotation using the impeller of stirred tank as electrode(s) and changing the speed of rotation of the electrode(s). While flocculation was produced by the rotation of the electrode, electrogenerated gas bubbles float the flocs.

Waste water from many metallurgical industries contain various metals and these could be effectively removed by electroflotation. The metal could be precipitated and subsequently separated by electroflotation. Minute quantities of metal value from plating baths can also be recovered by electroflotation. Using aluminium as an anode, waste water from cadmium plating bath has been clarified. The $Al(OH)_3$ formed at the anode has been found to be a strong flocculant. Processing of waste water from leather industry, textile industry, sugar industry and meat processing industry could also be done using electroflotation. These have been reviewed [23]. Matov [54] has described a consumable anode electroflotation cell. The influence of pH value on the effectiveness of electroflotation purification of effluents has also been reported.

Electroflotation behaviour of metal ions such as Cu, Ni, Zn and Cd using ammonium dodecyl dithiocarbamic acid ligand, the chelating surfactant, has been investigated [55]. The maximum recovery was as high as 95%, except for copper, for which it was from 85 to 90%. Since all the metal ions studied showed very high recovery values, without any selectivity, separation of these ions from one another is not possible by flotation.

SUMMARY AND CONCLUSIONS

Electroflotation is already being successfully used in the purification of municipal industrial waste waters. However, electroflotation studies on mineral systems have not progressed much from laboratory scale experimentation. In the processing of mineral fines, the amount of solids to be separated is small and also this removal has to be selective.

Although, the role of particle size in flotation has been discussed in considerable detail in flotation literature, not much work has been reported on the characterization of bubbles. Study of the effect of bubble size and charge on the various sub processes in flotation deserves immediate attention. Further, in view of the high energy requirements, the special advantages of electrogenerated bubbles in contrast to fine bubbles generated using other techniques have to be clearly brought out before electroflotation is applied to any mineral system.

REFERENCES

1. Sides, P.J.: In White, R.E. Bockris, J.O'M and Conway, B.E. (Eds): Modern Aspects of Electrochemistry, 18. New York : Plenum Press 1986, p. 303.
2. Vogt, H: In Comprehensive Treatise on Electro-chemistry 6, 1983, p. 445.
3. Westerheide, D.C. and Westwater, J.W.: Am. Inst. Chem. Engrs., J. 7 (1961) 357.
4. Janssen L.J.J. and Hoogland, J.G. : Electrochim. Acta 15 (1970) 1013.
5. Scriven, L.E. : Chem. Engg. Sci. 27 (1959) 1753.
6. Glas, J.P. and Westwater, J.W. : Int. J. Heat Mass Transfer 7 (1964) 1427.
7. Sides, P. and Tobias C.: J.Electrochem. Soc. 132 (1985) 583.
8. Kabanov, B.N. and Frumkin, A.N. : Z. Phys. Chem. 165 (1933) 539.
9. Venczel, J.: Electrochim Acta 15 (1970) 1909.
10. Mamakov A.A : In Nebera, P.V. (Ed). : Modern State and Perspective of Electrolytic Flotation. Kishinev 1975, pp. 3-66.
11. Janssen, L.J.J. and Hoogland, J.G.: Electrochim Acta 18 (1973) 543.
12. Glembotsky, V.A. Mamakov A.A., and Sorokina, V.N.: Electron. Obrab. Mater. 5 (1973) 66.
13. Ibl, N. and Venczel, J. : Metalloberflache 24 (1970) 365.
14. Ketkar, D.R., Mallikarjunan, R. and Venkatachalam, S. : J. Electrochem. Soc. India 37 (1988) 314.
15. Brandon, N.P. and Kelsall, G.H. : J Appl. Electrochem. 15 (1985) 475.
16. Somasundaran, P (Ed) : Fine Particle Processing, AIME, Vols. I & II, 1980.
17. Fuerstenau, D.W :In Somasundaran, P (Ed) : Fine Particle Processing, AIME, Vol. 1, 1980, p. 669.
18. Trahar, W.J. : Int. J. Min. Processing 8 (1981) 289.
19. Trahar, W.J. and Warren, L.J. : ibid 14 (1985) 195.
20. Chander, S. : Trans IIM, 31 (1) (1978) 12.
21. Jameson, G.J. : In Jones, M.H. and Woodcock, J.T. (Ed.) : Wark Symposium, Aust. Inst. Min. Metall. 1984, 215.
22. Derjaguin, B.V. and Dukhin, S.S. : In Laskowski, J (Ed): 13th Int. Min. Process.Congress 1979, pp. 21- 62.

23. Bhaskar Raju, G. and Khangaonkar, P.R. : Trans I.I.M, 37 (1) (1984) 59.
24. Kuhn, A.T : Chem. Process, 20, (June 1974) 9-12, (July 1974) 5-7.
25. Mallikarjunan R, and Venkatachalam, S : In Electrochemistry in Mineral and Metal Processing. Cincinnati, Ohio, 1984, 233.
26. Venkatachalam S: In Mineral Processing and Extractive Metallurgy Review Vol. 8, 1992, pp. 47-55.
27. Elmore, F. : British Patent No. 13578, 1904
28. Rivkin P.M, Olofinsky, N.F, Balabanov, E.V. and Borodin, P.P. : Author's certificate of USSR, No. 66558, 1946.
29. Zozulya, I.I.: Iogu, Akad, Nauk, SSR, Otdel, Nauk, 7 (1953) 946.
30. Bergmann, A.: Bergbau Wissenchaften, 7, No. 14, 1960.
31. Romanov, A.M., Nenno, E.S., Nenno, V.E. and Mamakov, A.A. : Electron. Obrab. Mater. 1 (1973) 54.
32. Malko : Electron. Obrab. Mater. 3 (1973) 33.
33. Glembotsky, V.A, Mamakov, A.A., Romanov, A.M. and Nenno, V.E. : In 11th International Mineral Processing Congress, 1975, 561.
34. Nebera, V.P. Zelentsov, V.I. and Kiselev, K.A. : In Somasundaran, P (Ed): Proc. Int. Symposium on Fine Particle Processing Vol. 1, 1980, 886.
35. Mamakov, A.A., Sorokina, V.N., Avvakumov, M.I. : Electron. Obrab. Mater. 4 (1969) 44.
36. Hogan, P, Kuhn A.J. and Turner J.F.: Trans, Inst. Min. Metall. 88 (1979) 93.
37. Bhaskar Raju, G and Khangaonkar, P.R.: Int.J.Miner Process. 9 (1982) 133.
38. Bhaskar Raju, G and Khangaonkar, P.R.: ibid 13 (1984) 211.
39. Ledesma, H and Guzman, M : In Plumton, A.J. (Ed) : Production and Processing of Fine Particles. Montreal : Canadian Institution of Mining and Metallurgy 1988, 195.
40. Rao G.V. : Erzmetall 40 (1987) 183.
41. Gorodetski, M.E., Yaschenk, N.N, Plaksa, N.E. Mekler, L.I and Golman, A.N. : In Janes, M.J. (Ed) : 10th International Mineral Processing Congress. Institution of Mining and Metallurgy, 1974, pp. 689-707.
42. Ketkar, D.R. Mallikarjunan, R. and Venkatachalam, S. : Int.J. Min. Process. 31 (1991) 127.
43. Khosla N.K., Venkatachalam, S. and Somasundaran, P. : J. Appl. Electrochem 21 (1991) 986.
44. Khosla, K, Khosla N.K., Venkatachalam, S. and Somasundaran, P. : Paper communicated for publication.
45. Setty, S.: Ph.D. Thesis submitted to IIT, Bombay 1993.
46. Collins, G.L., Motarjemi, M. and Jameson, G.J. : J. Collid Interface Sci. 63 (1) (1978) 69.
47. Usui, S. and Sasaki, H. : ibid 65 (1) (1978) 36.
48. Usui, S, Sasaki, H. and Matsukawa, H. : ibid 81 (1) (1981) 80.
49. Li, C, and Somasundaran, P. : ibid 146 (1) (1991) 215.
50. Li, C. and Somasundaran, P. : ibid 148 (2) (1992) 587.
51. Fukui, Y. and Yuu, S.: A.I. Ch. E. J. 31 (2) (1985) 201.

52. Fukui, Y. and Yuu, S.: Chem. Eng. Sci. 35 (1980) 1097.
53. Fukui, Y. and Yuu, S. : ibid 39 (6) (1984) 939.
54. Matov, B.M. : Electron. Obrab. Mater. 4 (1966) 94.
55. Srinivasan, V. and Subbaiyan, M, : Sep. Sci. Technol. 24 (1&2) (1989) 145-150.

Modelling Studies on Fine Coal Treatment Processes

T.C. RAO and B. GOVINDARAJAN
Regional Research Laboratory,
Bhopal - 462 026, India

ABSTRACT

Fine coal beneficiation processes such as water only cyclones, spirals, flotation and oil agglomeration are discussed in the paper. Simple mathematical models developed over the years for improving these processes are also reviewed. The need to fine tune the operating and design variables in order to achieve better performance in the plants is emphasised.

INTRODUCTION

The increasing demand for coking coal has forced the coal mining industry to adopt vigorous mechanised mining techniques resulting in increased amount of fines (–0.5 mm) production. Earlier these fines were mixed with the cleaned coarse coal, but with continuously degrading quality, these fines need to be washed before mixing with the cleaned coarse coal.

Cleaning coal at finer sizes is very difficult and it has been reported that loss of good coal in fine coal washing is 10 times more than that of coarse coal washing [1]. The major problems in fine coal cleaning may be attributed to

- strong influence of feed characteristics (which varies widely) on beneficiation performance of fine coal.
- the low price of end product which economically limits the use of sophisticated cleaning methods.

Within the limits of technical and economical feasibility, improvement in fine coal cleaning can be achieved by

- evolving simple and effective beneficiation methods
- existing process optimisation through systematic data collection and analysis.

In the area of process development much advance has been made especially using gravity concentration [2]. However, not much of attention has been paid so far to develop simple models for fine coal beneficiation processes which can be used to improve the efficiency of existing circuits. This paper reviews available mathematical models for fine coal beneficiation processes.

GRAVITY CONCENTRATION

The application of gravity concentration for fine coal cleaning has been very limited till recently due to the inefficiency of gravity concentration to treat the fines. However, the cost-factor in treating fine coal using the physico-chemical separation such as froth flotation and oil-agglomeration has made the revival of gravity methods resulting in development of novel approaches like Multi Gravity Separator [3] and Kelsey Jig [4]. These modern methods are still being tested only at pilot scale but bear good potential.

Water-only cyclones and spirals are widely used gravity concentration techniques to treat coal fines. Usage of heavy medium cyclones for fine coal cleaning has also been reported in literature [5-7]. Despite encouraging results on pilot plant testing of heavy medium cyclones, industry installations are not common.

Mathematical models available for fine coal cleaning gravity concentration methods are based on generalised distribution curve concept [8]. The products from the gravity concentration method i.e., concentrate and tailings are subjected to sink-float testing which gives the distribution of different specific gravity fractions in the feed to the products. These distribution points are plotted against specific gravity to get "Distribution curve" for any particular test. Fig. 1 shows a typical distribution curve of a gravity concentration operation as an

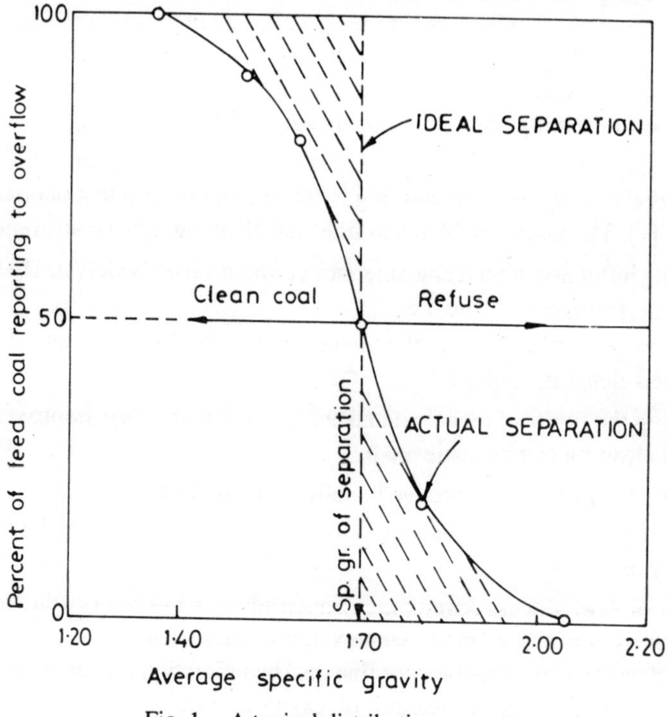

Fig. 1. A typical distribution curve

example. From the distribution curves, two important performance parameters, cut specific gravity (SG50) and probable error (E_p) are evaluated.

The specific gravity of separation represents the specific gravity of a particle which has 50% probability to either report to the overflow or underflow. The term probable error has been defined as

$$E_p = \frac{SG25 - SG75}{2} \tag{1}$$

where SG25 and SG75 are the specific gravities of the feed material that has reported to concentrate 25% and 75% respectively. The probable error values are used to compare the efficiency of a given operation. For an ideal separation the probable error value will be zero. Hence, the E_p is used as a selection criterion for operating level of variables.

The family of distribution curves obtained at different operating levels of variables of a coal cleaning equipment for the same feed coal can be merged together to form a single generalised distribution curve by plotting the distribution points against reduced specific gravity. The reduced specific gravity is the ratio of the median of the specific gravity fraction to the specific gravity of separation.

The generalised distribution curve provides the following advantages for the interpretation of the data obtained from a coal cleaning process.

(i) The generalised distribution curve is independent of process and design variables. Hence this can be used to characterise the equipment.

(ii) Since the specific gravity of separation is the only unknown in the generalised distribution curve, this can be used for modelling of the process.

The mathematical models available for fine coal gravity concentration operations using the generalised distribution curve concept are discussed in the following.

Water Only Cyclone

Presently, Water only cyclones are the only gravity concentration technique used in Indian coal washeries for treating coal fines. Water only cyclones have advantages such as high capacity, ability to treat oxidised coals and low production costs against demerits such as poor sharpness of separation and low ash reduction.

A typical distribution curve obtained from the water only cyclone is shown in Fig. 2. From this figure it may be noted that the short circuiting/by passing ratio is high in water only cyclone. Pillai et al. [9] suggested a correction factor for this type of distribution curves

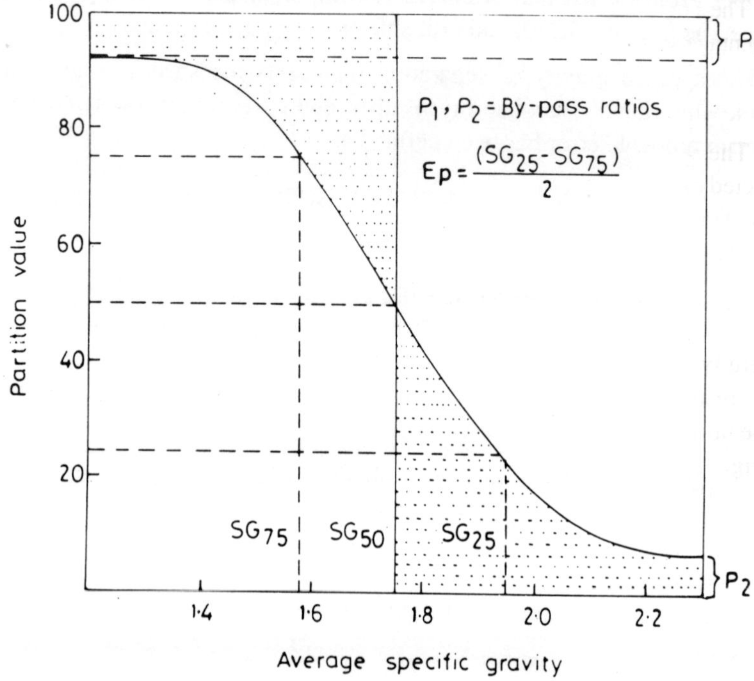

Fig. 2. Performance curve for water only cyclone

$$Y_c = \frac{Y_a - Y_1}{\{1 - (Y_1 + Y_2)\}} \tag{2}$$

where, Y_c and Y_a are corrected and actual distribution points and Y_1 and Y_2 are by pass ratios. Using the corrected distribution curve the values of $SG50_c$ (corrected cut specific gravity) are estimated. The generalised distribution curve obtained using the values of $SG50_c$ passes through the distribution points 0 and 100. Many research workers have quantified the generalised distribution curve mathematically using different equations. The following are some of the widely used equations for generalised distribution curve.

Plitt [10] $$Y = 100\{1 - \exp(-0.693\,X^M)\} \tag{3}$$

Gottfried [11] $$Y = 100\{\exp - (X - A)^B/C\} \tag{4}$$

Gottfried [12] $$Y = 100\,[E + F\{\exp - (X - A)^B/C\}] \tag{5}$$

where X is the reduced specific gravity (SG/SG50) and M, A, B, C, E and F are fitted constants.

The generalised distribution curve for the water-only cyclone has been mathematically defined by Suresh et al. [13]

$$Y = 100\exp^{-0.693\,(SG/SG50)^{4.559}} \tag{6}$$

The cut specific gravity (SG50) is related to the cone angle (CA) and the percentage slimes (–75 Micron) content (FC) in the feed by the following equation

$$SG50 = 1.574 + 0.021\ CA - 0.045\ FC \tag{7}$$

The water distribution in the water-only cyclone has been found to be affected only by the spigot diameter (SPD) and independent of other variables [14]. This phenomenon is confirmed with data from an industrial water-only cyclone also. The water split in water only cyclone is quantified as

$$(WOF/SPD) = 0.0144 + 0.9932(WF/SPD) \tag{8}$$

where WF and WOF are flowrates of water in tonnes per hour in feed and over flow respectively. Based on these studies Suresh et al. [14] have proposed a scale up strategy for water only cyclones using a generalised water plot shown in Fig. 3. This scale up strategy is similar to that of hydrocyclones [15].

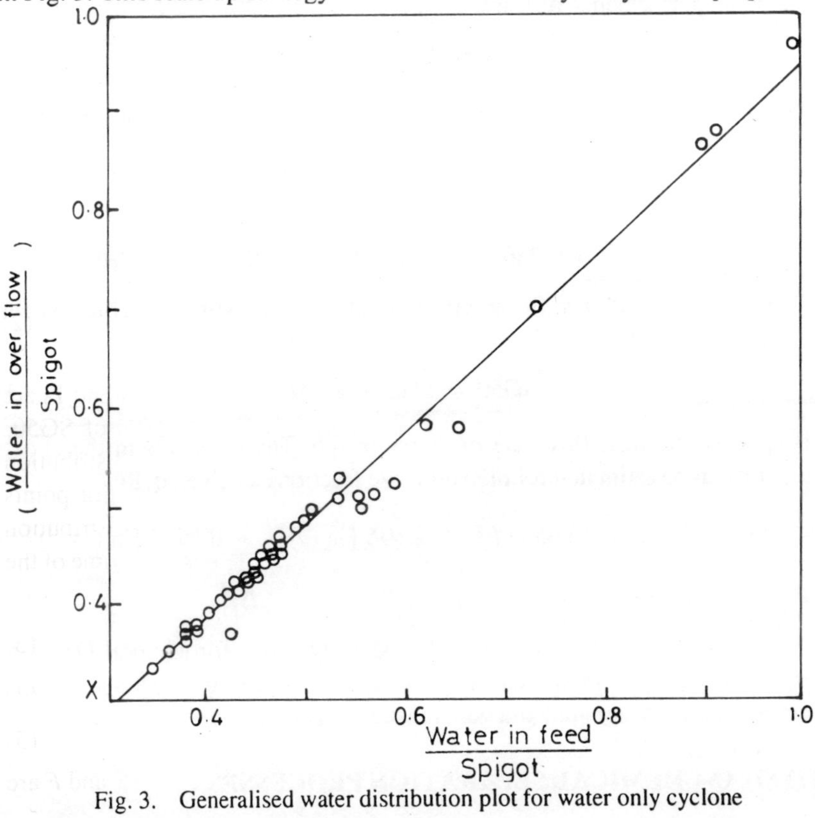

Fig. 3. Generalised water distribution plot for water only cyclone

Spirals

Application of spirals for coal washing has gained momentum in the recent times. The following are the major advantages of using spirals for fine coal cleaning [16].

(i) consistent separation of different size fractions, i.e., sharpness of separation does not drastically reduce with decrease in particle size as in other gravity concentration units.

(ii) ability to tolerate high amount of clay content in the feed.

However, capacity, floor space requirement and inability to treat slimes are the limitations of spirals.

Kelly [17] has proposed the following equation to quantify the generalised distribution curve of spiral

$$Y_d = \exp(-0.693\{(SG-1)/(SG50-1)\}^A) \quad (9)$$

Using the above equations, Subasinghe and Kelly [18] studied the variation in the values of A with different levels of operating varaible such as mass flow rates, feed slurry concentration, etc., and developed the following equations to predict the performance of spiral

$$A = A_0 + m \, . \, P \quad (10)$$

where P is feed slurry concentration % solids by mass. The constants A_0 and m are found to be size dependent and empirically related as

$$A_0 = 0.8537 \ln(d) - 2.145 \quad (11)$$

$$m = 0.2660 - 0.1227 \ln(d) + 0.01362\{\ln(d)\}^2 \quad (12)$$

Further, the cut specific gravity is related to operating variables by the following equations

$$SG50 = SG_0 + K \, . \, M \quad (13)$$

where M is the mass flow rate of slurry in tph. The constants in the above equation can be estimated for different size fractions as given below

$$SG_0 = 33.82 + 16.65 \ln(d) + 2.806\{\ln(d)\}^2 - 0.1553\{\ln(d)\}^3 \quad (14)$$

$$K = 1.600 - 0.4841 \ln(d) + 0.03715\{\ln(d)\}^2 \quad (15)$$

Kelly et al. [19] have developed model equations for Mineral Deposits spirals and Humphreys spirals by correlating operating parameters with cut specific gravity, water split and generalised distribution curve.

PHYSICO-CHEMICAL SEPARATION PROCESSES

The hydrophobic surface of coal fines makes them readily amenable to the processes such as froth flotation and coal oil agglomeration. Most of the clean fine coal produced in the country is obtained from the froth flotation process. Coal oil agglomeration is a very efficient process to treat coal fines and attempts are being made to reduce the cost of this process to make it economically viable.

Mathematical models available for froth flotation and oil agglomeration of coal fines are reviewed in the following sections.

Coal Flotation

Most of the models available for coal flotation in the literature are based on kinetic approach. Kinetic models for the flotation process have been developed by relating the recovery of solids with time. The models using kinetic approach are generally stated on a cumulative recovery basis. The classical first order model [20] is given by

$$R = RI\ (1 - e^{-kt}) \tag{16}$$

where RI is the equilibrium or maximum recovery obtained at infinite time. R is the cumulative recovery at time t and k is the rate constant.

Agar et al. [21] have suggested a time lag to be incorporated into the flotation model as

$$R = RI\,[1 - \exp\{-k\,(t + \phi)\}] \tag{17}$$

where ϕ is the correction factor for time lag.

Klimpel [22] modified the first order model assuming rectangular distribution of rate constants as

$$R = RI\ [1 - (1/kt)]\ [1 - \exp(-kt)] \tag{18}$$

The above model has been shown to have some what more flexibility to describe experimental flotation data than the simple first order model.

Lynch et al. [23] have modelled the effect of collector on coal flotation based on the results of batch tests. The increase in collector enhances the rate of flotation of combustable particles.

Lynch et al. [24] have also developed models for batch and continuous coal flotation using a three parameter kinetic equation. They have suggested an equation for predicting the recovery of solids based on the recovery of water and ultrafines (–38 μm) in the feed.

Sztaba and Tajchman [25] suggested that coal enrichment flotation can be described by a set of significant regression equations with the variable parameters: feed, concentrate and reject ash content, solids flow rate of the feed, consumption of reagents and concentration of solids in the feed. These equations can be used for predicting process results and introducing suitable parameter changes aimed at maintaining the co-efficients in the desired ranges.

Rao and his co-workers [26] have carried out extensive modelling studies on froth flotation of Indian coal fines. Vanangamudi and Rao [27] developed an empirical model based on kinetics to predict the recoveries of non-ash and ash material and also the yield and ash content of the products obtained at different flotation periods. They related the logarithm of k of non-ash and ash material with average size d and observed a straight line relationship. Under different operating conditions, slope (m) of the straight line remained constant and only the intercept value (N) changed. Based on these observations, they have developed the following modelling methodology

$$\ln k = N + m\,d \tag{19}$$

The intercept (N) has been related to the operating variables such as collector and frother dosages and pulp density using a regression equation.

The above model has been further refined by Rao et al. [28] by incorporating the feed coal characteristics also. They found that the rate constant k for any size fraction is dependent not only on the average size of the size fraction but also on the proportion of ash and non-ash fractions within that size range and in the fines below 75 micron in the feed.

Arnold and Aplan [29, 30] have fitted a dimensionless residence time distribution model for continuous coal flotation banks as

$$Q\,(t') = a\,(t')^n\ \exp\,(-\,bt') \tag{20}$$

where $Q\,(t')$ is dimensionless concentration, t' is ratio of actual time to mean residence time and a, n and b are constants. They have also compared the performance of batch and continuous flotation systems and concluded that using rate equations in conjunction with residence time distribution is a good approach for scale up.

The performance of froth flotation is affected by entrainment of slimes to the froth alongwith water which has been studied by many workers [31-33]. Vanangamudi and Rao [34] have developed a model to predict the slimes and water recovery in batch coal flotation. Govindarajan et al. [35] have shown that the relationship between the recovery of different size fractions and water recovery is also influenced by collector dosage during batch coal flotation tests. Based on this observation, the characteristic non-ash and ash recovery curves for a given coal have been obtained as shown in Figs. 4 and 5. These curves are independent of operating variables and once the water recovery is known then the recoveries of non-ash and ash material can be predicted at any other levels of variables.

The water recovery in batch flotation has been related to the collector (C) and frother (F) dosages and time (T) through "Water Recovery Factor" which is defined as

$$\text{WRF} = (1 + C)\ .\ F\ .\ T \tag{21}$$

The typical relationship between WRF and water recovery is shown in Fig. 6. Equations have been proposed to quantify the curves shown in Figs. 4-6 to develop the model.

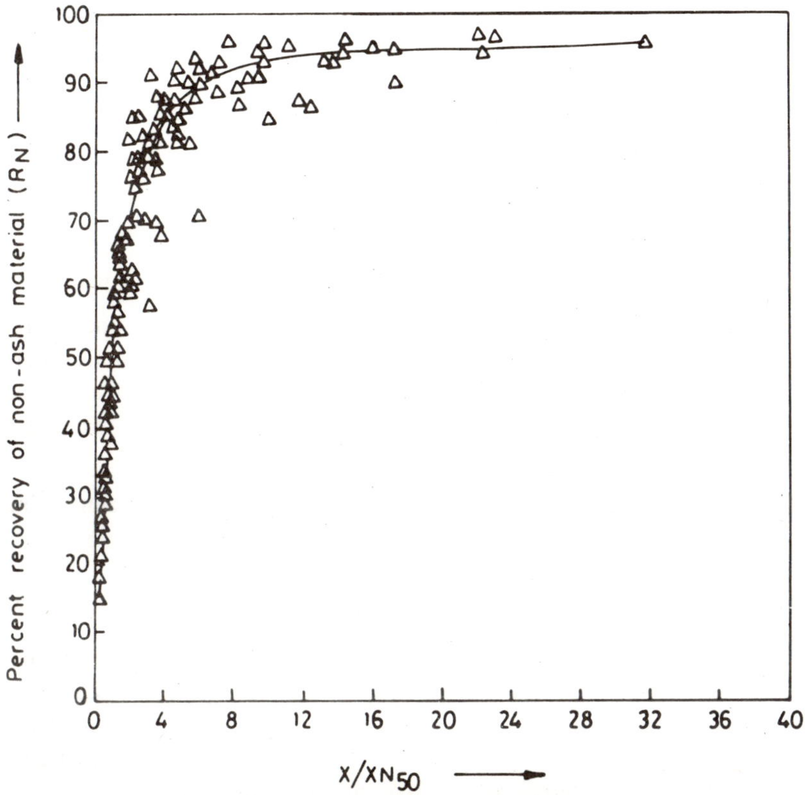

Fig 4. Characteristic non-ash recovery curve for batch coal flotation

Coal Oil Agglomeration

The coal-oil agglomeration process is the most efficient method to treat coal fines below 0.5 mm. This process has the following advantages:

(i) Capable of treating extreme fines and oxidized coal
(ii) Dewatering of the product is easy
(iii) The product has a low and consistant level of impurity

The modelling methodologies available for spherical agglomeration of different mineral fines [36-38] can be adopted for the coal oil agglomeration process also. Swanson et al. [39] have studied the kinetics of coal oil agglomeration in terms of inversion time, i.e, the time at which power consumption is maximum due to phase inversion.

The modelling of the coal-oil agglomeration process proposed by Vanangamudi [40] involves prediction of (i) the size distribution of the agglomerates and (ii) the yield of the agglomerates. The size distribution plots obtained for a given top feed size of coal have been used to make dimenisonless size distribution plot by plotting the cumulative weight fraction passing through size

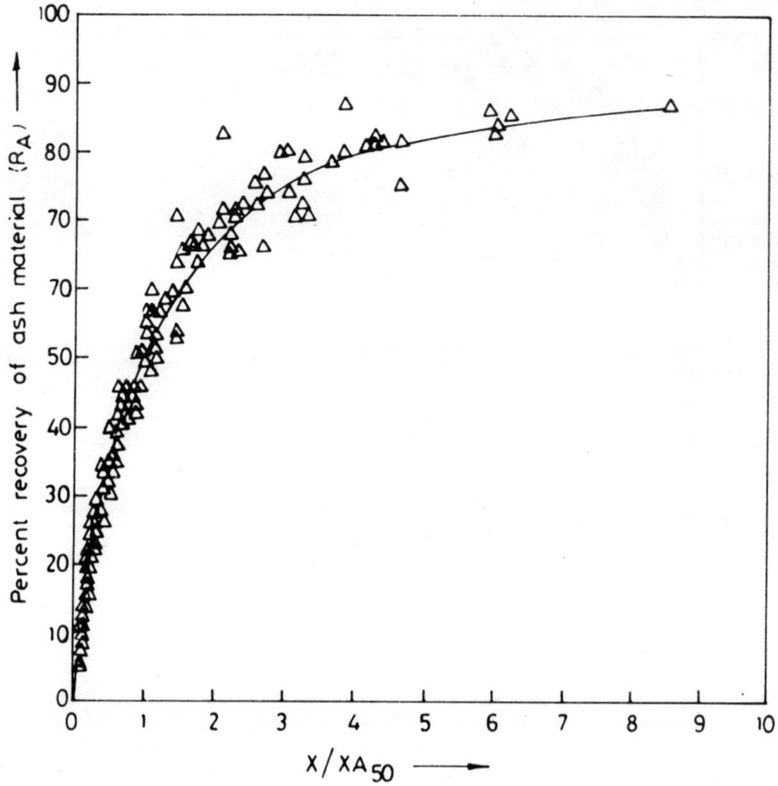

Fig 5. Characteristic ash recovery curve for batch coal flotation

d against $d/d50$. These curves do not pass the origin hence the x-axis absscia is modified as [41]

$$X = \frac{d - d_f}{d50 - d_f} \tag{22}$$

The modified plot is shown in Fig. 7 which is designated as the "Characteristic curve" for a given coal sample and top feed size. The characteristic curve of coal-oil agglomeration is independent of process variables and it can be used for modelling of the process. The characteristic curve is quantified as

$$YD = 1 - \exp(-0.693\ X^{3.05}) \tag{23}$$

Further, the $d50$ values have been related to the process variables such as time (t), furnace oil dosage (FO), feed pulp density (PD) and degree of agitation (RPM) by the following equation

$$d50 = 1.384\ t^{0.202}\ \mathrm{FO}^{2.022}\ \mathrm{PD}^{0.415}\ \mathrm{RPM}^{-0.273} \tag{24}$$

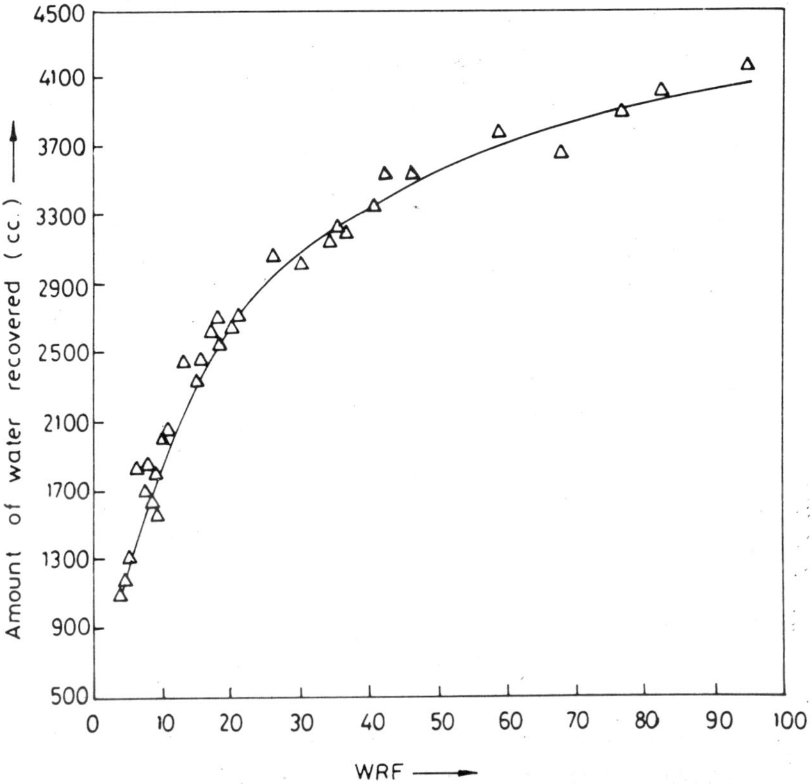

Fig 6. Characteristic water recovery curve for batch coal flotation

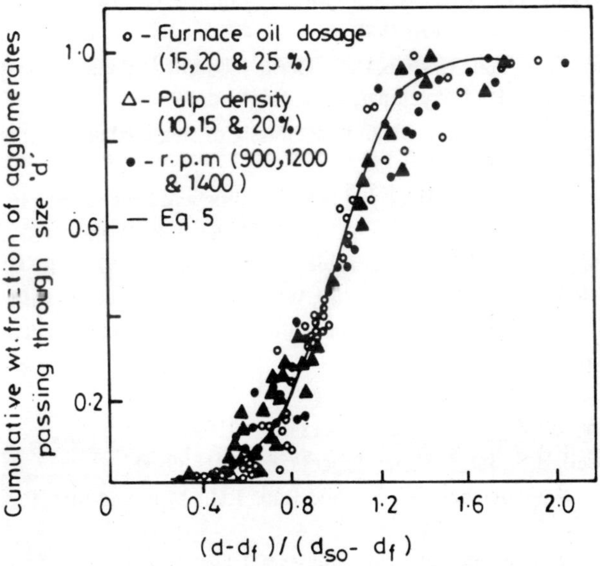

Fig 7. Characteristic curve for coal-oil agglomeration

Now, Eqs. (23) and (24) can be used predict the size distribution of agglomerates. The yield of the agglomerates (Y) is related to the $d50$ as given below

$$Y = 95\,\{1 - \exp(-0.083\,(d50 - 130)^{0.837})\} \tag{25}$$

Haldar et al. [42] have used dimensional analysis to quantify the effect of conditioning on the yield of agglomerates. They have concluded that the recovery of coal is highly influenced by the variables of conditioning system.

CONCLUSIONS

The overview of literature on modelling of fine coal treatment processes reveals that simple models for fine coal gravity concentration units are limited in scope and require estimation of cut specific gravity through time consuming sink-float tests. Though many models are available for coal flotation, most of them are based on laboratory batch tests only. The importance of process performance improvement through systematic data collection and analysis to fine tune operating and design variables is yet to be recognised.

REFERENCES

1. Mikhail, M.W., Picard, J.L., and Humenuik, O.E.: Performance evaluation of gravity separators in Canadian washeries. Proc. CIM Second Technical Conference on West Canadian Coals, Edmonton 1982.
2. Turner, J.F.: Minerals Engg. 4 (3/4) (1991) 213-224.
3. Chan, S.K., and Mozley, R.H.: Enhanced gravity separation for the beneficiation of fines and ultra-fines. Richard Mozley Limited, Cornwall, UK, 1987.
4. Anon.: The Miner, April 1992.
5. Lathior, R.A., and Osborne, D.G.: Dense medium cyclone cleaning of fine coal. 2nd Int. Conf. Hydrocyclones, Bath, England, 1984.
6. Killmeyer, R.P.: Dense medium cycloning of fine coal at low specific gravity. US DOE/PETC/TT 83-1, 1982.
7. Mengelers, J.H., and Absil, J.H.: Coal Mining and Processing, (1976) 62-64.
8. Gottfried, B.S., and Jacobson, P.S.: Generalised distribution curve for charactersing the performance of coal clening Equipment. USBM, RI 8238, 1977.
9. Pillai, K.J., Spottiswood, D.J., Bull, W.R. and Kelley, E.G.: A preliminary model of a water only cyclone and its application to product quality control. 2nd, IFAC, Australia, 1985.
10. Plitt, L.R.: CIM Bull. 69 (1976) 114-123.
11. Gottfried, B.S.: Int. J. Miner. Process. 5 (1978) 1-20.
12. Gottfried, B.S.: Int. J. Miner. Process. 8 (1981) 89-91.
13. Suresh, N., Vanangamudi, M., and Rao, T.C.: Min. and Met. Process. (1989) 130-137.
14. Suresh, N., Vanangamudi, M., and Rao, T.C.: Minerals Eng. 3(5) (1990) 537-541.

15. Lynch, A.J., and Rao, T.C.: Modelling and scale-up of hydrocyclone classifiers. Proc. 11th Int. Min. Proc. Congr., Cagliari, paper 9, 1975.
16. Mikhail, M.W., Salama, A.I.A., Parsons, I.S., and Humenuik, O.E.: Coal Prep. 6 (1/2) (1988) 53-78.
17. Kelly, E.G., et al.: Trans SME 276 (1985) 1943-1949.
18. Subasinghe, G.K.N.S., and Kelly, E.G.: Coal Prep. 9 (1/2) (1991) 1-12.
19. Kelly, E.G., et al.: The development and application of mathematical models for concentrating spirals. XIV Int. Mineral Proc. Cong., Stockholm, Sweden, 1988.
20. Arbiter, N., and Harris, C.C.: In Fuerstenau D.W. (Ed) : Froth Flotation, 50th Ann. Vol. New York : AIME, 1962
21. Agar, G.E., Stratton-Gawley, R., and Bruce, T.J.: CIM Bulletin 73 (1980) 173-181.
22. Klimpel, R.R.: In Klimpel,R.R. and Luckie, P.T. (Eds.) : Industrial Practice of Fine Coal Processing. Littleton, Colorodo, (1988) 113-127.
23. Lynch, A.J., McKenzie, C.K., Leach, K.R., and Bateman, K.W.: Modelling of Central Queensland coal flotation circuits for control purposes. Aus. Inst. Min. Met. Confr., North Queensland, 1978.
24. Lynch, A.J., Johnson, N.W., Manlapig, E.V., and Thorne, C.G.: Mineral & Coal Flotation Circuits Their Simulation and Control. New York : Elsevier Publ. Comp. 1981.
25. Sztaba, K., and Tajchman, Z.: Regression model of coal flotation process. 7th Intl. Coal preparation Congress, Paper No. J-1
26. Rao, T.C., Vanangamudi, M., and Barnwal, J.P.: Modelling of coal flotation operation. Progress Report, Dhanbad : Indian School of Mines, 1989.
27. Vanangamûdi, M. and Rao T.C.: Intl. J. Miner. Proces. 16 (1986) 231-243.
28. Rao, T.C., Govindarajan, B. and Vanangamudi, M.: Minerals Engineering 2 (3) (1989) 403-414.
29. Arnold, B.J., and Aplan, F.F.: In Chander, S. and Klimpel, R.R., (Eds) : Advances in Coal and Mineral Processing Using Flotation. ME, Littleton, Colorado, 1989, pp. 130-145.
30. Arnold, B.J. and Aplan F.F.: The application of rate models to pilot and commercial scale coal Froth Flotation circuits. Presented at the 1989 SME Annual Meeting, Las Vegas NV, March 1989.
31. Bishop, J.P. and White, M.E.: Trans. Inst. Min. Metall. 85 (1976) C191-C194.
32. Trahar, W.J.: Intl. J. Miner. Proces. 8 (1981) 289-327.
33. Warren, L.J.: Intl. J. Miner. Proces. 14 (1985) 33-44.
34. Vanangamudi, M. and Rao, T.C.: Minerals Engineering 2 (2) (1989) 185-198.
35. Govindarajan, B., Vanangamudi, M., and Rao, T.C.: Minerals and Metallurgical Processes May (1991) 110-115.
36. Kawashima, Y., and Capes, C.E.: J. Powder & Bulk Solids Technol. 2 (1978) 53-57
37. Kawashima, Y., and Capes, C.E.: Ind. Eng. Chem. Fundamentals 19 (1980), 312-314.
38. Bemer, G.G., and Zuiderweg, P.J.: In Somasundaran, P. (Ed.) : Fine particle processing. New York : AIME 1980, pp.1524-1546.

39. Swanson, A.R., Bensley, C.N., and Nicol, S.K.: Some fundamental aspects of the selective agglomeration of fine coal. Proc. 2nd Int. Symp. on Agglomeration, Atlanta, 1977.
40. Vanangamudi, M.: Modelling and kinetic studies on coal flotation and coal oil agglomeration processes. Ph. D Thesis. Indian School of Mines 1983.
41. Rao, T.C., Vanangamudi, M., and Hanumantha Rao, K.: Int. J. Min. Proc. 9 (1982) 235-243.
42. Haldar, D.D., Hazra, S.K., Vanangamudi, M., and Rao, T.C.: Coal Prep. 9 (1/2) (1991) 27-36.

Role of Surface Chemistry in High Gradient Magnetic Separation

SHARAD G. DIXIT
Department of Chemical Technology,
University of Bombay, Matunga, Bombay-400 019, India.

ABSTRACT

Surface phenomenon plays an important role in the High Gradient Magnetic Separation (HGMS) . It affects the particle-particle interaction as well as the particle-filament adhesion. Various factors affecting the heterocoagulation and dispersion of particles have been discussed with reference to HGMS of clays, mineral beneficiation, magnetic flocculation, magnetic coating and magnetic surfactants. It has been shown that both selectivity and grade can be improved by controlling the surface chemistry.

INTRODUCTION

High Gradient Magnetic Separation (HGMS), a versatile separation technique, achieves separation of particles by the combination of a magnetic field and a magnetic field gradient. It is probably the only technique which can accomplish particle separation at a micron or even submicron level (Fig. 1). In order to be able to cause separation through HGMS, one or some of the constituents of the particulate system must be at least weakly paramagnetic. The technique found its first commercial application in removing the coloring impurities from Kaolin clays thus improving dramatically the brightness of clays [2]. Several other applications were soon demonstrated at a laboratory level. These include deashing and desulphurising of coal [3], mineral processing [4] and removal of soluble phosphates from water [5]. Some of these were subsequently commercialised. While the process parameters affecting the separation and its grade and recovery were experimentally investigated, significant effort went into developing the mathematical models describing the operation and functioning of HGMS [6, 7, 8, 9]. Another significant aspect on which the success of the process depends namely, the conditions leading to dispersion of the multicomponent particulate system has received only scant attention. Furthermore, the enhancement of selectivity by modulating the susceptibility of particle has also been largely neglected. Both these aspects involve a great deal of surface chemistry. It is the purpose of this overview to highlight the surface chemical aspects of HGMS. It is hoped that the manipulation of the surface chemical

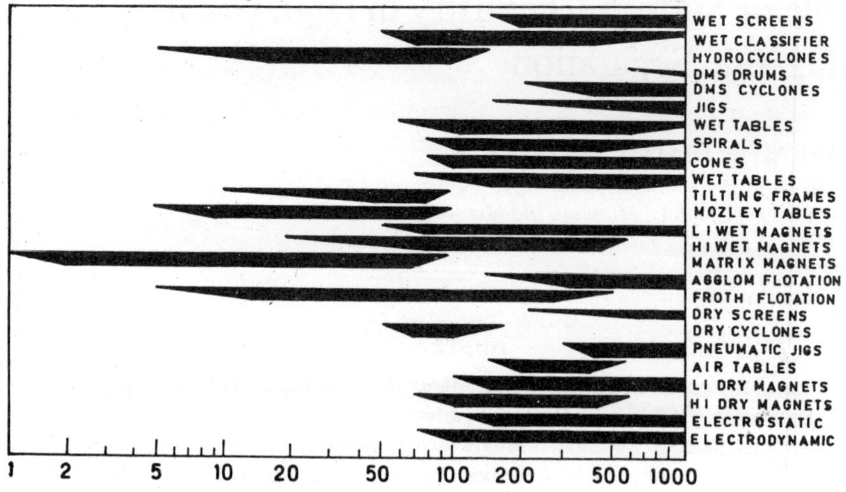

Fig. 1 Effective range of application of mineral processing techniques (after Ref. [1])

factors may significantly broaden the application of HGMS and open up interesting areas for investigation.

PRINCIPLE OF HGMS

It is necessary to know the principle on which the HGMS works in order to appreciate the role of surface chemistry. Solid particles behave differently when placed in the externally applied magnetic field H as shown in Fig. 2. The ferromagnetic solid achieves saturation magnetization of high magnitude as the field is increased. In case of paramagnetic material magnetisation continuously increases though at a much lower level than ferromagnetic materials. On the other hand, a diamagnetic (non-magnetic) material has a negative value of magnetic susceptibility. Thus, these materials are repelled from regions of higher magnetic field towards regions of lower magnetic field.

The equation giving the force on a particle in a magnetic field may be written as follows,

$$F_m = X \,.V \,.\, H \,.\, dH/dX \tag{1}$$

The force on the particle in a magnetic field therefore can be optimised in three different ways.

- Increasing the field H:

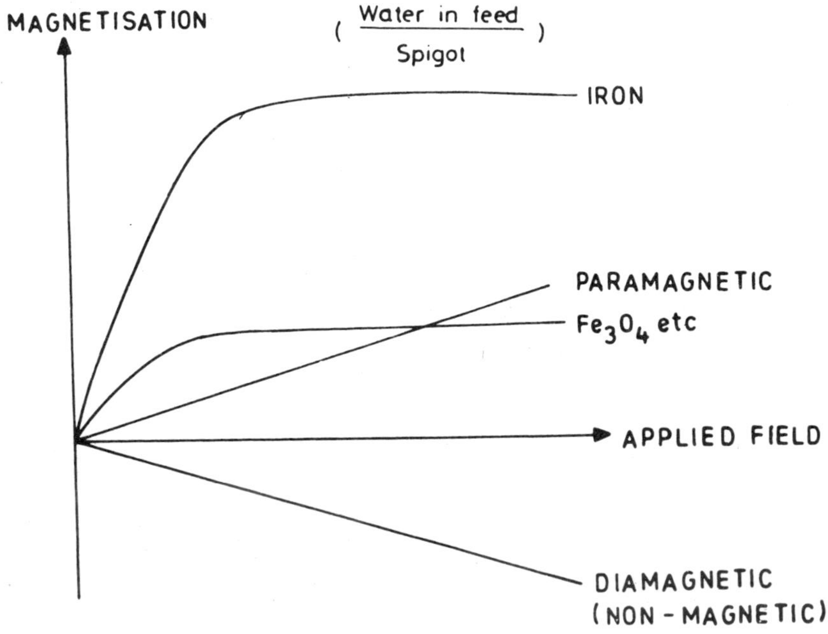

Fig. 2 Magnetisation curves for ferromagnetic, paramagnetic and dimagnetic materials

In HGMS it can be increased upto 20 kGauss using powerful electromagnet. It can be further increased upto 50 to 70 kG using superconducting magnet. Both these involve higher capital costs and also higher power consumption in case of electromagnets.

- Increasing the field gradient (*dH/dX*):

  It can be simply and inexpensively achieved by incorporating the filamentary matrix in the canister and by reducing the diameter of the filaments. Values of field gradient for different types of matrices are given in Table 1. Such high gradients obviously increase the force on the particle significantly.

- Increasing or introducing magnetic susceptibility in the particles:

  It can be achieved easily and inexpensively by using a ferromagnetic coating material such as magnetite. Thus, in case of a coated particle the field requirement is drastically reduced. The coating can be done selectively in a mixture of particles and thus selectivity of HGMS can be greatly improved. It is here that surface chemistry can play an important role.

Whatever be the conditions, the suspension has got to be completely dispersed for efficient separation. This involves the use of dispersants and close control of the surface chemistry of the system.

Table 1: Field gradient as a function of matrix size

Type of matrix	Size of matrix (d)	*B* Gauss	Field Gradient *dH/dX* Gauss/cm
Ball	0.95	20,000	26,536
Grooved Plate	0.60	20,000	42,017
Grooved Plate	0.30	20,000	84,034
Filamentary	0.02	20,000	1,260,507
Filamentary	0.01	20,000	2,521,014
Filamentary	0.003	20,000	8,403,381
Filamentary	0.001	20,000	25,210,143

HGMS OF CLAYS

The HGMS technology found its first commercial application in improving the brightness of clays by selectively removing the coloring paramagnetic contaminants such as iron minerals and ferruginous anatase in the magnetic fraction. The improvement obtained in the brightness of clays using HGMS has been well established as seen from the work of Ianicelli [2] and Maurya and Dixit [10, 11]. It was demonstrated that field, flow rate, pH, dispersant and its type and concentration significantly affect the separation. Whereas the effect of field and flowrate have been investigated for several clays, the role of physico-chemical parameters such as pH and dispersant has not received adequate attention.

Effect of pH

Maurya and Dixit [12] investigated in detail the role of pH on HGMS of clays. Typical results for the Dhrangadhra and Chotanagpur clays are shown in Fig. 3 and Fig. 4 respectively. The variation of percent brightness and percent removal of acid-soluble Fe have been shown in the figures. The curves exhibit a maxima in the pH range 8.5-9.5. The variation of the percent iron removal and percent brightness curves are similar. Similar correlation was obtained in respect of other clays [13]. It is therefore obvious that a common explanation for the behavior should be possible. Clays are hydrous aluminosilicates having a layer structure consisting of silica and alumina sheet forming a stack [14, 15]. It may be considered as a combination of oxides, namely silica and gibbsite. The surface charge on the kaolinite when suspended in water is pH dependent because of protonation and deprotonation involving H^+ and OH^- ions. The following units may be present on the clay surface.

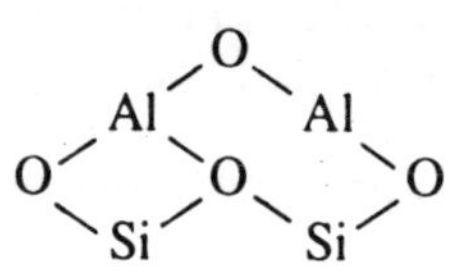

(a)

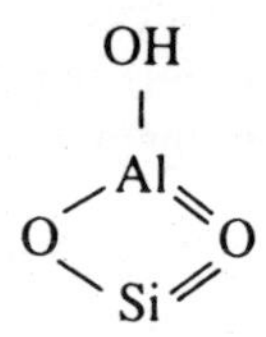

(b)

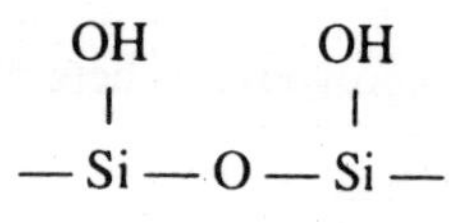

(c)

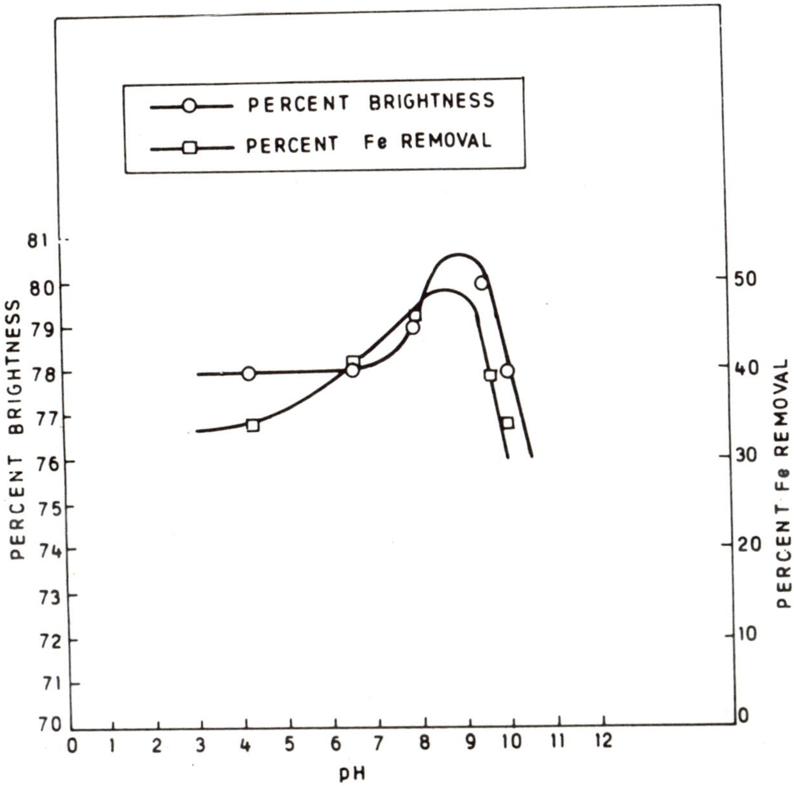

Fig. 3 Effect of pH on change in brightness and iron removal during HGMS of Dhrangadra clay (after Ref. [13])

These react with H^+ ions as follows

$$AlOH + H^+ ==== AlOH_2^+ \tag{2}$$

$$AlOH ==== AlO^- + H^+ \tag{3}$$

$$SiOH ==== SiO^- + H^+ \tag{4}$$

Thus, variation of the pH influences the sign and magnitude of charge. The clays present two types of surfaces, the face surface and the edge surface as can be seen from the typical clay structure of kaolinite shown in Fig. 5. It is known that the face surface charge on the kaolinite remains negative over a wide range of pH from 3-11 [16]. However, the protonation and deprotonation mostly takes place on the edge surface where aluminum and silicon sites are exposed. It has been reported [17] that edge surface undergoes charge reversal and the pzc has been estimated to be around pH 7. The edge surface charge reversal above pH 7 may lead to increased dispersion because in the alkaline range both the surface and edge charges are negative. Aggregation due to edge-surface interaction is thus prevented under these conditions.

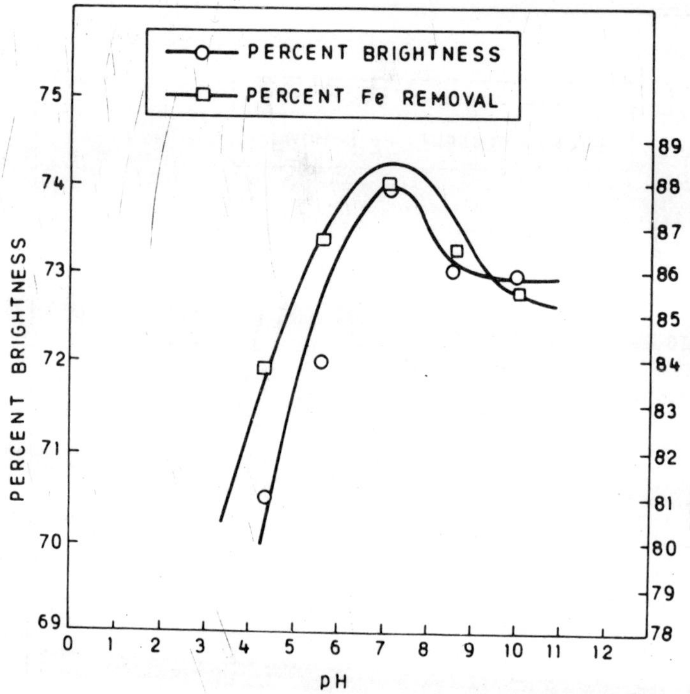

Fig. 4 Effect of pH on change in brightness and iron removal during HGMS of Chotanagpur crude clay (after Ref.[13])

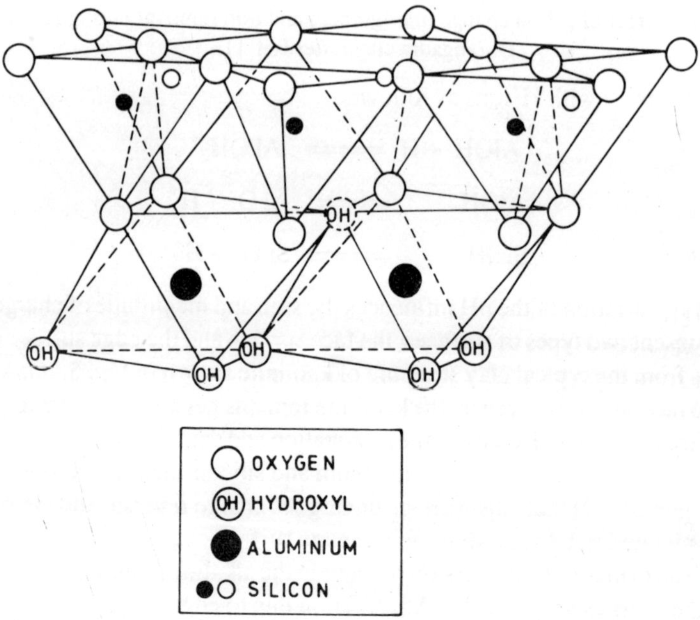

Fig. 5 Atomic arrangement of kaolinite crystal (after Ref.[13])

pH - Drift Experiments

The role of pH was further investigated by Maurya and Dixit [12] using simple pH drift experiments. The initial pH value of the suspension was kept in the range of pH 2-11 and it was allowed to drift as a function of time over a period of 24 hours. Typical results are shown in Fig. 6 for Chotanagpur crude clay. It will be seen from the figure that the natural variation of pH from different initial values is typical of such surfaces. It was found to be the same for all the clays

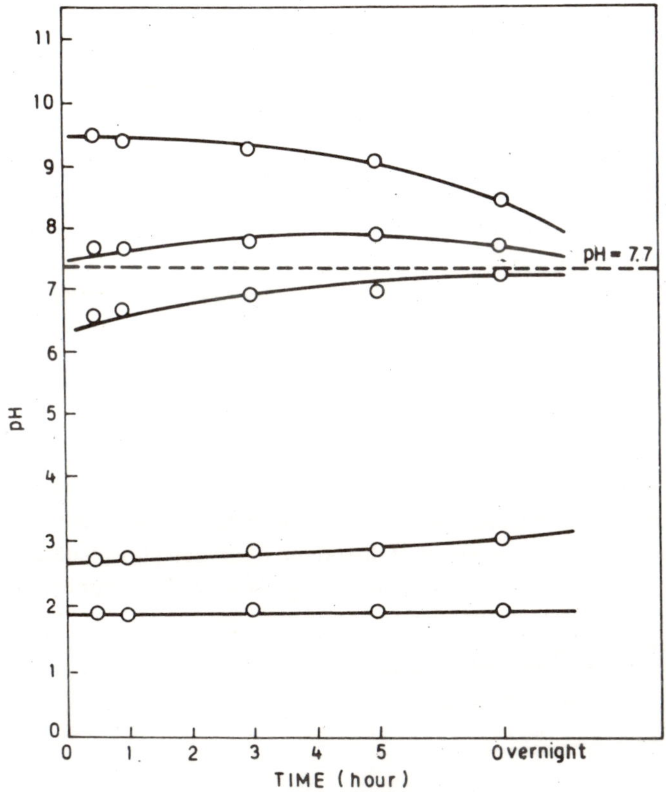

Fig. 6 pH drift as a function of time: Chotanagpur crude clay (after Ref. [12])

studied [12]. If the initial pH is very acidic (pH 2-3) there is not much variation of pH over a period of 24 hours. The final pH values always remained between 2.2 and 2.7 for all the clays studied and was in the vicinity of the point of zero charge (pzc) of silica i.e. 2 –2.3 [18]. This is obviously due to a predominant role of –SiOH silanol group in controlling the conditions at the clay surface.

In the alkaline region the drift curves converge towards a common pH value, which for Udaipur clays occurs at pH 7.7. Similar results were obtained for other clays the steady pH value falling in the range of 7.5-8.5 [12]. It may be noted that pzc for aluminum hydroxide has been reported to be in the range of pH 7–9.2 [18]. The steady pH values for different clays fall in this region.

Thus,the improvement in percent brightness in HGMS experiments in the alkaline pH region is mostly controlled by aluminol group. Ferris and Jepson [19] have presented evidence that kaolinite is coated by polymeric Al–O–OH species and the surface behaviour as a function of pH is governed by these coatings. The pH drift results clearly support such a view. The clays also contain colouring paramagnetic impurities which must remain fully dispersed in order to get removed in the magnetic fraction. The pzc for common colouring minerals occurring in clay are: sepecular hematite (6.6 ± 0.2), Fe_2O_3 (6.5) and rutile, TiO_2 (4.7) [18]. It has been shown by Matijevic [20] that if the zeta potential value is near the pzc i.e. within ± 15 mV, rapid coagulation is observed. The particles having zeta potential outside this range remain dispersed. If the suspension contains more than one type of mineral, pzc of the other minerals has also to be taken into account. For improving the brightness of clays, it may be noted that pzc for both iron oxides (5.5– 6.5) and TiO_2 (4.7) are removed far away from that of gibbsite (pH 9). In the alkaline pH range the clays are also negatively charged. Hence, tendency towards heterocoagulation among impurity and clay particles is minimum. This would result in better capture of the dispersed coloring species during HGMS and greater improvement in percent brightness.

Effect of Dispersant Addition

In HGMS complete dispersion of the particles in suspension is necessary for selective separation. Therefore, dispersing reagents such as sodium silicate, disodium hydrogen phosphate and sodium hexameta-phosphate are commonly used. Typical results of the effect of di-sodium hydrogen phosphate are shown in Fig. 7 for Chotanagpur crude clay and in Fig. 8 for Chotanagpur washed clay.

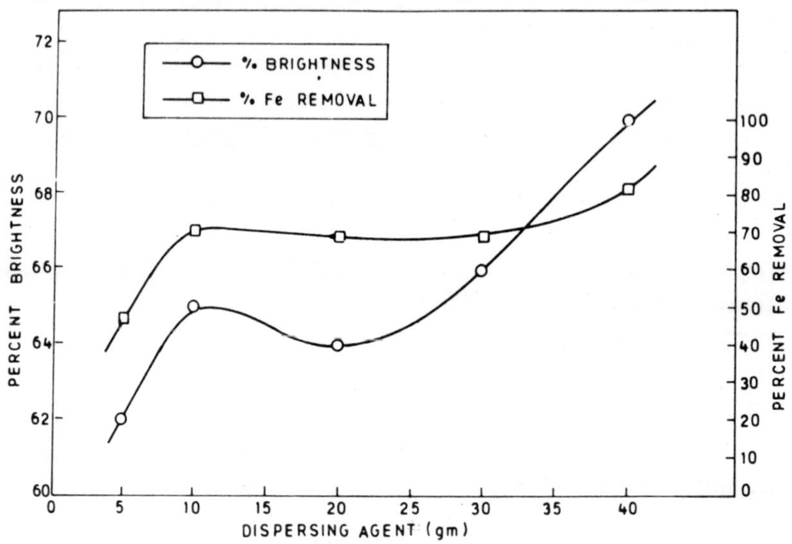

Fig. 7 Effect of addition of di-sodium hydrogen orthophosphate as dispersant: HGMS of Chotanagpur crude clay (after Ref. [12])

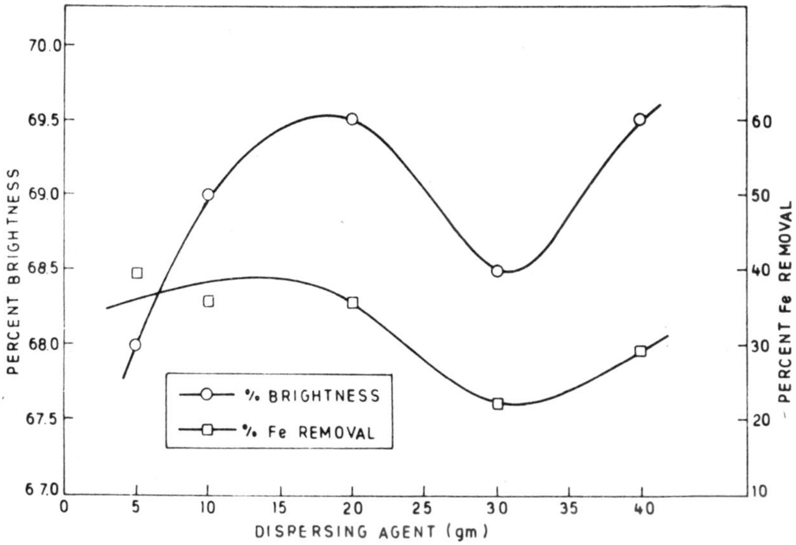

Fig. 8 Effect of addition of di-sodium hydrogen orthophosphate as dispersant : HGMS of Chotanagpur washed clay (after Ref. [12])

It is seen from both these figures that maximum improvement in brightness (as also percent removal) first occurs at an addition of 15 g/l of the dispersant. As the amount of dispersant is increased beyond maxima the improvement in brightness first decreases and then again tends to increase at an addition of 35 g/l of the dispersant. It will also be noted that the percent brightness as well as percent Fe removal curves correlate well with each other.

Effect of sodium silicate addition was studied at lower addition (0.05 to 1 g/l) and higher addition (5-40 g/l). Typical results are shown in Fig. 9 for Chotanagpur crude clay and Fig. 10 for Chotanagpur washed clay. It will be

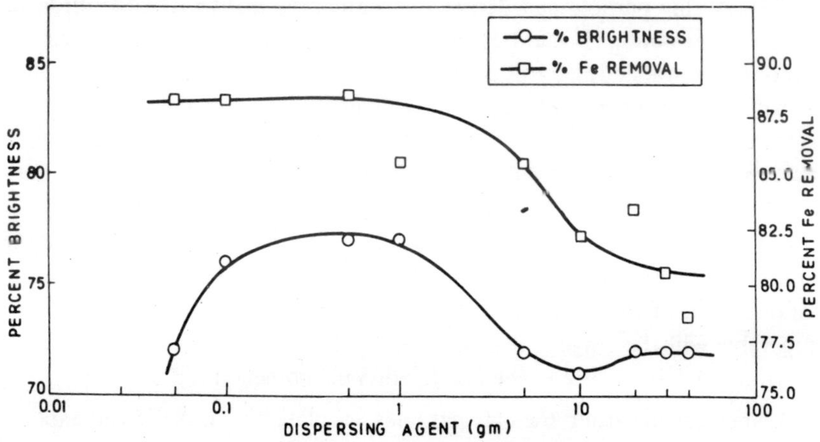

Fig. 9 Effect of addition of sodium metasilicate as dispersant : HGMS of Chotanagpur crude clay (after Ref. [12])

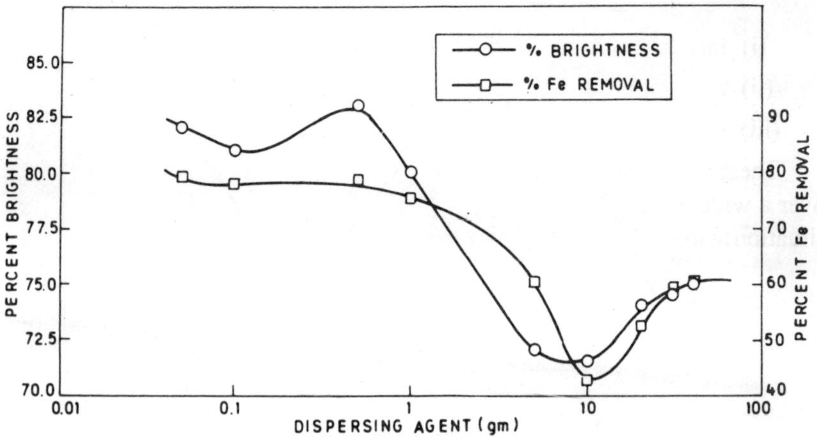

Fig. 10 Effect of addition of sodium metasilicate as dispersant : HGMS of Chotanagpur washed clay (after Ref. [12])

seen from Fig. 9 that best results are obtained at a lower addition of 0.5 g/l of silicate. At higher addition the brightness again tends to increase. In case of washed clay also significant improvement in brightness is obtained at lower addition of 0.5 g/l of sodium silicate whereas at higher addition of 25 g/l brightness curve again shows another maxima though less pronounced. The Fe removal curves correlate with the brightness curves in case of both crude and washed clays.

Similar results were obtained in respect of several other clays. Sodium metasilicate gave better results as compared to phosphate as dispersant. Settling studies [13] further confirmed this behaviour.

The major force affecting the dispersion/flocculation behaviour are the electrical double layer and the van der Waals attractive forces between particles [21, 22]. A clay particle can derive surface charge due to (i) adsorption of potential determining H^+ and OH^- ions. (ii) preferential adsorption of a cation or anion from an added electrolyte and (iii) imbalance of charge within the solid. When placed in water the diffused double layer of the two particles interact and this results in repulsion. The addition of electrolytes leads to compression of the double layer and as a result the range of repulsive forces is reduced. It is this effect which is responsible for the decreased stability of the system.

The nonspecific van der Waals-London attractive dispersion force between two atoms decays rapidly with distance. However, the sum of all forces between all atom pairs of the particles has a much longer range in practice and is comparable with the range of double layer repulsion. Therefore, van der Waals attraction is a major cause of the flocculation phenomenon [23, 24, 25].

As mentioned earlier, the clay particles are plate like in shape and expose two crystallographical surfaces, face surfaces and edge surfaces. Therefore, the following interactions are possible.

(i) face to face interaction

(ii) edge to face interaction and

(iii) edge to edge interaction.

These are depicted in Fig. 11. The face surfaces are negatively charged over a wide range of pH between 3–11 because of cationic deficiencies. The situation at the edge surfaces is different as it is created due to breakage of bonds

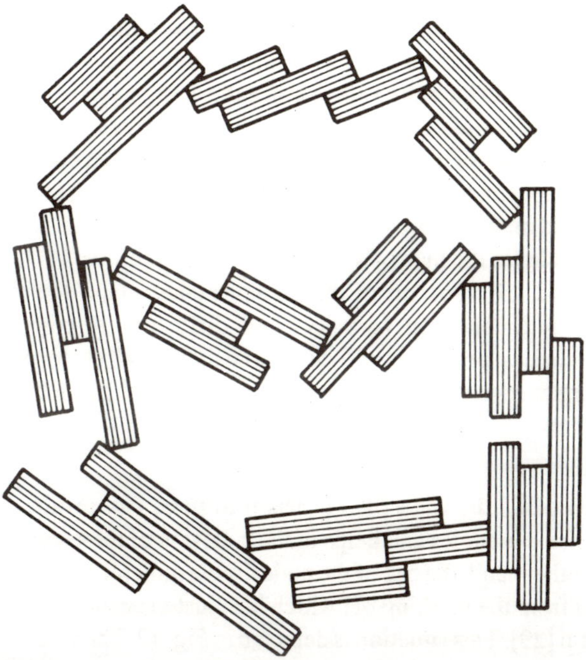

Fig. 11 Representation of structural network of (i) edge to face (ii) edge to face and (iii) face to face aggregate of multilayered platelets of kaolinite (after Ref. [17])

at the surface and there are both positive and negative charges present at the surfaces depending on pH. The point of zero charge (pzc) for the edge surface has been estimated to occur near pH 7.0. Since below pzc face surface is negatively charged while edge surface is positively charged, edge to face flocculation can occur due to electrostatic attraction. It can be broken by reversing the charge of the edge surface to negative charge either by increasing pH or by addition of dispersant containing anions which are chemisorbed on the edge surface. Since the surface area of the edge surface is very small as compared to face surfaces, the quantity of the dispersant required to reverse the edge surface charge is also small. It is believed that the increased dispersion and increased brightness observed at low sodium metasilicate concentrations (0.5 - 1 g/l) is due to the edge charge reversal.

Alternatively the edge to face interaction can be destroyed by reversing face surface charges due to the adsorption of sodium cations on the face surfaces. However, the requirement of the dispersant will be quite large. The improvement in brightness at higher additions of sodium metasilicate addition (25 - 35 g/l) may be attributed to such a process. The experimental results (Fig. 9 and 10) indicate that this enhancement is not very pronounced.

HGMS OF MINERALS

Svoboda [27] and Arellano [28] have established that if the particle size is small (25 μm) the inertial forces on particles such as magnetic, gravity and hydrodynamic drag have to compete with surface forces between particles and between the particle and the matrix. In practice the particles should be colloidally unstable i.e. they should not repel each other and prevent contact. On the other hand, the particles of undesirable gangue minerals must not heterocoagulate with the valuable mineral particles. The surface forces such as electrostatic, van der Waals and magnetic dipolar interactions determine whether the attachment of the particles to the matrix will occur and also the strength of adhesion. On the other hand, the transport of the particles takes place due to magnetic traction force, sedimentation (gravity) and diffusion. A combination of transport and adhesion processes will determine the overall performance of the high gradient magnetic separator. The surface potential is given by

$$\psi = (kT/2.303e)\,(pH_{pzc} - pH) \qquad (2)$$

When $pH = pH_{pzc}$ the value of ψ is nearer to zero. This results in unstable conditions for these particles and helps in the adhesion of these particles on the matrix. On the other hand, the pzc of the undesirable mineral component should be far removed from the working pH which is adjusted nearer to the pzc of the valuable mineral [29]. This situation is depicted in Fig. 12. The valuable mineral is denoted by *A*. It is necessary to have the working pH to be higher than but closer to pzc of (*A*) because in this region the particles of dissimilar minerals repel one another. In the region pzc (*B*) < pH working < pzc (*A*), the separation will be non selective hence excellent grade can not be achieved. At pH working < pzc (*B*) the process is selective but inefficient since it does not promote the coagulation of valuable mineral *A* [29].

The experimental results of HGMS of uranium obtained by Svoboda et al. [29] on the uranium-gold tailings from Witwatersrand-Klerksdorp are presented in Table 2. It will be seen that both recovery and grade depend on pH. The maximum in grade and recovery occurs in the pH range 1.2-1.8. Mackenzie and Lovell [30] determined the zeta potential of uranium rich and uranium poor (quartz) ore from the same source. The pzc for U-rich uraninite occurs at pH 3-3.2. On the other hand the curve for U-poor ore is similar to that of quartz and shows a pzc at pH 2-2.3 . The pH of maximum grade and recovery for uraninite occurs at 1.2-1.8 . This would be expected according to the theory suggested and discussed earlier in the preceding paragraph.

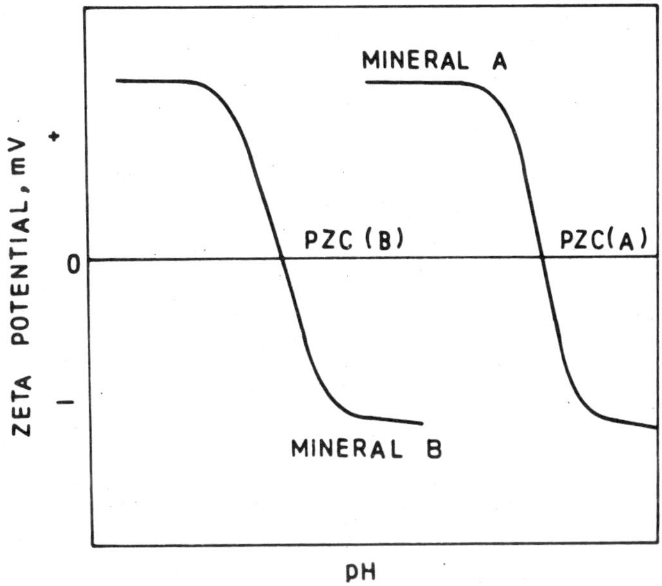

Fig. 12 General zeta potential curves for mineral *A* and *B* as a function of pH

Table 2: HGMS tests on uranium-gold tailings from Witwatersrand-Klerksdrop (after Ref. [29])

pH	Mass yield into mags		U_3O_8 grade of mags (g/t)		Recovery of U_3O_8 (%)	
	Balls	Mesh	Balls	Mesh	Balls	Mesh
1.2	6.0	8.0	997	952	47.2	60.3
1.8	6.6	9.3	973	857	50.1	62.3
3.0	7.1	10.2	963	791	46.6	56.4
5.0	7.5	9.9	936	835	47.3	55.9
8.9	8.2	11.1	821	765	49.7	55.7
11.5	8.3	11.4	840	732	46.0	56.7
13.0	7.9	12.6	842	683	45.1	57.0

Enhancement of Magnetic Response

The magnetic force on the particle is dependent on the magnetic susceptibility of the particle as given by Eqn (1). A number of processes have been described to increase the magnetic response of selected constituent in a mixture. In general a strongly magnetic substance such as magnetite is incorporated into a particle of otherwise nonmagnetic or very weakly paramagnetic material. This results in increasing the susceptibility which in turn significantly improves magnetic response shown by particle containing even small amount (0.01-01 % by volume) of a highly ferromagnetic material such as magnetite due to the greater permeability of magnetite as compared with the weakly paramagnetic material. Parsonage [31] has reviewed the literature. Hazen Research developed the magnex process [32] in which decomposition of iron pentacarbonyl took place selectively onto pyrite and ash surface in the presence of coal. This allowed their selective removal from organic coal substance by magnetic means.

If hydrophobic particles are contacted with oil laden with magnetite particles the suspension spreads on the surface of the hydrophobic particles. This can form the basis of selective magnetisation and separation. The Murex Process described by Taggart [33] is based on the above principle. Schubert [34] used kerosene based magnetic fluid for separation of chalcocite from silica and coal from ash.

Selective Coflocculation With Magnetite

Coflocculation of a mineral with fine magnetite particle may incorporate these particles in the floc. The flocs are then amenable to magnetic separation [35, 36]. Alternatively, dissolved phosphates can be precipitated in the presence of magnetite particles suspended in water and subsequently the precipitate removed by HGMS [37]. Selective coflocculation of one mineral with magnetite is possible. Thus, gibbsite from quartz [38], alunite from quartz [39], and hematite from silica [40] were separated.

According to Parsonage [31] coflocculation with magnetite followed by magnetic separation has several technical advantages over the conventional process of selective flocculation followed by settling.

- It is not important to produce large, fast settling flocs. Therefore, higher shear rates during conditioning stage can be used.
- High shear mixing may help to reduce entrainment of gangue in the flocs and thus improve grades.
- The problem of contamination of the flocculated concentrate by coarse gangue particles settling with the floc is avoided.
- The process is much faster, requires less space and little control.

Selective Magnetic Coating

In this process, the surface adsorption of individual magnetite particles onto the selected grains forms the mechanism of magnetisation. Frangiskes et al. [41] were the first to use this approach to separate silica and calcite in the magnetic fraction from magnesite which was uncoated and hence reported as the non-magnetic fraction. ARQUAD-ZC, diesel oil and Flotol B were used as reagents to coat magnetite.These reagents are the same as those used for selective flotation but the quantities required for magnetic coating are much smaller. Separation of calcite and dolomite from phosphate minerals by selective coating of carbonate was reported by Parsonage [31]. Addition of a surfactant such as sodium oleate reduced the quantity of magnetite required for coating. More examples of selective separation using selective coating investigated by the above author are given in Table 3.

Table 3: Separation using selective magnetic coating (after Ref. [31])

(1) Cassiterite/quartz, –20 microns, 5.10^{-4} M Na oleate, pH 4.5, 2.10^{-3} M $NaClO_4$

	Wt%	Cassiterite	Recovery (%)
Magnetics	3.7	74.9	71.4
Non-magnetics	96.3	1.1	28.6
Feed	100.0	3.8	100.0

(2) Beryl/glass/quartz, – 150 + 75 microns, 0.1% Aeropromoter 825, pH 2.5

	Wt%	Beryl(%)	Recovery (%)
Magnetics	27.4	93	95
Non-magnetics	72.6	2	5
Feed	100.0	27	100

(3) Barite/calcite, – 150 + 75 microns, 10^{-3} M Na_2SO_4, pH 9, 2.10^{-3} M $NaClO_4$

	Wt%	Barite (%)	Calcite (%)	Barite recovery (%)	Calcite recovery (%)
Magnetics	39.2	13.9	86.1	8.6	91.6
Non-magnetics	60.3	94.9	5.1	91.4	8.4
Feed	100.0	63.1	36.9	100.0	100.0

(4) Scheelite/calcite, – 150 + 75 microns, 10^{-4} Na oleate pH 9, 2.10^{-3} M $NaClO_4$

	Wt%	Scheelite (%)	Recovery (%)
Magnetite	59.1	0.39	0.6
Non-magnetics	40.9	97.2	99.4
Feed	100.0	40.0	60.0

Recently, Shaikh and Dixit [42] used the coating technique to beneficiate phosphate ore using the selective coating followed by two stage high gradient magnetic separation. Two flowsheets were proposed for Rajasthan phosphate ore and Sauger (M.P) ore. These are shown in Figs. 13 and 14 respectively. The HGMS results are shown in Table 4. In case of Rajasthan phosphate, a product containing 30.6% P_2O_5 was obtained at an overall recovery of 55.6%. In case of Saugar (M.P) ore, a product containing 31.5% P_2O_5, 2.59% iron and 8.7% silica was obtained at an overall recovery of 65.04%. It may be noted that knowledge of flotation surface chemistry can be utilized fruitfully to achieve selectivity in HGMS as also to devise suitable reagent scheme.

Table 4: Results of HGMS investigation on phosphate ore (after Ref. [42])

Rajasthan rock phosphate

	% P_2O_5	% Calcite	% Silica	% Recovery
Feed	26.3	11.62	21.4	–
1st Stage	27.66	1.59	21.72	75
2nd Stage	30.6	negligible	8.8	(overall)

Saugar (Madhya Pradesh) rock phosphate

	% P_2O_5	% Iron	% Silica	% Recovery
Feed	27.2	8.035	20.6	–
1st Stage	29.034	0.9486	20.8	88.5
2nd Stage	31.55	2.54	8.7	65.04 (overall)

Surface Phenomenon in Coating Process

To understand the factors responsible for the coating of fine magnetite particle on the mineral surface it is necessary to consider various forces acting between magnetite and mineral particles. Parsonage [31, 43] has listed these forces as follows:

- Born repulsion due to overlapping electron clouds of two nearest atoms, V_B.
- Van der Waals interaction due to dipoles and London forces, V_A.
- Electrical interaction due to charges on the particles and overlap of double layers, V_R.
- Steric effects, due to interaction between adsorbed species on the mineral surfaces, V_{ST}.
- Hydrocarbon chain association due to interaction of hydrocarbon chains on opposing surfaces, V_{assoc}.
- Hydration effect due to changes in water structure near the hydrated surface, V_{HDN}.
- Hydrophobic effect, long range attractive forces between hydrophobic surface, V_{HPB}.

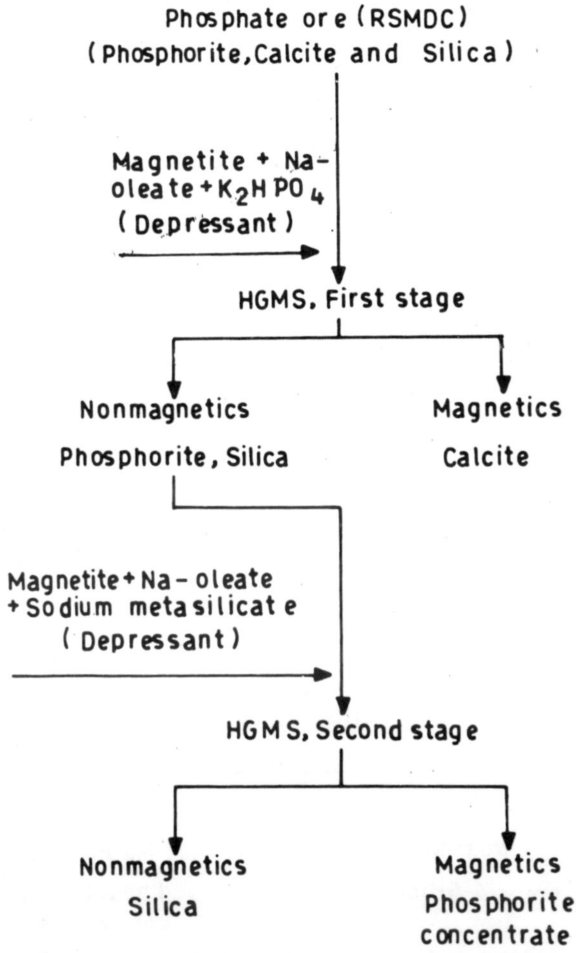

Fig. 13 Flowsheet for Rajasthan phosphate (after Ref. [42])

The total interaction energy, V_T, is given by the sum of all the above mentioned energies. A net repulsive energy or high energy barrier will tend to prevent contact of the mineral and magnetite particles and will inhibit formation of coating. On the other hand, a net attractive energy of interaction will favour contact of the magnetite and the mineral finally resulting in coating. A typical potential energy curve is shown in Fig. 15. For coating to take place ideally there should be no potential energy barrier and with a deep primary minimum to ensure strong adhesion. On the other hand, a high barrier and shallow minimum will prevent coating formation.

Though theoretical aspects of adhesion leading to coating are reasonably well known there are only a few experimental studies in relation to the relative importance of various factors in specific cases. The present author [44] has studied experimentally the role of magnetite and sodium oleate in the HGMS

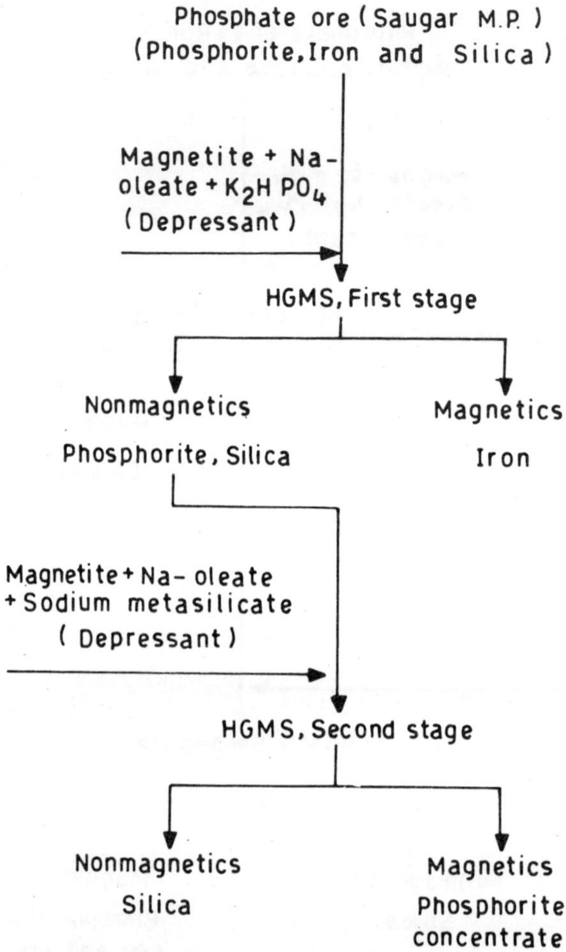

Fig. 14 Flowsheet for Saugar (M.P.) rock phosphate (after Ref.[42])

of calcite using magnetic coating. The effect of magnetite addition in the absence and presence of sodium oleate are shown in Fig. 16 and Fig. 17 respectively. In the presence of sodium oleate (25 mg/l), only 100 mg of magnetite are required to get 100% recovery of calcite in the magnetic fraction. The SEM photograph of calcite coated with magnetite is shown in Fig. 18a, whereas the X-ray mapping of the same particle for iron is shown in Fig. 18b to identify the magnetite particles. The white spot in Fig. 18b indicate magnetite particles. The corresponding particles in the SEM photograph have been encircled. It is seen that the magnetite particles do not coat the entire surface of calcite particles; rather a few of them are anchored to the calcite surface and this is enough to impart necessary magnetic characteristics to the calcite particle. It is, therefore, not surprising that a very small amount of magnetite (0.4%) is required.

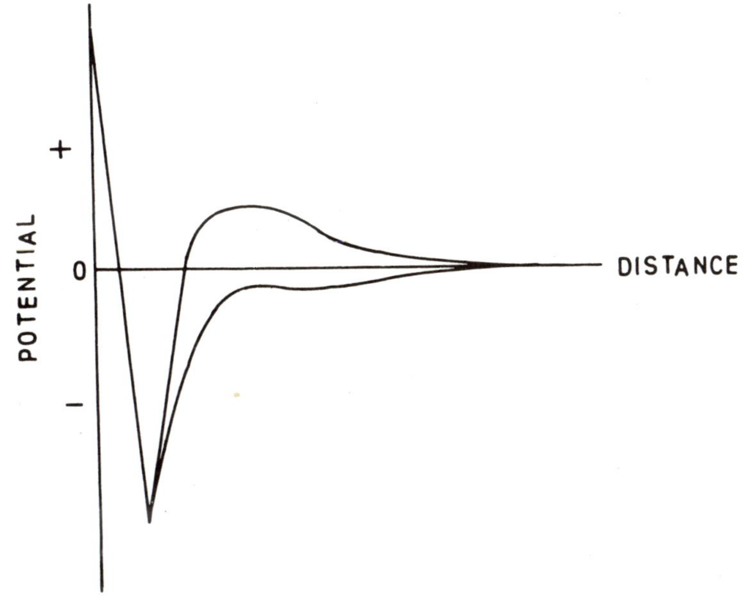

Fig. 15 Typical potential energy profile

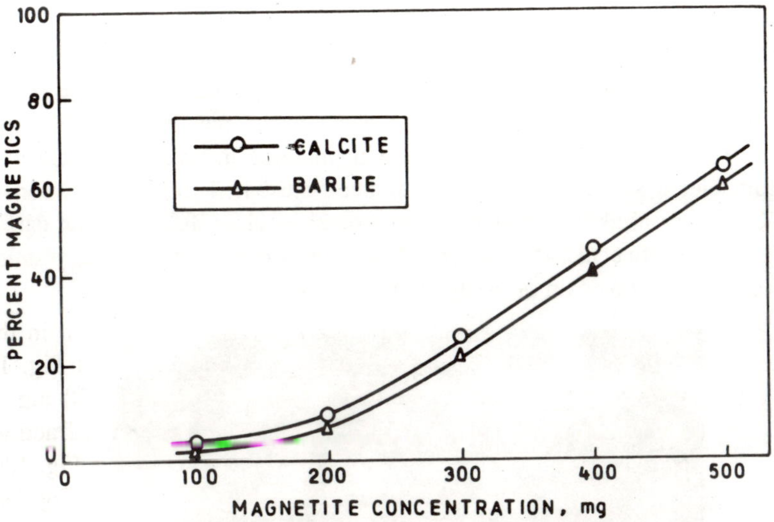

Fig. 16 Effect of magnetite addition in the absence sodium oleate on HGMS of calcite and barite (after Ref. [44])

The zeta potential studies of calcite and magnetite in the presence of sodium oleate shows that in case of magnetite, the zeta potential is negative above pH 4 and as high as 120 mV. For calcite also the values are negative and have a large magnitude (40-120 mV). Thus, the role of electrostatic attraction in the

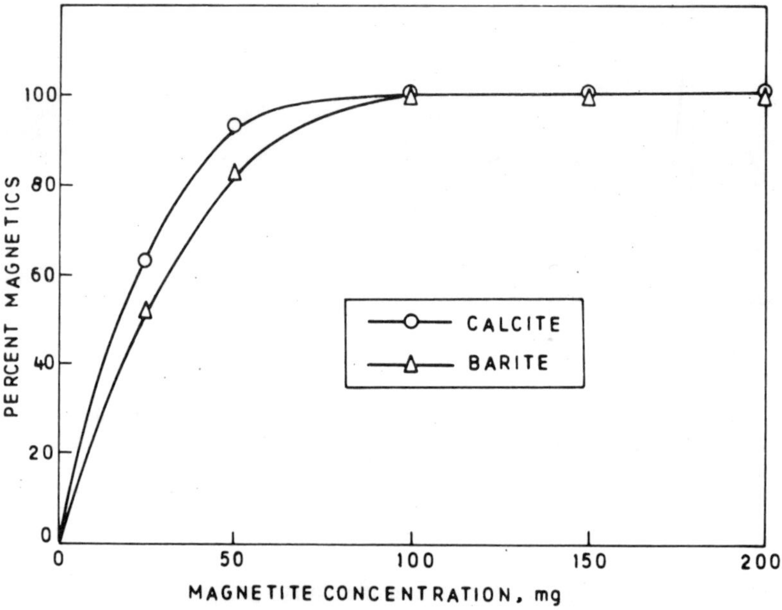

Fig. 17 Effect of magnetite addition in the presence of sodium oleate on HGMS of calcite and barite (Na oleate-25 mg) (after Ref. [44])

coating process appears to be negligible. The FTIR spectra of calcite coated with magnetite in presence of sodium oleate shows a new peak at 1095 cm^{-1} (Fig. 19). The peak becomes intense with increase in time of contact with sodium oleate solution. The peak has been attributed to the formation of –C–O–C– crosslinking between the hydrocarbon chain following an earlier suggestion by Hu et al. [45]. The crosslinking may result in the polymerisation of the hydrocarbon chain of oleate as shown below.

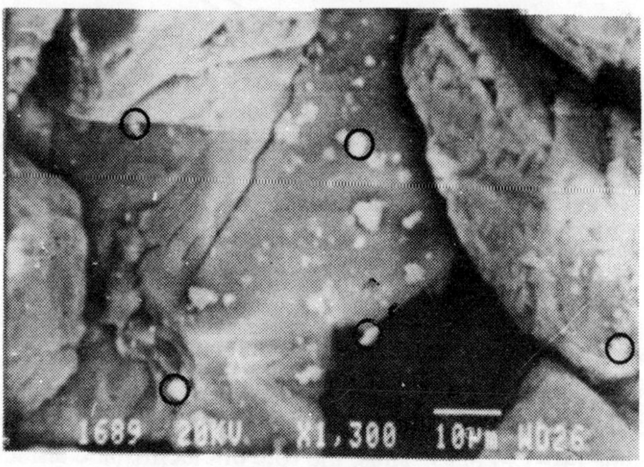

Fig. 18 (a) SEM of calcite particle coated with magnetite

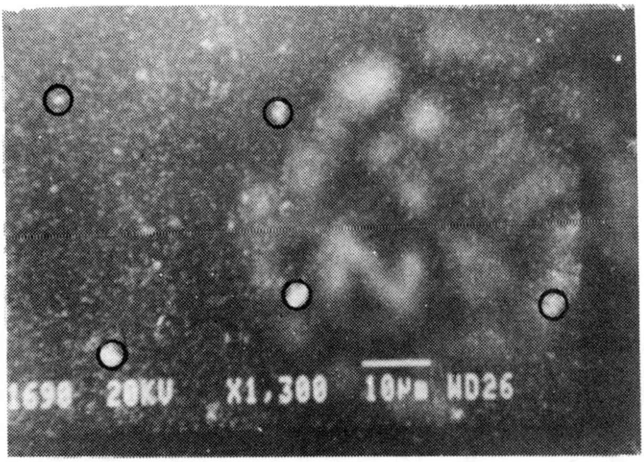

Fig. 18 (b) X-ray mapping of calcite particle shown in (a) (after Ref. [44])

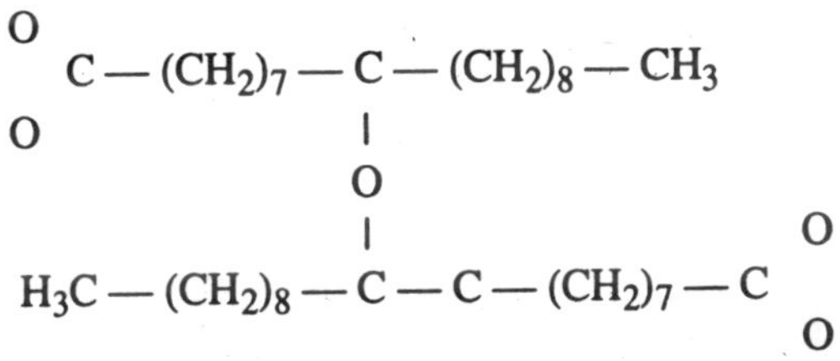

The surfaces develop more hydrophobicity which increases the attractive forces between the particles leading to more coating. Thus, sodium oleate plays a significant role in the coating process.

Magnetic Surfactants

Many times it may not be desirable to introduce impurities like magnetite as it may introduce undesirable colour and create clogging problems. In a bid to overcome these difficulties an effort was made to impart magnetic characteristics by using surfactant with magnetic functionality [46]. The magnetic functionality is introduced by using a cation with known magnetic properties to form a salt with the surfactant such as manganese oleate and stearate. Such a technique has several advantages.

- It will not impart an intense colour to a mineral.
- The adsorption of the surfactant on the nonmagnetic mineral surface is expected to be more uniform.
- It is possible to modulate the field and flow rate as well as use additives to enhance selectivity and recovery.
- A wide variety of reagents can be used to suit a specific ore and the vast knowledge of flotation chemistry could be profitably used.

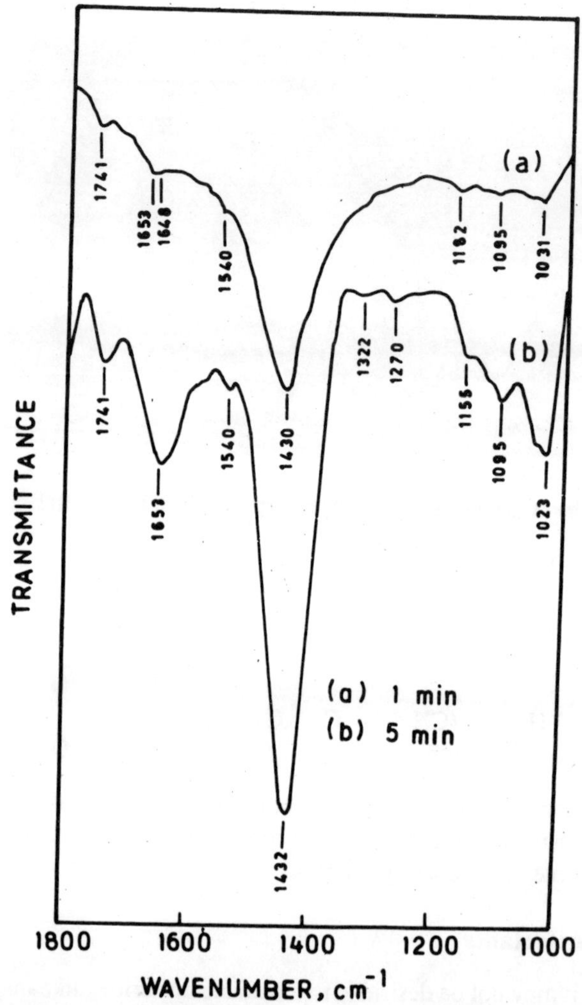

Fig. 19 FTIR spectra of calcite coated with magnetite in the presence of sodium oleate (a) 1 min and (b) 5 min contact (after Ref. [44])

The present author studied the HGMS of pure calcite and barite using manganese oleate and stearate. The results are depicted in Fig. 20. It will be seen that 100% recovery of the minerals in the magnetic fraction was possible. Manganese stearate alone is not very effective. But in the presence of sodium oleate, it was possible to achieve adequate recovery. This was attributed to lower solubility [47] of manganese stearate (pKsp = 17.5) as compared to manganese oleate (pKsp = 12.3). Comparable results were obtained by the addition of manganese sulfate along with sodium oleate.

The effectiveness of manganese soap and salt can be understood if the values of magnetic susceptibility are compared. The values for soaps are not available but $MnSO_4$.n H_2O has a susceptibility of 14200-14700 (10^6 CGS

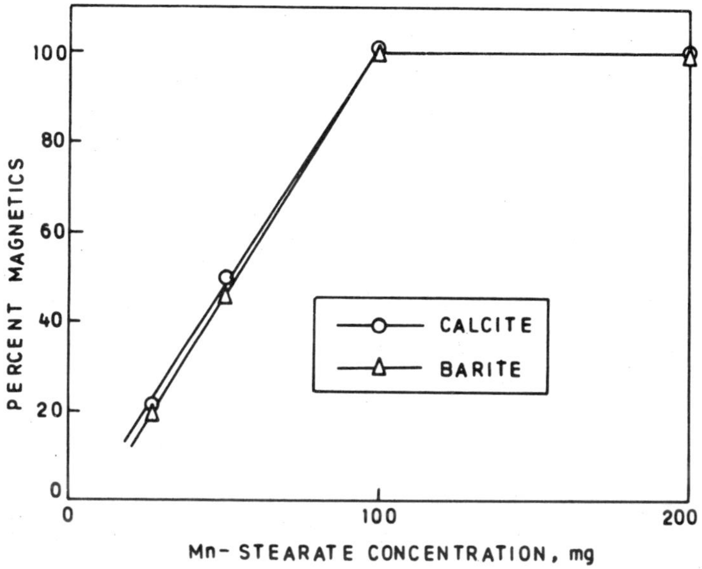

Fig. 20 Effect of manganese stearate addition in the presence of Na oleate (25 mg) on the HGMS of calcite and barite (after Ref. [46])

units) whereas FeO and Fe_2O_3 have a susceptibility of 7250 (293 K) and 3568 (1033 K) respectively Thus, manganese salts are more magnetic as compared to even FeO and Fe_2O_3 [48]. Also, if the substances are arranged in order of their affinity to iron metal having an arbitrary value of 100, then Fe_3O_4 will have an affinity of 40,Fe_2O_3 1.32 and $MnSO_4$ 5 [49]. Therefore, it is not surprising that Mn-salt and soaps impart sufficient magnetism to otherwise non-magnetic material.

The use of magnetic surfactant in HGMS open up several possibilities for research because here the surface chemistry is interfaced with magnetic separation.

CONCLUDING REMARKS

The role of surface chemistry in HGMS is illustrated with appropriate examples. The surface chemistry plays an important role by influencing flocculation, hetero-flocculation, dispersion, magnetic coating and magnetic surfactants. Surface chemistry is responsible for selectivity in several cases. It can also be utilized to broaden the applications area of HGMS.

ACKNOWLEDGEMENTS

The author gratefully acknowledges the financial support given by Department of Science and Technology, Government of India. He appreciates the hard work done by his students C.B.Maurya, A.M.H.Shaikh and S.S.Banerjee during the experimental investigation.

NOMENCLATURE

F_m – force on particle

– magnetic susceptibility

V – volume of the particle

H – background field

dH/dX – magnetic field gradient

– surface potential

pH_{pzc} – point of zero charge

pH – experimental value

k – Boltzman constant

T – temperature, K

e – electronic charge

REFERENCES

1. Wills, B.A.: Mineral processing technology, 2nd Edn; Oxford, England: Pergamon Press 1981.
2. Iannicelli, J.: Clays and Clay Min 24, (1976) 64-68.
3. Lin, C.J., Liu, Y.A., Vives, D.L., Oak, M.J., and Crow, G.E.,: IEEE Trans on Magnetics Mag-12 (5) (1976) 513.
4. Murray, M.H.: IEEE Trans on Magnetics Mag-12 (5) (1976) 498.
5. Keland, D., Kolm, H., Dehatour, C., Maxwell, E., and Oberteuffen, J.: High gradient magnetic separation: industrial applications of magnetism. New York: NATO Adv. Study Inst 1974.
6. Watson, J.H.P.: J. Appl. Phys 44 (1973) 4209-4213.
7. Lubrosky, F.E., and Drummond, B.J.: IEEE Trans. on magnetics, Mag-11 (1975) 1696.
8. Lawson, W.F., Simons, W.H., and Treat, R.P.: J. Appl. Phys. 48 (1977) 3213-3224.
9. Liu, Y.A., and Oak, M.J.: AIChE Journal 25 (5)(1983) 771-779.
10. Maurya, C.B., and Dixit, S.G.: Trans IIM 41 (1988) 161-168.
11. Maurya, C.B., and Dixit, S.G.: Bull. Mater. Sc 10 (5) (1988) 471-475.
12. Maurya, C.B., and Dixit, S.G.: Int'l. J. Mineral Processing 28 (1990) 199-207.
13. Maurya, C.B.: High gradient magnetic separation and improvement of brightness of clays. Ph.D. Thesis, University of Bombay, 1989
14. Grim, R.E.: Clay mineralogy. New York : McGraw Hill 1953.
15. Marshall, C.E.: Soil materials. New York : Wiley 1964

16. Pradip: In Mehrotra, S.P., and Ramchandran, T.R; (Ed) : Progress in metallurgical research: Fundamental and applied aspects. Tata-McGraw Hill 1985 pp 45-50.
17. Williams, D.J.A., and Williams, K.P.: J. Colloid and Interface Science 65 (1) (1978) 79-87.
18. Parks, G.A.: Chem. Rev. 65 (1965) 177-198.
19. Ferris, A.P., and Jepsin, W.B.: J. Colloid and Interface Science 51 (1975) 245-259.
20. Matijevic, E.: J Colloid and Interface Science 58 (2) (1977) 374-389.
21. Derjaguin, B., and Landau, L.D.: J. Exp. Theor. Phys. USSR 15 (1945) 662.
22. Verwey, E.J.W., and Overbeek, J.Th.G.: Theory of the stability of lyophobic colloids. Elsevier 1948.
23. Hamaker, H.C.: Physica 4 (1937) 1058-1072.
24. Vold, M.J.: J. Colloid Science 9 (1954) 451-459.
25. Vold, M.J. : J. Colloid Science 16 (1961) 1-12
26. Mahanty, J. and Ninham, B.W. : Dispersion forces. New York : Academic Press 1977.
27. Svoboda. J.: Aufbereitungstechnik 24 (1983) 520.
28. Arellano, M.E. : Digest of the Inter Mag conference, Canada 1984.
29. Svoboda, J., Corrans, I.J., and Spitze, M.H.E.: Int'l. J. Mineral Processing 17 (1986) 83-98.
30. Mackenzie, J.M.W., and Lovell, V.M.: The application of double layer properties to the upgrading of uranium in Witwaterarand ores by selective flocculation. Rep.1413, National Institute for Metallurgy, Randburg, 1972.
31. Parsonage, R.: Int'l. J. Mineral Processing 24 (1988) 269- 293.
32. Kindig, J., and Turner, R. : AIME Fall Meeting Denver USA. Preprint No 76-F-306 (1976).
33. Taggart, A.F.: Handbook of mineral dressing. New York : John Wiley and Sons 1945.
34. Shubert, R.H.: Magnetic separation of particle mixtures. US Patent 30, 360, 1980
35. Delatour, C. IEEE Trans. Magnetics, MAG-9 (3) (1973) 314-316.
36. Kurinobu, S. and Uchiyama, S.: IEEE Trans. Magnetics MAG- 18 (6) (1982) 1526-1538.
37. Shaikh, A.M.H., and Dixit, S.G. : Wat. Res 26(6) (1992) 845-852.
38. Hwang, J.Y., Kulluerud, G., Takayasu, M., Friedlaender, F.J., and Wankat, P.C. : IEEE Trans. Magnetics. MAG-18 (6) (1982) 1689-1691.
39. Hwang, J.Y., Kulluerud, G., Frielaender, F.J., and Takayasu, M.: SME-AIME annual meeting, New York 1985.
40. Iwasaki, I.: U.S. Patent 4,298,169; 1981.
41. Frangiskos, A.Z., and Gambopoulos, T.: Proceedings XII International Mineral Processing Congress. Sao Paulo Brazil 1977.
42. Shaikh, A.M.H., and Dixit, S.G. :Int'l. J. Mineral Processing 37 (1993) 149-162.
43. Parsonage, P.: Particle interactions in colloidal suspensions. Rep. Warren Spring Lab No. LR 600 (MM) 1987.

44. Shaikh, A.M.H., and Dixit, S.G.: J.Colloid. Interface Science 155 (1993) 340-346.
45. Hu, J.S., Mishra, M., and Miller, J.D. : Int'l. J. Mineral Processing 18 (1986) 73-84.
46. Shaikh, A.M.H., Banerjee, S.S., and Dixit, S.G. : Separation Technology 3 (1994) 36.
47. Du Rietz, C.,: Trans IV Int'l Mineral Processing Congress Stockholm, Almquist and Wiksell Publ. 1957 pp 417-433.
48. Weast, R.C.,:CRC handbook of chemistry and physics CRC Press. Inc. 1983-84.
49. Gaudin, A.M.: Principles of Mineral Dressing. New York : McGraw-Hill Book Co. Inc 1939.

Subject Index